Proceedings of the International Symposium on

Gold Metallurgy

Pergamon Titles of Related Interest

Kachaniwsky INT. SYMP. IMPACT ON PRODUCTIVITY OF NON-FERROUS METALLURGICAL PROCESSES

Macmillan INT. SYMP. QUALITY AND PROCESS CONTROL IN REDUCTION AND CASTING OF ALUMINUM AND OTHER LIGHT METALS

Rigaud INT. SYMP. ADVANCES IN REFRACTORIES FOR THE METALLURGICAL INDUSTRIES

Ruddle INT. SYMP. ON ACCELERATED COOLING OF ROLLED STEEL

Saimoto SOLUTE-DEFECT INTERACTION Theory and Experiment

Tyson INT. SYMP. FRACTURE MECHANICS

Wills MINERAL PROCESSING TECHNOLOGY, 3rd Edition

Related Journals

(Free sample copies available upon request.)

ACTA METALLURGICA
CANADIAN METALLURGICAL QUARTERLY
GEOCHIMICA ET COSMOCHIMICA ACTA
SCRIPTA METALLURGICA

Proceedings of the International Symposium on

Gold Metallurgy

Winnipeg, Canada
August 23-26, 1987

Co-Sponsored by Canadian Mineral Processors

Vol. 1 Proceedings of the Metallurgical Society of the
Canadian Institute of Mining and Metallurgy

Edited by

R.S. SALTER

General Manager, Lakefield Research, Lakefield, Ontario
Canada

D.M. WYSLOUZIL

Consulting Metallurgist, Lakefield, Ontario
Canada

G.W. McDONALD

Assistant General Manager, Selco Division
BP Canada Limited, Toronto, Ontario
Canada

PERGAMON PRESS

New York Oxford Beijing Frankfurt São Paulo Sydney Tokyo Toronto

Pergamon Press Offices:

U.S.A.	Pergamon Press, Maxwell House, Fairview Park, Elmsford, New York 10523, U.S.A.
U.K.	Pergamon Press, Headington Hill Hall, Oxford OX3 0BW, England
PEOPLE'S REPUBLIC OF CHINA	Pergamon Press, Qianmen Hotel, Beijing, People's Republic of China
FEDERAL REPUBLIC OF GERMANY	Pergamon Press, Hammerweg 6, D-6242 Kronberg, Federal Republic of Germany
BRAZIL	Pergamon Editora, Rua Eça de Queiros, 346, CEP 04011, São Paulo, Brazil
AUSTRALIA	Pergamon Press (Aust.) Pty., P.O. Box 544, Potts Point, NSW 2011, Australia
JAPAN	Pergamon Press, 8th Floor, Matsuoka Central Building, 1-7-1 Nishishinjuku, Shinjuku-ku, Tokyo 160, Japan
CANADA	Pergamon Press Canada, Suite 271, 253 College Street, Toronto, Ontario M5T 1R5, Canada

---

First printing 1987

Library of Congress Cataloging in Publication Data

Library of Congress Number: 87-18997

In order to make this volume available as economically and as rapidly as possible, the authors' typescripts have been reproduced in their original forms. This method unfortunately has its typographical limitations but it is hoped that they in no way distract the reader.

Printed in the United States of America

Preface

The International Symposium on Gold Metallurgy was conceived at the 25th Annual Conference of Metallurgists in Toronto in August 1986. The rapid expansion of the gold mining industry in North America had resulted in a shortage of metallurgists experienced in gold milling. Although several short courses and international symposia had been held in recent years on gold metallurgy, none had been hosted in Canada. The Hydrometallurgical Section and the executive of the Metallurgical Society of CIM agreed that a forum for the presentation and discussion of present plant practice and of new work being undertaken both in the laboratory and in the plant was required. Such a meeting was perceived as a service to the gold mining industry, worldwide, and in particular to the North American industry, to be an opportunity for training and exposure to international experience for its junior metallurgists.

The selection of the 26th Annual Conference of Metallurgists as venue and the agreement of the Canadian Mineral Processors to join in the effort allowed the symposium organizing committee to concentrate on the technical program, the assembly and editing of the proceedings, some specialized publicity, and the development of the short course on carbon-in-pulp and Hemlo-Red Lake tour to be associated with the symposium. All arrangements, registration, general publicity and agreements with the publisher were well advanced by the conference organizing committee. The response to the requests for papers was excellent and papers were attracted from around the world confirming that the theme of the Symposium was both timely and apposite.

To set the stage, the keynote speaker places gold metallurgy into perspective by discussing the historic evolution of gold metallurgy and the technologies and challenges facing metallurgists today. This theme is continued into the first technical session where the problems with obtaining meaningful measurements in gold processing are investigated from sampling through to desorption from carbon. In the following sessions, existing plant practices are extensively reviewed; new research, development and engineering techniques are discussed; emerging processes which will be important in the future are revealed; and, finally, some environmental control aspects of gold processing are considered. This Symposium contributes significantly to the current state of knowledge and should provide the background and stimulus for metallurgists to continue with the advance of technology in the treatment of gold ores.

The editors express their appreciation to the authors, session chairmen, institutes, universities, and companies who, through their willingness to participate and contribute, made this Symposium possible.

The assistance of Joanne Munro and Shelley Kocjan are gratefully acknowledged. Their attendance to the abundance of secretarial detail was a significant factor in the success of this Symposium.

R.S. Salter, Lakefield
D.M. Wyslouzil, Lakefield
G.W. McDonald, Toronto

June 1987

SYMPOSIUM AND CONFERENCE ORGANIZERS

SPONSORING SOCIETIES

The Metallurgical Society of CIM
The Canadian Mineral Processors of CIM
The Hydrometallurgical Section of the Metallurgical Society of CIM

SYMPOSIUM CO-CHAIRMEN

R.S. Salter
Lakefield Research
Lakefield, Ontario, Canada

G.W. McDonald
Selco Division - BP Canada Limited
Toronto, Ontario, Canada

SYMPOSIUM ORGANIZING COMMITTEE

G.W. McDonald
Selco Division - BP Canada Limited
Toronto, Ontario, Canada

R.S. Salter
Lakefield Research
Lakefield, Ontario, Canada

K.W. Sarbutt
Lakefield Research
Lakefield, Ontario, Canada

D.M. Wyslouzil
Consulting Metallurgist
Lakefield, Ontario, Canada

SESSION CHAIRMEN

Technical Session No. 1
Measurements in Gold Processing

D.M. Wyslouzil
Consulting Metallurgist
Lakefield, Ontario, Canada

B. Thorndycraft
Consulting Metallurgist
Winnemucca, Nevada, U.S.A.

Technical Session No. 2
Plant Practice 1

K.G. Thomas
American Barrick Resources
Toronto, Ontario, Canada

C. Fleming
MINTEK
Randburg, South Africa

Technical Session No. 3
Effect of Mineralogy on Flowsheet Development

V. Lakshmanan
Ontario Research Foundation
Mississauga, Ontario, Canada

R. Bhappu
Mountain States Research and Development
Tucson, Arizona, U.S.A.

Technical Session No. 4
Plant Practice 2

K. Wright
Dome Mines
Toronto, Ontario, Canada

G.W. McDonald
Selco Division - BP Canada
Toronto, Ontario, Canada

Technical Session No. 5
Refractory Gold

K.W. Sarbutt
Lakefield Research
Lakefield, Ontario, Canada

J.A. Eisele
USBM Reno Research Station
Reno, Nevada, U.S.A.

Technical Session No. 6
New Developments

P.R. Mainwaring
CANMET
Ottawa, Ontario, Canada

D.M. Doyle
CANMET
Ottawa, Ontario, Canada

CONFERENCE CO-CHAIRMEN

L.E. Seeley
Falconbridge Limited
Falconbridge, Ontario, Canada

M.N. Bassim
University of Manitoba
Winnipeg, Manitoba, Canada

CONFERENCE ORGANIZING COMMITTEE

M.N. Bassim
University of Manitoba
Winnipeg, Manitoba, Canada

J.R. Cahoon
University of Manitoba
Winnipeg, Manitoba, Canada

M.C. Chatervedi
University of Manitoba
Winnipeg, Manitoba, Canada

M. Goble
Inco Thompson
Thompson, Manitoba, Canada

L.E. Seeley
Falconbridge Limited
Falconbridge, Ontario, Canada

D. Smith
Falconbridge Limited
Falconbridge, Ontario, Canada

Table of Contents

KEY NOTE ADDRESS

MEASUREMENTS IN GOLD PROCESSING

PLANT PRACTICE 1

EFFECT OF MINERALOGY ON FLOWSHEET DEVELOPMENT

PLANT PRACTICE II

REFRACTORY GOLD

NEW DEVELOPMENTS

Keynote Address

Judith A. Eisele
Research Supervisor, USBM Reno Research Station
Reno, Nevada

The introduction of cyanide leaching almost 100 years ago revolutionized the treatment of gold ores and allowed gold to be recovered from ores that could not be amalgamated or concentrated by gravity. Major milestones since then have been: (1) the use of activated carbon to collect gold from solutions, (2) electrowinning of gold from pregnant solutions, (3) carbon-in-pulp technique, (4) oxidation pre-treatment to make carbonaceous gold ores amenable to cyanidation, and (5) the application of heap leaching to low-grade ores. These techniques are common practice and widely used today, but 35 years ago if the gold ore was not high-grade and amenable to treatment in a standard countercurrent decantation cyanide mill with zinc precipitation, there were no processing options.

Areas of active research today include: (1) the use of acidic thiourea as an alternative to alkaline cyanide leaching, (2) bacterial leaching as a pretreatment for sulphide ores to make them leachable by cyanidation, and (3) method for preventing heaps from freezing so that heap leaching can be used in northern climates and throughout the winter months.

Technical Sessions

Session No. 1

Measurements in Gold Processing

SAMPLING AND ASSAYING OF GOLD ORES

G.W. Poling and A.J. Reed
Dept. of Mining and Mineral Process Engineering
University of British Columbia
Vancouver, B.C.
V6T 1W5

EXTENDED ABSTRACT

This paper reviews major factors affecting the reliability of assay results of gold ores such as:

(1) the natural variability of the gold distribution in the ore
(2) the size of the sample collected
(3) the numbers of samples collected
(4) the precision of the analytical technique

Sampling and assaying practices for gold ore reviewed with comments upon problems such as "nugget effects", "metallics" and "hidden gold".

INTRODUCTION

Sampling and analyses are the first critical elements in project evaluation involving geochemical exploration, ore reserve estimation, mine planning or mill performance optimization. The main problems are to ensure that the "sample" is representative of a larger population and then to obtain a reliable estimate of the gold content of this sample. Professional geologists and engineers need to understand sampling and analyses requirements in order to avoid many pitfalls. These professionals should not abrogate their responsibilities to service people, such as samplers and assayers, to make decisions which might critically affect subsequent evaluations.

Extremely low grade exploration samples will require some form of preconcentration prior to analyses in order to achieve statistical significance. The professional requires knowledge of the natural variability of the gold distribution within the deposit, the gold particle size distribution present and the approximate abundance of gold in order to specify a sampling program and minimum sample sizes for analyses.

SAMPLING AND SAMPLE SIZE

Relatively simple methods to calculate adequate sample size have been presented by Gy (1968), Clifton and colleagues (1969) and Ingamells (1980). These techniques are reviewed and compared in this paper. In sampling gold deposits containing relatively coarse gold, these techniques will all three dictate sometimes alarmingly large sample weights are required to obtain a representative sample, particularly of a low grade deposit. For example, in order to sample an alluvial gold deposit containing of the order of 250 mgm Au/m^3 with the coarsest gold being 1 mm diameter, at a 95% confidence limit of being within ± 25 mgm/m^3 would require a minimum sample size of 1.5m^3 (or around 5000 lbs). A general conclusion of Clifton's work is that "the number of gold particles in the sample is the only factor controlling the precision of chemical analyses". A precision of ± 50% at the 95% confidence limit is achieved when samples for analyses each contain a minimum of 20 particles of gold. Although these relatively simple sampling theories depend on the gold particles being randomly distributed (which seldom occurs) they certainly provide useful indications of adequate sample size. Implications of these sampling statistics have led at least one author to conclude that there are deficiencies in sampling placer gold deposits that are "perhaps insurmountable at reasonable cost" Fricker (1976).

Natural distributions of gold values in many deposits appear to be lognormal Krige (1978) (a plot of log cumulative frequency versus log of the gold content should be a straight line). This means that typical histograms of gold values show distributions of gold assays highly skewed with large numbers of samples near zero in gold content and only a very few high assay values. Koch and Link (1970, 1971) found that gold values in the Getchell Mine and Homestake Mine were skewed from lognormal. Geographically, gold values were highly clustered in ore shoots which meant that more samples were needed to provide a specified precision than that predicted by standard error-of-the-mean statistics. Also the variance of gold values were then not inversely proportional to sample volumes as would be predicted by Gy, Clifton and Ingamells-theories. Koch and Link therefore conclude that smaller diameter drill core (i.e. XRT) might provide nearly as precise results as the more expensive larger diameter (EX) core. They also concluded that 5-foot sample intervals provided just as precise results as the usual one-foot increments. For the submicron-Getchell-type gold ore body, they found that less expensive drilling producing "cuttings" yielded as good or better results than the more expensive diamond drilling technique.

Geostatistics presents a powerful tool for studying the continuity of gold values within a pay streak or ore shoot or an entire ore body. Examples of experimental semi-variograms clearly demonstrating directional anisotropy in the Witswatersrand reefs has been reviewed recently by Janisch (1986).

Geostatistical analyses of spatial relationships of gold values enables reasonable weighting of specific samples for improved grade estimations. This method can also assist in the design of more rational and more economical drilling programs and mining strategies. Using other methods of statistics can result in biasing estimates such that they could not be weighted or adjusted reliably. The detailed paper will include examples of these problems.

GOLD ANALYSES

Total variance of gold sampling and assaying data sets is the sum of: (i) the natural variance; (ii) the sampling variance, (iii) the processing variance and (iv) the analytical variance. Difficulties often arise because neither geologist-prospectors nor assayer-analysts take sufficient interest in the origin, collection and processing of each sample for analysis. Since realistically, the analytical chemist bears little responsibility for the way his assays are used, the major responsibility for understanding all aspects lies with the geologist-prospector or engineer who submits the samples. The analyst should certainly be told about the nature of the samples submitted, their anticipated gold contents and the anticipated size distribution of gold in the sample. With this information in hand the most appropriate analytical technique can be selected. A preliminary "orientation study" as described by Burn (1981) is perhaps the best way to begin design of a detailed sampling program.

Analytical methods for determining trace amounts of gold were reviewed recently by Hosking (1982). Detection limits of 0.1 ppb have been achieved by combining fire assay fusion and cupellation techniques for preconcentration followed by neutron activation analyses of the dore head. Sample preparation technique should always consider the need of preconcentration and separation of "metallics" at 150 μm. Attempts to pulverize or "grind" coarse gold to improve sampling statistics or the reliability of sample splits should be avoided due to the probability of smearing of gold onto the grinding mechanism. Arbitrary "cutting" of high gold assay values should not be permitted. The assay should be considered real and perhaps an indicator of relatively coarse gold with implications of inadequate sample size. Perhaps the high assays indicate a bonanza or at least an economic deposit.

The classical fire assay technique with a normal lower limit of detection of around 0.01 troy oz/ton has been shown capable of collecting essentially all the gold in a wide variety of sample types. Hosking shows loss of gold: to the slag, to the crucible, in dressing the button, to the cupel, and during parting might total up to 2%. This magnitude of error is generally not significant when compared to other errors such as in sampling. The paper will provide more detail on comparison of the several techniques available in a modern analytical facility.

In sampling and assaying, with appropriate knowledge high precision can often be attained with existing technologies. A major question is what level of acceptable precision can be attained at a reasonable cost?

REFERENCES

Burn, R.G., (1981). Data reliability in ore reserve assessments, Mining Mag., October, 289-299.

Clifton, H.E., Hunter, R.E., Swanson, F.J. and Phillips, R.L., (1969). Sample size and meaningful gold analysis, USGS Professional Papers, 623-627 A-F, C1-C15.

Fricker, A.G., (1976). Placer gold measurement and recovery, Aus. I.M.M. Melbourne Branch, Sampling Symposium, September, 115-127.

Gy, P., (1968). Theory and practice of sampling broken ores, Ore Reserve Estimation and Grade Control, CIM, 5-10.

Hosking, J.W., (1982). Review of analytical methods for determining trace amounts of gold in ores and process streams, Aus. I.M.M. Carbon in Pulp Seminar, July, 351-367.

Ingamells, C.O., (1980). Sampling demonstration, Int. Prec. Met. Inst., 2nd Sym. on Sampling, March.

Janisch, P.R., (1986). Gold in South Africa, J.S.A. Inst. Min. and Met., 86, 273-316.

Koch, G.S. Jr. and Link, R.F., (1970). A statistical interpretation of sample assay data from the Getchell Mine, U.S. Dept. of the Interior, R.I., 7383.

Koch, G.S. Jr. and Link, R.F., (1971). Sampling gold ore by diamond-drilling in the Homestake Mine, U.S. Dept. of the Interior, R.I., 7508.

Krige, D.G., (1978). Lognormal-de Wijsian geostatistics for ore evaluation, S.A. Inst. Min. and Met. Monograph Series.

MATERIALS BALANCES IN GOLD PROCESSING PLANTS

D. Cimon, G. Barbery, F. Flament and D. Hodouin

GRAIIM, Département Mines et Métallurgie, Université Laval, Québec, G1K 7P4, Canada

ABSTRACT

The problems encountered in the application of conventional methods of materials accounting in gold processing plants are presented. The values of statistically coherent materials balancing programmes are shown to be of importance for gold processing plants, in which the experimental data is both noisy and redundant. The existing programmes, such as BILMAT, are not directly applicable to gold mills, since the valuable element is transferred from the solid phase to an aqueous phase. A modification of the programme, called BILMATOR, is presented and illustrated for the case of the DOYON gold mill in Québec, Canada.

KEYWORDS

Gold ore processing, metallurgical accounting, materials balance.

INTRODUCTION

The use of materials balance programmes in the minerals industry has developed slowly since the introduction of generalized software in the early '80s. It is generally accepted that these programmes are of outstanding value in order to extract the statistically most reliable information from the noisy experimental data which characterize plant data. The corrected values that are provided satisfy the mass balance equations, and correction of experimental data minimizes a target value, usually a least square criterion. These programmes, such as BILMAT (Hodouin and Everell, 1980, Hodouin, Gelpe and Everell, 1982) are of necessity when experimental data is redundant, and when sampling errors are large. All these methods have been developed for solid-solid or solid-liquid separations, that is under conditions when there is no dissolution of an element in the aqueous phases. Gold processing plants, although they are similar in a number of respects to other mineral processing plants, have major differences when considering the application of materials balance programmes: gold, the valuable element is transferred from the solid ore into aqueous solutions as an aurocyanide complex; it is then cemented on zinc powder in the conventional Merrill-Crowe process or adsorbed on activated charcoal. It is thus moving from one solid phase to aqueous solutions and then to another solid phase, along the process plant. Since materials balance programmes are able to reduce the uncertainty in experimental data, through the use of redundant information and mass balance equations, it has seemed a useful exercice to try and adapt an existing general programme, BILMAT, to the case of conventional gold processing plants. The paper presents the development of the programme, BILMATOR, and its application to the case of the DOYON gold ore processing plant in Québec. Indications for the use of the programme are given.

CONVENTIONAL METALLURGICAL ACCOUNTING

The example of the Doyon gold processing plant will be used extensively in the paper; the plant flowsheet is given on Fig. 1. The operations have been described by Hope (1984) and McMullen (1984). The process used is typical of plants built in Canada before 1983, and incorporates the following features: grinding in closed circuit with hydrocyclones; use of cyanide in grinding, hydrocyclone overflow is thickened; the clear solution is sent to the Merrill-Crowe process; thickener underflow is leached in a series of agitated tanks and then filtered on drum filters; washing of filter cakes is carried out with barren solution; pregnant solution from the filters is directed to the thickener feed; final plant effluent is filter cake from the last drum filter diluted with barren or process water.

In such plants, it is a routine practice to measure the following parameters:

- tonnage of ore (wet and dry)
- analysis of the run of mine ore
- solid tailings analysis
- analysis of the solution with the tailings
- pregnant solution flowrate
- analysis of the pregnant solution
- solids analysis in hydrocyclone overflow
- solids analysis in tank #1
- solids analysis in primary filters feed
- solids analysis in secondary leaching tanks #1 and #4
- solids analysis in secondary filters feed
- solids analysis in tertiary filter feed
- flowrate of tailings solution.

With these elements, the plant metallurgist is able to quantify recoveries at various stages in the process. The only information used is gold content and solids concentration. Feed analysis is not taken into account in practice; the so called "calculated" head replaces it for balancing all production figures. The following calculations are performed:

net production = pregnant solution assay * pregnant solution flowrate
gold in tailings = solids gold assay * dry ore flowrate
+ tailings solution assay * tailings solution flowrate

$$\text{calculated gold grade} = \frac{\text{net production + gold in tailings}}{\text{dry ore tonnage}}$$

recovery = net production/(calculated gold grade * dry ore flowrate)
% extraction = 100*(1- measured solid grade/calculated ore grade)

On this basis, a metallurgical accounting software, CALMET, has been developed for IBM-PCs, to prepare the daily, monthly and yearly report in a convenient manner.

The use of uncorrected data in the analysis of complex plants is difficult to justify; the method does not use all the data available, in particular the solids content in the various streams. As an example of the poor quality of data available, and of the need to apply mass balancing techniques Table 1 compares the results obtained in the calculation of dilutions at some points in the circuit. Dilution calculated with solution and solids analyses was obtained by:

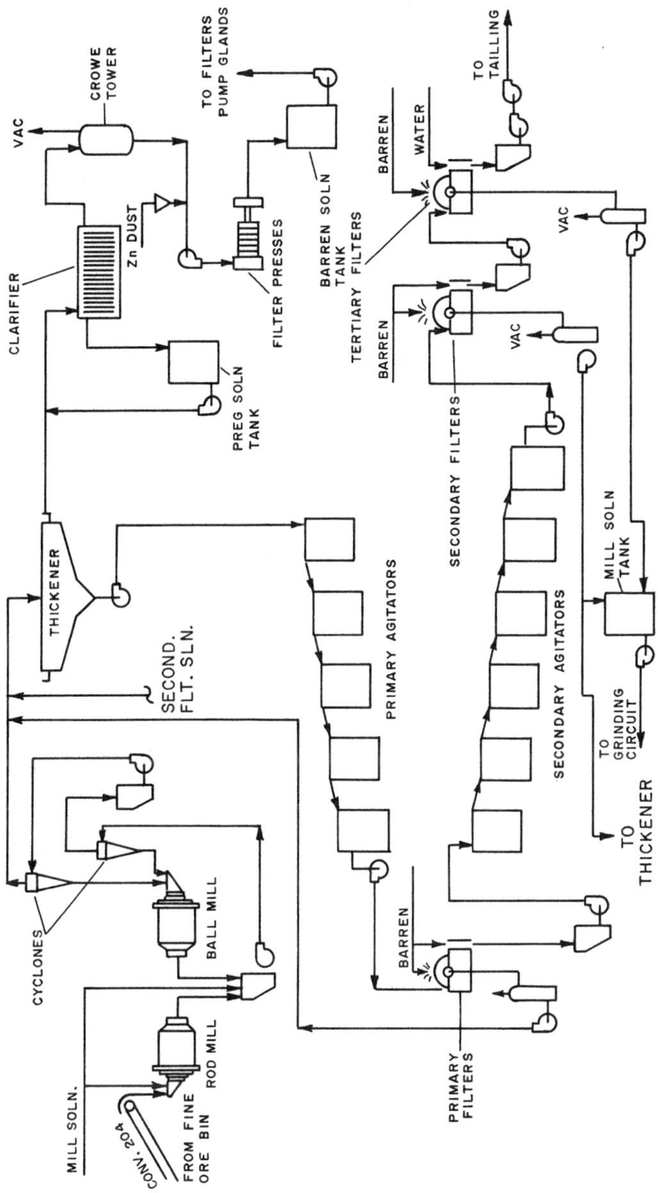

Fig. 1. Flowsheet of the DOYON gold processing plant, Québec, Canada

$D = (X_{in} - X_{out})/(Y_{out} - Y_{in})$ where X and Y stand for solids and solutions analysis.

TABLE 1 Sampling campaign of October 1985. Examples of the inadequacy of conventional methods in the calculation of dilution with various uncorrected experimental data.

		Dilution calculated from solids and solutions assays		
Sampling #1	Measured dilution	for Au	for Cu	for Zn
in primary leach tanks	0.966	1.256	2.811	2.667
in secondary leach tanks	0.965	0.370	-86.667	1.000
Sampling #2				
in primary leach tanks	0.893	0.956		-9.0
in secondary leach tanks	0.854	4.706	3.704	
Sampling #3				
in secondary leach tanks	0.925	0.842	14.0	2.222

It is based on the type of problems illustrated by the incompatibility of results obtained for the various constituents (solids, gold, copper, zinc), that the present work was developed.

SAMPLING GOLD PROCESSING PLANTS

The basis for all materials balance calculation is the extraction of data from the process. Some of the data can be obtained as solids or liquids flowrates, through the use of instruments in the plant, but most information is derived from the extraction of samples and their analysis. Sampling in mineral processing plants has been the object of many studies, in particular by Gy (1979). During the sampling campaigns that were used for the testing of the materials balance programme, care was taken to follow the good practice recommended by Gy:

- assessment of the fundamental error, as a function of sample mass, was made through the use of Gy sampling formula
- sampling was done, whenever possible, at points for which minimization of sample extraction errors are possible, and using the recommanded cutter geometry and speed (Gy, 1979)
- preparation and analysis errors were minimized by due care in sample handling, filtration, drying...

In the course of the work, it became apparent that drum filters had to be sampled in a particular manner, in order to extract the relevant information for the gold and water balance in the circuit. Filter cakes were sampled at various points on the discharge from the filter. Wash water added to the filters proved to be more difficult to assess. Since there was no flowmeter on the wash water lines, it was necessary to fabricate a sampling implement, through which the flowrate could be estimated. Wash water is added on the drum filters from a horizontal 20 feet long pipe, in which many small holes are drilled to release the wash water along the length of the filter. Sampling this pipe to quantify wash water flow was made through a sampling trough, 2 feet long and 6 inches wide, to which a 12 feet long, 2 inches diameter copper pipe was fitted. A rubber hose enabled to collect the wash water sample from the copper pipe into a bucket. Timed samples taken along the length of the filter gave the primary data.

APPLICATIONS OF MATERIALS BALANCING TO GOLD PROCESSING PLANTS

The main characteristics of materials balancing programmes have been outlined by various authors. A major feature is the materials conservation equation, which states that there is no accumulation in the circuit, in other words that the circuits are operated under stable conditions. For gold plants, the following remark can be made, that the solids flowrate is constant in the plant, in other words, solids dissolution is considered to be negligible compared to the flow of ore. The second characteristics of gold plants is the dissolution of the valuable element. The mass balance equation along any unit operation will be described as

$$(\text{gold in solids})_{FEED} + (\text{gold in solution})_{FEED} =$$
$$(\text{gold in solids})_{OUTLET} + (\text{gold in solution})_{OUTLET}$$

The main variables used in materials balance equations are the following:

S: solids feed flowrate
L: solution flowrate
D: dilution $D = L/S$ or $D = (100 - \%\ \text{solids})/\%\ \text{solids}$
$\%\ \text{solids} = 100\ S/(L + S)$
X: solids analysis
Y: solution analysis

As an example, Fig. 2 gives the mass balance equations for a filter with wash water and repulping. Similar equations have been prepared for the thickener and the leaching tanks.

Flowsheet Description

According to the terminology used in BILMAT, the flowsheet must be described under two matrices, one (M_S) describing the solids, and the other (M_L) the solution. Element M_{ik} of the matrices is 1 if stream k is going to unit i, -1 if it is coming from unit i, and 0 otherwise. Use is made of the relative solutions flows (d), with respect to solids flow taken as 1. A diagonal matrix ($\underline{d}$) containing the relative solution flows is thus introduced. Under these conditions the materials balance equations can be written as:

$M_L d = 0$ for solution mass balance

$M_L \underline{d} Y + M_S X = 0$ for species conservation (gold...)

Minimization Criterion

In order to find a solution to the mass balance equations given above, use must be made of experimental measurements, which contain errors. The criteria to be minimized are the following:

- for solids analysis: $J_1 = \Sigma[(X - X_M)^T V_X^{-1} (X - X_M)]$

which represents the sum of squares of corrected values X minus experimental values X_M, each being weighted by the inverse of the measurement variance V_X to take into account the sampling and measurement error.

- for solutions analysis: $J_2 = \Sigma[(Y - Y_M)^T V_Y^{-1} (Y - Y_M)]$
- for solution flowrates: $J_3 = \Sigma[(L - L_M)^T V_L^{-1} (L - L_M)]$
- for feed solid flowrate: $J_4 = (S - S_M)^2 V_S^{-1}$

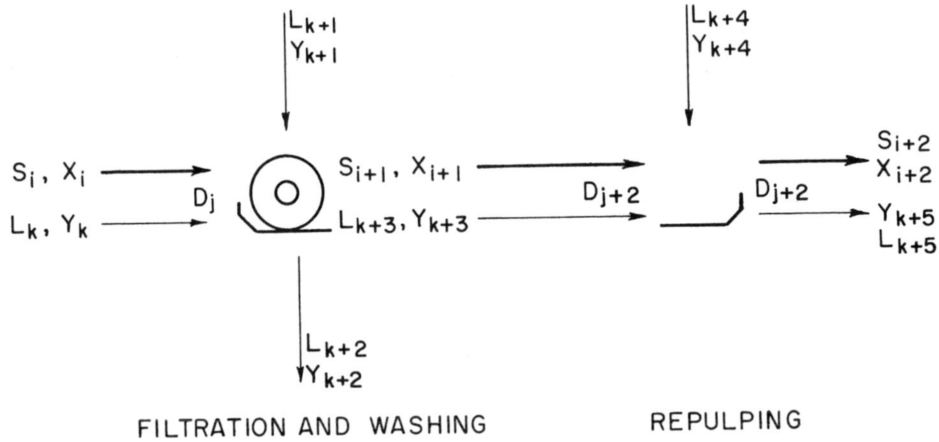

Fig. 2. Examples of mass balance equations for a drum filter with a filtration and washing section, and a repulping section.

In the filtration and washing section:

Mass conservation: $S_i = S_{i+1}$; $L_k + L_{k+1} = L_{k+2} + L_{k+3}$

Dilution: $D_j = L_k/S_i$; $D_{j+1} = L_{k+3}/S_{i+1}$; $L_{k+1} = D_{j+1}S_{i+1} + L_{k+2} - D_jS_i$

Conservation of species: $S_iX_i + L_kY_k + L_{k+1}Y_{k+1} = S_{i+1}X_{i+1} + L_{k+3}Y_{k+3} + L_{k+2}Y_{k+2}$

Which can also be written as: $X_i - X_{i+1} + D_j(Y_k - Y_{k+1}) + D_{j+1}(Y_{k+1} - Y_{k+3}) +$

$$(L_{k+2}/S_i)(Y_{k+1} - Y_{k+2}) = 0$$

In the repulping section:

Mass conservation: $S_{i+1} = S_{i+2}$; $L_{k+3} + L_{k+4} = L_{k+5}$

Dilutions: $D_{j+1} = L_{k+3}/S_{i+1}$; $D_{j+2} = L_{k+5}/S_{i+2}$

Conservation of species: $S_{i+1}X_{i+1} + L_{k+3}Y_{k+3} + L_{k+4}Y_{k+4} = S_{i+2}X_{i+2} + L_{k+5}Y_{k+5}$

Which can also be written as: $X_{i+1} - X_{i+2} + D_{j+1}Y_{k+3} - D_{j+2}Y_{k+5} + (L_{k+4}/S_{i+1})Y_{k+4} = 0$

- for dilutions: $J_5 = \Sigma[(D - D_M)^T V_D^{-1} (D - D_M)]$

The overall criterion to minimize is $J_1 + J_2 + J_3 + J_4 + J_5 = J_{tot}$; minimization must be performed under the constraint of the mass balance equations.

Overall Method Used in BILMATOR

The first step is the selection of a group of independant relative solution streams, θ, from which all the other streams can be calculated:

$d = T*\theta$, where d is the solution vector, θ is the independant solution vector and T is a Gauss elimination matrix consisting of 1, -1 and 0.

1. A first estimate of θ is required to start the procedure
2. The relative solution flowrates are calculated with $d = T*\theta$
3. A computation of corrected analyses criteria is made by a minimization of $J_1 + J_2$, under the constraint $M_{\underline{L}}dY + M_SX = 0$.

Minimization under constraint is done with the Lagrangian procedure, which provides a direct analytical solution:

$$Z = Z_M - V_{\underline{Z}}\underline{dM}_C^T (M_{\underline{C}}dV_{\underline{Z}}\underline{dM}_C^T)^{-1} M_{\underline{C}}dZ_M$$

where Z and Z_M are matrices corresponding to corrected values and experimental values, V_Z is the matrix of variances of the measures, M_C is a calculation matrix, equal to $[M_{\underline{L}}d;M_S]$.

4. A computation of S and D by the minimization of $J_3 + J_4 + J_5$ under the constraint $M_T d = 0$. The solution in this case is fairly simple:

$$S = (d^T V_L L_M + V_S S_M)/(d^T V_L d + V_S)$$

Solutions flowrates and dilutions are calculated by:

$L = dS$
$D = d$

5. A minimization of the overall criterion J_{tot} with respect to the subset θ of independant relative flowrates. A search procedure involving the Powell algorithm is used in the minimization.

When a minimum has been found at step 5, the procedure is restarted at step 2 until an overall minimum is found.

Special Case of Solution Splitting

Due to the particular nature of gold processing plants, it is common practice to divide a solution stream into several subsidiaries. Such an example can be seen on Fig. 1 for the case of barren solution, which is sent to six different points, as wash water or dilution water. The composition of all the streams must be identical, and a special treatment of the mass balance equations must be incorporated. A constraint network is created, which forces the equality in composition to the various streams. For a stream divided into N subsidiaries, N-1 constraint equations are added to the overall materials balance equations.

APPLICATION TO THE DOYON GOLD PROCESSING PLANT

During the sampling campaigns that were carried out in the Summer of 1984 and in the fall of 1985, various strategies were tested in order to obtain the most efficient information. The complete results are presented by Cimon (1987). The results presented here have been selected in order to indicate the type of problems and solutions that were encountered and developed. Fig. 3 presents the conceptual flowsheet for the application of BILMATOR.

The following comments can be made. Slurries streams are given two numbers, one for solution and the other for solids. For example, hydrocyclone overflow is stream 1 for solution and stream 28 for solids. For convenience, all solutions streams are numbered before solids, so that solutions streams number 1 to 26, whereas solids streams number 27 to 38. Stream 24 is introduced to assess water losses in the circuit, streams 25 and 26 represent process water added to the final repulping and to the solution head tank, respectively. Stream 38, taken as a dry solid flow is gold cement leaving the circuit. In order to limit the number of search variables, this solid stream has been affected the same mass flowrate as the feed. Examples of solution divisions are units XIV, wash water split, and XII, solution head tank.

Data was available from the plant metallurgical records for: ore dry flowrate (dry tonnes per day), solution flowrates for streams 10, 17 and 18. Solution flowrates were estimated for streams 19, 20 and 21 (filters wash water) with the method described above. Solids concentration data was measured on all slurry streams (solution stream numbers 1 to 10), and gold concentrations were measured from samples collected on solution streams 1 to 24, and solid streams 27 to 37. Process water gold concentration was taken as 0; for gold in stream 38, gold cement composition and flowrate was assumed to be unknown.

Weighting Factors for the Various Measurements

As it has been shown in the previous section, the use of materials balance programmes requires estimates of weighting factors for all measured variables. In the BILMAT family of programmes, the basic information required is the relative standard deviation (RSD) for all measurements. The programme computes and uses the inverse of the absolute variance in the calculations. In the application of the programme, measured variables with a high RSD are more corrected than those with a low RSD. The estimation of RSDs for experimental data is not an easy task, especially when comparing measures which have little in common, for example the ore dry mass flowrate compared to, say the gold content in filter cake solution. Users of the materials balance programme should realize that the weighting factors, taken as RSDs, have the same influence on the overall results as the actual measured values they put in.

For the purpose of the application to the DOYON gold processing plant, the following considerations have been taken into account:

- in gold plants, operators are very efficient at maintaining slurry solids concentration at stable values. These measurements have been taken as the most reliable, with RSD of 0.05 or 0.08, depending on the ease of sampling,

- solid mass flowrate has been taken as relatively precise (RSD = 0.1); solutions mass flowrates measured by the mill operators have been given the same precision. Solution flowrates measured by the technique developed was given a low priority (RSD = 0.3),

- gold content in solution was estimated to be known with a higher precision that in solids, due to the simpler sampling problems encountered. RSD values of 0.1, 0.15 and 0.2 were given to the streams, according to the ease of sampling. For solids, RSDs of 0.15, 0.20, 0.25 and 0.50 were given. The lowest precision, following Gy (1979) fundamental error value, was given to crushed ore; other

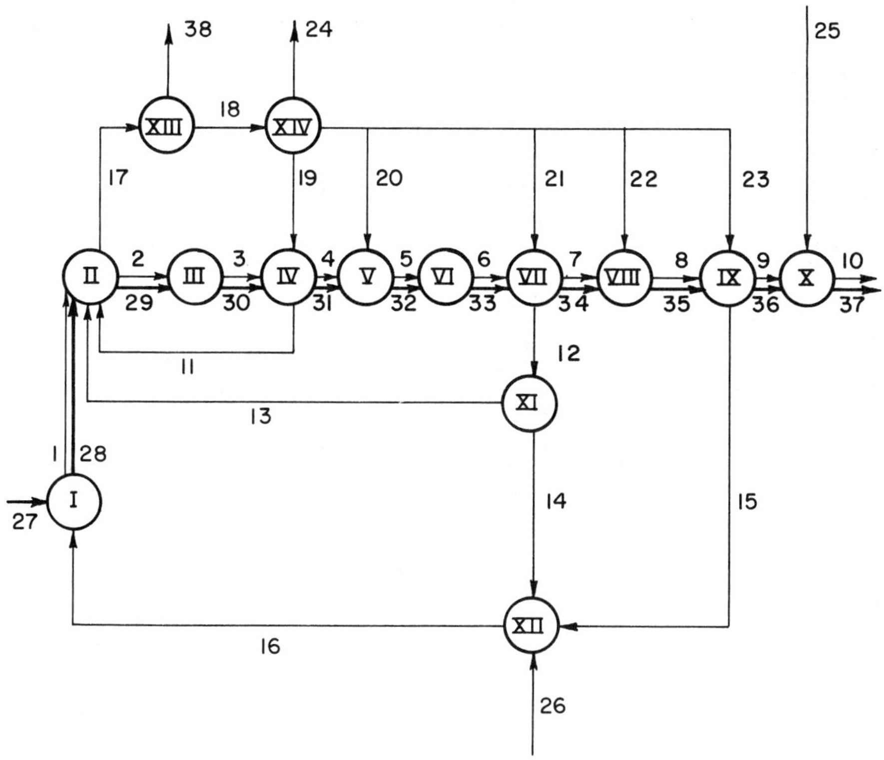

Fig. 3. Conceptual flowsheet used for the sampling campaign of October 1985

Description of equipments

I	Grinding and classification
II	Thickener
III	Primary leaching tanks
IV	Primary filter (washing section)
V	Primary filter (repulping section)
VI	Secondary leaching tanks
VII	Secondary filter (washing section)
VIII	Secondary filter (repulping section)
IX	Tertiary filter (washing section)
X	Tertiary filter (repulping section)
XI	Solution division
XII	Mill head tank
XIII	Gold precipitation
XIV	Barren solution tank

values were attributed to streams depending on the sampling problem encountered.

Table 2 presents the measured data, the RSD for the stream characteristic, and the corrected values according to BILMATOR.

Results

At the present time (May 1987), BILMATOR is available as an APL programme, and is devoted to mainframes, since APL is not an efficient language for computing speed in microcomputers. A Fortran version for IBM-PCs is being developed and should be available to interested organizations in the Fall of 1987. In terms of calculations, the application of BILMATOR to a conventional gold processing plant such as the DOYON mill is relatively large, considering especially the number of streams. The results of Table 2 enable to assess the corrections that were applied to the experimental data. It should be pointed out that in the calculations, no constraint was placed on positive values for flowrates or gold contents. No negative value was found. Although analyses for elements other than gold were available, they were not included in the present work, since the behaviour of the elements did not contain any information useful for materials balance; at best their value was corrected to agree with the mass balance equations, at worse, they had a negative influence on gold contents.

TABLE 2 Measured values, relative standard deviations and corrected values for the various streams in Fig. 2 . Sampling campaign of October 1985.

Stream #	Measured tmpd	RSD %	Corrected tmpd	Adjustment %
Solids flowrates				
27	1367.5	10	1551.0	6.51
Solutions flowrates				
10	1437.7	10	1315.13	8.51
17	3718.7	10	3534.68	4.94
18	3718.7	10	3534.68	4.94
19	400.0	30	527.22	31.80
20	300.0	30	420.90	40.30
21	480.0	30	429.64	10.49
Solids concentrations				
Stream #	Measured % solids	RSD %	Corrected % solids	Adjustment %
1	29.00	8	31.20	7.58
2	55.29	5	52.38	5.26
3	52.65	5	52.38	0.51
4	69.46	8	86.50	18.77
5	52.37	5	53.47	2.10
6	51.51	5	53.47	3.80
7	71.63	8	91.32	27.48
8	52.31	5	53.19	1.68
9	66.49	5	72.15	8.51
10	51.43	8	52.54	2.15

TABLE 2 (end)

Gold content (g/t)

Stream #	Measured gold g/t	RSD %	Corrected gold g/t	Adjustment %	Relative Mass Flowrate d	Stream #
1	1.7685	15	1.56	11.79	2.205	1
2	1.8603	15	1.76	5.39	0.909	2
3	3.9791	10	3.98	0.00	0.909	3
4	1.1604	10	1.15	0.90	0.212	4
5	0.4672	15	0.45	3.68	0.870	5
6	1.3097	10	1.07	18.30	0.870	6
7	0.3931	10	0.39	0.79	0.095	7
8	0.2086	10	0.23	10.26	0.880	8
9	0.1145	10	0.11	3.94	0.386	9
10	0.0459	15	0.05	8.93	0.903	10
11	2.7946	10	2.91	4.13	1.059	11
12	0.5506	10	0.60	8.97	1.064	12
13	0.5506	10	0.60	8.97	0.071	13
14	0.5506	10	0.60	8.97	0.993	14
15	0.0000	10*	0.61	-	0.789	15
16	0.8175	10	0.49	40.07	2.205	16
17	1.7351	10	1.97	13.54	2.427	17
18	0.0792	20	0.08	1.01	2.427	18
19	0.0792	20	0.08	1.01	0.362	19
20	0.0792	20	0.08	1.01	0.659	20
21	0.0792	20	0.08	1.01	0.289	21
22	0.0792	20	0.08	1.01	0.785	22
23	0.0792	20	0.08	1.01	0.295	23
24	0.0792	20	0.08	1.01	0.037	24
25	0.0000	0.001*	0.0000	-	0.517	25
26	0.0000	0.001*	0.0000	-	0.424	26
27	16.4590	50	4.98	69.75	1	27
28	3.7330	25	2.61	30.08	1	28
29	3.5036	20	2.80	20.08	1	29
30	0.0000	10*	0.78	-	1	30
31	1.2930	20	1.10	15.03	1	31
32	1.2930	20	1.01	15.03	1	32
33	0.5839	15	0.46	21.22	1	33
34	1.3764	20	0.74	46.24	1	34
35	0.4171	15	0.64	53.44	1	35
36	0.4171	15	0.34	18.48	1	36
37	0.3545	25	0.34	4.09	1	37
38	0.0000	10*	4.59	-	1	38

* Absolute standard deviation

A few points are worth mentioning in the results. The very high correction applied to run of mine ore grade is typical of the problems encountered in sampling coarse gold ores. The sample mass was too small to provide a reliable figure. Water loss from the plant (as barren solution) was small (3.7% of ore flowrate), indicating that the data, overall, was of fairly good quality.

Application to Metallurgical Performance Assessment

The production of corrected, balanced values for all solutions and solids flowrates, as well as for gold analyses, is a valuable tool in the assessment of a gold processing plants. Efficiencies of various equipments, recoveries in various parts of the circuit can be evaluated in a safer fashion than for raw data, since the materials balance programme takes into account all data, in a systematic manner. In the case of the DOYON plant, for example, the overall plant recovery is estimated, from the result of Table 2, at 0.922, whereas the accounting method used by the plant operators is 0.921, and a proper balance to take into account actual production is 0.918. Such differences are small, and in the course of the work that has been done for DOYON, it has been found that the differences in overall recovery for the conventional procedure and the results of BILMATOR (six sampling campaigns), are small. Considering the economic impact of such differences, they are worth investigating, since a 0.5% difference in recovery for a 1000 tpd plant processing a .15 oz/t ore means US $ 120 000 per year!

Detailed and precise information can be obtained on the behaviour of filters. For example it could be claimed, from the comparison in the gold content in solids, as sampled, before and after filtration, that the filters act as a good dissolution devices. The corrected values enable to obtain a better picture of the efficiency of the filters. White (1976) has suggested to use the following criteria for filter washing efficiency, and for filter cyanidation efficiency:

$$\text{Washing efficiency} = (C_A + C_L - C_S)/C_A$$

where the C stand for concentration in solutions for: A, feed to the filter, L in the wash water and S in the filtercake solution. The experimental values are found to be: 72.8, 76.0 and 83.0, respectively for the first, second and third filter. The corrected efficiencies are 73.1, 81.3 and 87.0. The value of these efficiencies should be assessed by looking at the proper load of each filter (which is obtained from the materials balance results), and the flowrate of wash water. It provides a very good tool for plant metallurgists to decide on the actions to take for any piece of equipment.

Comments

During the course of the development of the BILMATOR programme, various sampling strategies were tested in order to minimize the number of streams to sample and analyze. The most significant parameter to measure, apart from the measurements made on a routine basis and listed above, is the wash water flowrates on filters. All sampling campaigns in which this measurement was omitted gave poorly reliable results. The importance of experimental errors on the estimation of wash water flowrate, when not measured, can be seen from Fig. 3, in which a working equation is presented for the direct evaluation. The authors would like to stress the need, in the application of any materials balance programme, to have flowrate measurements; this requirement is of particular importance for gold processing plants in the filtration areas.

CONCLUSIONS

A materials balance programme has been developed for gold processing plants; it provides corrected values for all experimental measurements, which satisfy the mass balance equations, and in which experimental values are weighted according as a function of their reliability. The application of the programme enables metallurgists to have access to a systematic procedure for the production of statistically coherent data, and thus to prepare metallurgical reports which have more reliable figures. It also provides estimates of all streams characteristics, and is a valuable tool for the identification of unit operation efficiencies.

REFERENCES

Cimon, D., (1987), Bilans matières dans les usines de traitement des minerais d'or, M.Sc. Thesis, Université Laval, July 1987.

Hodouin, D. and Everell, M.D., (1980), Int. J. Mineral Processing, 7, 91-116.

Hodouin, D., Gelpe, T. and Everell, M.D., (1982), Powder Technology, 32, 139-153.
Hope, G., (1984), Proc. 16th Ann. Meet. Canad. Miner. Proc, Ottawa.
Gy, P., (1979), Sampling of particulate materials - theory and practice, Elsevier, Amsterdam
McMullen, J., (1984), Traitement des minerais d'or, CRM, Québec, pp. 115-129.
White, D.A., (1976), Chem. Eng. Sci., 31, 419-425.

PROCESS MINERALOGY OF GOLD ORES - A REVIEW

Paul R. Mainwaring
Mineral Science Laboratories
CANMET
Ottawa, Ontario
Canada, K1A 0G1

Process mineralogical studies of gold deposits focus attention on those factors which will influence the ultimate recoveries of gold from the ores. The availability of automated image analysis systems interfaced to electron microprobes has made it possible to perform systematic mineralogical examinations of a deposit before any extensive metallurgical testwork is undertaken. This can be particularly significant for those deposits in which mineralogical zoning is thought to occur. These instruments can determine bulk mineralogy, mineral proportions, grain size of valuable minerals and their distribution. In addition, data on mineral associations can be provided to the metallurgist.

Refractory gold ores present unique problems for both the process mineralogist and the extractive metallurgist due to specific minerals or mineral assemblages present in the ore or due to fine grain size and distribution of the valuable minerals. The latter case is now coming under intense study in order to assess the role of solid solution of gold in pyrite and arsenopyrite. Indications are that both these minerals are capable of carrying small but volumetrically important amounts of gold in solid solution and should be considered in the development of pretreatment steps in the processing of gold ores.

OPTIMISING CARBON PROPERTIES FOR USE IN CIP

Authors: D. McArthur, C.G. Schmidt and J.A. Tumilty

Anglo American Research Laboratories, Johannesburg, South Africa.

ABSTRACT

Properties of activated carbon were investigated with respect to their effect on the use of carbon in a carbon-in-pulp circuit. Particular regard was paid to determining the properties to be used for specification of virgin carbon for use in a circuit. The roles of these properties in determining the efficiency of thermal regeneration were also studied.

It was found that activated carbon is not a homogeneous product. Carbon activity is decreased by attrition. This is due to the differing resistance to attrition of carbon particles of differing activity. That is, the higher the activity of the carbon, the softer it is, hence the carbon lost first due to attrition, is the softer, more active portion.

Thus, if high activity carbon is used in a CIP circuit, its activity is degraded by attrition in the circuit to an extent such that it has the activity of the harder component of the carbon. Therefore, when selecting carbons, the activity is measured after a period of attrition and the selection criterion is such that the attrition resistance should be equivalent to that of a plant carbon. When selecting amongst carbons which have passed the attrition test, the carbon with higher activity after attrition is selected. The aim of such specifications is reduced carbon losses due to attrition and to base carbon purchases on activity values which will have relevance to plant use.

Thermal regeneration efficiency is usually represented as a percentage restoration of virgin carbon activity, with 100 per cent as the target. Regenerating to this extent causes softening of the carbon and increased attritional losses. Evaluation of regeneration by restoring activity to the maximum possible without causing structural damage that causes increased attrition losses, is recommended.

KEYWORDS

Activated-carbon; carbon-in-pulp; carbon-seletion; thermal-regeneration.

INTRODUCTION

When selecting a proprietary brand of activated carbon for use in a CIP or CIL circuit two, conflicting, properties of the carbon are of importance, namely activity and hardness. Activity is required as this is necessary to allow the most efficient recovery of gold from solution. Hardness is required as the carbon has to withstand the severe abrasive conditions of the adsorption circuit. In order to be able to select the type of carbon most suited to use in the adsorption circuit it is necessary to have testing procedures and purchasing specifications to allow the comparison of carbons from different suppliers. At the Anglo American Research Laboratories (AARL) specification tests for selection of activated carbons have been in use for many years. When these tests were formulated they were based on the then available knowledge of the conditions to be expected on CIP plants. In the intervening years much more information regarding the operation of these plants has been acquired.

Testwork was initiated to allow the formulation of carbon specifications which are related to the carbons performance in CIP circuits and which are designed to provide efficient gold adsorption at minimal carbon consumptions. It has also indicated that thermal regeneration of plant carbons should be evaluated on the same basis.

MEASUREMENT OF CARBON PROPERTIES

Testing methods have been developed at AARL to measure some of the important physical and chemical properties of activated carbons. The relevant tests are listed below and described in detail in Appendix I.

Gold Adsorption Capacity

This gives the gold loading on the carbon in equilibrium with a 1 ppm gold solution after 24 hours of contact.

Gold Adsorption Rate

The rate of gold adsorption is measured by agitating carbon in a 10 ppm gold solution for 60 minutes. The percentage gold adsorption in this period is used as the gold adsorption rate.

Attrition Resistance

This is determined by attriting the carbon by agitation in a tumbling bottle in a water suspension. The amount of undersize material generated (i.e. for an 18 x 16 mesh carbon the amount of minus 16 mesh) is used to express the attrition resistance.

PREVIOUS CARBON SPECIFICATIONS

The carbon specifications used for selection of activated carbons for purchase were based on the tests described above. The tests were conducted on the carbon samples as received from the suppliers (after washing etc.). The specified values were:

gold loading capacity : 25 kg Au/t C
gold adsorption rate : 60% in 1 hour
attrition resistance : less than 2% loss in 24 hours

These specification values were estimates, as insufficient knowledge was available of the behaviour of activated carbons in gold recovery circuits. Experience in the operation of CIP plants has since indicated that applying these values did not provide for the purchase of carbons most suited to the operating conditions. The emphasis on having high activity values was leading to the selection of carbons which were too soft for use in adsorption circuits. New testing procedures and specification values were thus required.

DESIGN OF NEW TESTING PROCEDURES

As a first step in the design of new testing procedures it was necessary to investigate the chemical and physical properties which may affect the carbon's behaviour in a CIP circuit, particularly, the relationship between activity and hardness. It is well known that activity and hardness are inversely related, that is the more active a carbon is made the softer it becomes. It was also noticed that the hardness of carbon particles within a particular batch of virgin carbon varied. This was possibly due to small variations in the manufacturing process, for example differences in the carbon's residence time in the activation kiln. Thus an investigation of variations in the properties of carbon within a single batch was initiated.

The variations in carbon hardness could be explained by differences in the pore columes within the carbon. This is, the softer carbon particles would have had a greater proportion of their structure burned away during the activation process. Thus the softer particles should be less dense than the harder particles. Hence it was proposed to determine the physical and chemical properties of different density fractions from the carbon. This could be achieved by means of elutriation. However elutriation also separates according to shape and size and it was therefore necessary to first remove small and flat particles from the carbon.

Size Separation

This was readily achieved by means of screening the carbon on a square mesh screen. All of the minus 1,19 mm material was removed by screening.

Shape Separation

To separate flat particles out of the carbon it was necessary to develop a method which could separate the carbon according to only one axis. This was done by using a pair of slightly divergent rotating rollers (i.e. with a V-shaped gap between them), Fig. 1. The two rollers are set at about 4 degrees to the horizontal, such that the gap between them varies from 0 to 2 mm. The rollers roll in opposite directions outwards. Carbon is fed onto the rollers where the gap is zero and moves down the rollers towards the 2 mm gap. The action of the rollers is such that it separates the carbon according to its smallest cross-sectional axis. Therefore the carbon falling through the first part of the gap consists of flat carbon particles or 'platelets'. All of the carbon particles which had one axis less than 1 mm were removed using these rollers.

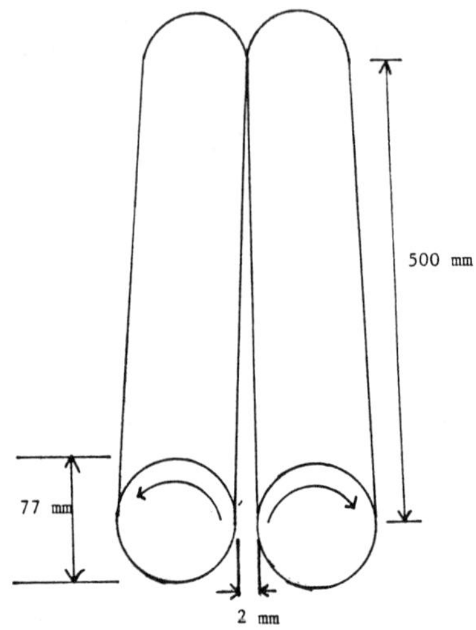

Fig. 1. Shape Sorting Rollers

Density Separation

A suitable device is an air elutriator, Fig. 2, in which particles either rise or descend in a rising airstream according to their mass and size. Subsequent to the above separating stages the carbon was fed into the air elutriator. The air flowrate was set such that about 10 per cent (by weight) of the carbon was collected as a light fraction i.e. that carbon which rose in the airstream. The sink fraction was again fed into the elutriator but at a higher air flowrate such that about 10 per cent of the carbon was collected as a heavy fraction. The physical and chemical properties of these heavy and light fractions were then compared.

EFFECT OF VARIATIONS IN DENSITY ON CARBON PROPERTIES

The properties compared were attrition resistance, gold adsorption rate and gold loading capacity. These were measured using the procedure described below. Results are given in Table 1.

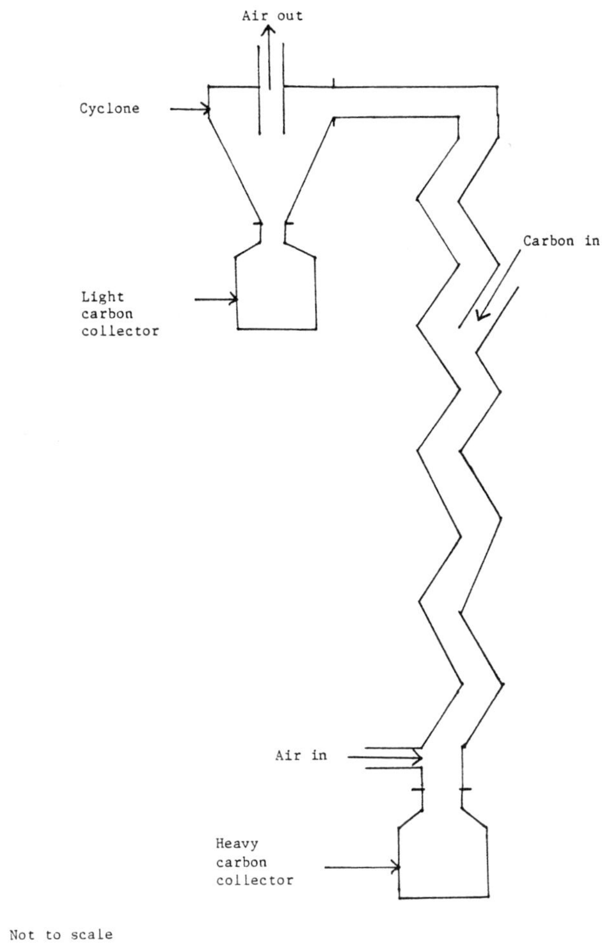

Fig. 2. Cross Section of an Air Separation

TABLE 1. Specification Tests on Light and Heavy Carbon Fractions

Fraction	Capacity kg Au/t C	Adsorption Rate % in 1 hour	Attrition Loss %
Light	25,4	75,8	4,1
Heavy	24,8	35,4	2,1

This showed the light carbon to have a much higher gold adsorption rate but a much poorer resistance to attrition i.e. it is softer, than the heavy carbon fraction. However, the gold loading capacities are similar which shows that these different fractions have similar surface areas available for gold adsorption. As the higher gold adsorption rate for the light fraction indicates a greater proportion of macro and meso pores but also a similar proportion of micropores it is indicated that for the light carbon a greater portion of the carbon's structure has been burned away during the activation process. It is this removal of more of the carbon structure during activation which weakens the carbon particles.

The standard attrition resistance test used at AARL employed a 24 hour contact in a tumbling bottle at 40 rpm. As there was such a big difference in the attrition losses over a 24 hour period an extended attrition test was conducted on the light and heavy fractions described above. The results of these tests are shown in Fig. 3.

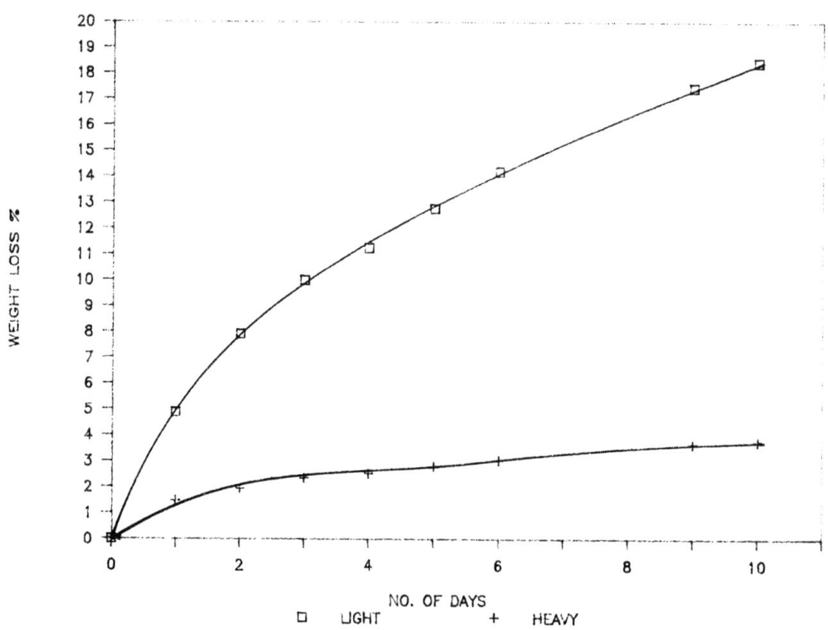

Fig. 3. Attrition of Light and Heavy Carbon Fractions

This shows the marked difference in the attrition resistance (or hardness) of the light, active carbon and the heavy, less active carbon. Whilst there is about 3 per cent less mass loss over 24 hours for the heavy carbon, after 10 days of attrition there is 14 per cent less mass loss for the heavy carbon. It is significant that the rate of loss for the light carbon is much greater at 10 days than that for the heavy carbon thus indicating that if the test had been extended the difference in mass loss would have been even greater.

As the more active component of activated carbon has a lower attrition resistance it follows that the carbon lost first due to attrition is the more active component. Thus if the activity of a sample of virgin carbon is measured before and after attrition there should be a marked reduction in activity. This was demonstrated for several brands of activated carbon. The gold adsorption rate and loading capacity were measured for the as received carbon and after the 10 day attrition resistance test described above. The results, see Table 2, demonstrate a significant reduction in both rate and capacity.

TABLE 2. Activity Measurements Before and After Attrition

	Capacity kg Au/t C	Ads. Rate % in 1 hour	Capacity kg Au/t C	Ads. Rate % in 1 hour
Carbon A	23,2	70,9	19,9	54,6
B	23,3	68,8	21,4	49,6
C	24,2	67,5	21,3	42,2
D	24,9	65,4	23,2	51,3
E	22,1	64,2	20,5	47,8
F	25,7	60,8	23,0	49,4
G	25,1	60,5	24,5	53,5
H	25,4	59,6	24,5	38,0
I	21,2	54,0	17,3	43,8
J	26,9	53,9	22,7	47,3

These results implied that when fresh activated carbon is added to a CIP circuit it is the lighter, more active carbon that is lost first due to attrition. Losing this more active component means that in use the activity of the carbon will drop due to physical losses. When selecting carbons for CIP circuits the activity of the carbon should be determined after attrition. Thus when comparing different carbons the activity measurement after attrition will have more relation to the activity of the carbon in use and by removing the softer component will generate a more homogeneous product for the activity tests.

It is also important to note that the ranking of the carbons change i.e. those carbons with the highest activity before attrition no longer have the highest activity after attrition. This is particularly marked as far as adsorption rate is concerned with some very marked decreases in rate being observed.

NEW CARBON SPECIFICATIONS

As it had been demonstrated that the activity of activated carbons is reduced by attrition, it indicated that the carbons which the existing specifications were selecting had activity values which were too high. Hence a premium was being paid for high activity but this was being lost shortly after introducing this carbon into a CIP circuit. New specifications would have to be designed leading to the purchase of carbons which would not lose such a high proportion of their activity when introduced into the circuit. The basis for this selection was designed on the basis that the attrition resistance of the carbon should be as close as possible to that of the carbon in CIP circuits on the supposition that if softer carbons are added they will lose their softest components first and tend towards having an attrition resistance the same as the carbon already in the circuit.

However, this move towards using harder carbons should as far as possible not cause large reductions in the activity of the carbons used. High activity is still a desirable quality. Hence the criteria for activity should be based on that of the currently available carbons after attrition.

The required value for attrition resistance will be derived by using eluted carbons from operating CIP plants in the Anglo American Group. The attrition resistances of these carbons will be measured using the method described above except that the time of agitation will be increased to 5 days to make the attrition more intensive. From this target value for attrition resistance will be set.

The target values for activity will be based on the activity of virgin carbons with similar attrition resistances to that of the plant carbons. Virgin carbons from several manufacturers will be put through the same test and the activity values will be measured for those carbons which come close to the target value. Those carbons which had higher losses will be resubmitted to the attrition resistance test. Carbons with an acceptable attrition loss on the second attrition will be submitted for activity testing.

In future when adjudicating carbon samples from suppliers the routine followed will be;

An attrition resistance testing, using more severe conditions than previously (more severe either in terms of time or rate of rotation of the tumbling bottle) will first be done.

The gold adsorption rate and loading capacity will then be measured for the attrited carbon.

The data in Table 2 indicate that when selecting carbons on the basis of adsorption rate the proposed new specification test will select different carbons from the old method. This is because carbons which satisfactorily met the previous specification will, after attrition, have much lower adsorption rates.

SIGNIFICANCE FOR THERMAL REGENERATION

It has been a common practice for the efficiency of thermal regeneration of carbons to be evaluated by comparing its activity to that of the original carbons added to the circuit. However, it follows from the above argument that if carbon with too high an activity and hence too low an attrition resistance is used then it is also undesirable to regenerate back up to this degree. If regeneration is conducted to this extent it would become necessary to create new pore structures to replace those lost due to attrition. Such a highly regenerated carbon will lose its high activity when re-introduced into the CIP circuit. This was shown by regenerating a plant carbon to a high degree, measuring its activity, submitting it to an attrition test and measuring the activity of the attrited carbon. Results are shown in Table 3.

TABLE 3. Tests on Virgin, Eluted and Regenerated Carbons

Carbon Type	Capacity kg Au/t C	Ads. Rate % in 1 hour	Attrition Loss %
Virgin	25,7	60,8	5,7
Eluted	11,9	29,1	1,5
Regenerated	14,6	48,0	3,7

This shows that measuring regeneration efficiency by means of activity only, could lead to high carbon losses, as has been reported by some South African CIP plants.

Eluted carbon from a CIP plant was regenerated under a variety of temperatures, ranging from 700 to 900°C. The activity of the regenerated carbon was measured before and after attrition, Table 4. Marked reductions in activity were measured after attrition again showing that on attrition the softer, more active portion of the carbon is lost first due to attrition. Thus whilst high activities can be achieved by the use of more aggressive regeneration conditions this high activity is rapidly degraded by attrition.

TABLE 4. Tests on Regenerated Carbon Before and After Attrition

Regenration Temp. °C	Before Attrition Capacity kg Au/t C	Before Attrition Ads. Rate % in 1 h	After Attrition Capacity kg Au/t C	After Attrition Ads. Rate % in 1 h
725	19,9	54,4	17,0	42,6
750	19,8	52,0	17,8	43,2
800	19,7	54,4	18,0	47,4
850	19,1	56,1	18,6	50,5
900	19,1	58,4	16,8	52,5

It would be necessary to use other criteria as well as the restoration of activity to judge regeneration efficiency. At the AARL it has become routine to not only measure gold adsorption rates for regenerated carbons but the attrition resistance as well. This allows the regeneration conditions to be set such that restoration of activity is gained without causing undue softening of the carbon.

CONCLUSION

It has been found that the activity of activated carbon, as measured by gold adsorption rate and loading capacity, is degraded due to attritional losses. This indicates that it is not an effective strategy to add activated carbons of a very high activity to a CIP circuit. Selection of carbons should be based on hardness values obtained for carbons from an operating plant and the activity of virgin carbons should be measured after removal of the most active portion by attrition. Regeneration of carbon should not be carried out to such an extent that it causes softening of the carbon.

ACKNOWLEDGEMENTS

The work described in this paper was sponsored by the Gold and Uranium Division of Anglo American Corporation of S.A. Ltd. (AAC) and is published with their permission.

APPENDIX I

A DETERMINATION OF ATTRITION RESISTANCE OF ACTIVATED CARBON MESH SIZE 8 x 16

PROCEDURE

1.1 Screen through 1,19-2,38 mm sieves, 150 g of dry carbon.

1.2 Thoroughly wash the carbon collected on the 1,19 mm sieve to remove dust and small particles.

1.3 Dry at 105°C to constant mass. Cool in a desiccator.

1.4 Measure 50,00 g of the dried sample. Place in a "Consol" jar and add 200 mℓ of water. Seal and wedge the jar in a plastic beaker in a top/bottom, side/side manner to ensure an end-over-end rotation.

1.5 Place the sample on a set of rolls. Agitate at a rate of 40 rpm for a period of 24 hours.

1.6 Place a sieve of 1,19 mm mesh size over a 2 litre beaker and transfer the wet carbon, from the jar, on to the sieve. Wash the carbon thoroughly with water. Collect the washings.

1.7 Filter the washings, collected in the 2 ℓ beaker, through a GF/C filter membrane.

1.8 Transfer the GF/C membrane, with the fines of carbon, to the oven and dry, to constant mass, at 105°C.

1.9 Cool in a desiccator and measure the mass.

CALCULATION

Calculate attrition resistance, as carbon loss (fines);

$$\text{Carbon fines (\%)} = \frac{A-B}{C} \times 100$$

Where:

A is the mass of filter membrane with carbon fines (g)
B is the mass of filter membrane (g)
and C is the mass of the dry carbon sample (50 g).

B. DETERMINATION OF GOLD ADSORPTION CAPACITY (K-VALUE) OF ACTIVATED CARBON

PRINCIPLE

The material is ground, and the carbon sieved through 0,43 mm mesh size is collected and tested for gold adsorption capacity.

The actived carbon, after acid washing, is contacted with 100 mg/ℓ Au solution for 20 hours at 25°C.

A five point isotherm, in the range of 0,1 to 0,5 grammes of carbon per 100 mℓ of a 100 mg/ℓ gold solution in the presence of cyanide is plotted.

The gold equilibrium constant is calculated, using the Freundlich isotherm relationship: X/M=kC.

The X/M value defines the quantity of Au (mg) adsorbed on the carbon (g). The X/M value at 1 mg/ℓ equilibrium gold concentration, determined from the isotherm plot, is the loading capacity.

REAGENTS

All reagents are of analytical grade or equivalent. Where water is mentioned, deionized water is understood.

2.1 Sodium hydroxide solutions : 1 mol NaOH/ℓ. Transfer 40,0 g of sodium hydroxide to a 500 mℓ beaker. Add 400 mℓ of water and stir to dissolve. Allow to cool. Transfer to a 1 ℓ volumetric flask and dilute to volume.

2.2 Borate buffer solution : pH 10. Transfer 6,184 g of boric acid (H BO) to a 2 ℓ beaker. Dissolve in 1500 mℓ water. Adjust the pH to a value of 10,0 with the 1 mol/ℓ NaOH solution. Transfer to a 2 ℓ volumetric flask and dilute with water.

2.3 Gold standard stock solution : 100 mg Au/ℓ. To a 1 ℓ volumetric flask, add 0,1463 g of K Au(CN), previously dried to constant mass. Dissolve and dilute to volume with the buffer solution.

2.4 Potassium aurocyanide solution : 100 mg Au/ℓ. To a 1 ℓ volumetric flask, add 0,1463 g of K Au(CN). Dissolve in the buffer solution. Immediately prior to use, add 200 mg of sodium cyanide. Mix to dissolve and dilute to volume.

2.5 Hydrochloric acid solution : 3% (v/v) HCl. Measure 30 mℓ of concentrated HCl into a 1 ℓ volumetric flask and dilute to volume with water.

PROCEDURE

3.1 Using a rotary sample splitter, select a mass of about 100 g of the activated carbon.

3.2 Grind the carbon, using a mechanical mortar and pestle.

3.3 Screen the sample through a 0,43 mm sieve, and collect on a receiving pan.

3.4 Transfer the collected carbon to a 400 mℓ beaker, washing the adhering material from the sieve with water. Add 100 mℓ of the hydrochloric acid solution. Boil for 10 minutes. Decant, and wash the carbon three times with water.

3.5 Transfer the carbon to an oven and dry at 105°C to constant mass. Cool in a desiccator.

3.6 Accurately measure 0,1000, 0,2000, 0,3000, 0,4000 and 0,5000 g of the sample and transfer each increment to a 250 mℓ Erlenmeyer flask.

3.7 To each flask, add 100 mℓ of the potassium aurocyanide solution. Stopper the flasks.

3.8 Place the Erlenmeyer flasks on a linear shaker-incubator. Adjust the temperature to 25°C and the speed to 130 oscillations/minute. Commence shaking.

3.9 Continue the agitation of the flasks for 20 hours.

3.10 Filter the contents of each flask through a Whatman No. 542 filter membrane, into a 100 mℓ glass bottle. Discard the first 10 mℓ of the filtrate.

3.11 Determine the equilibrium concentration of gold in each solution by direct automatic absorption measurement.

3.12 Determine the Au concentration of the potassium aurocyanide solution with each batch of carbon samples.

CALCULATION

To determine the gold adsorption capacity of the activated carbon use the empirical. Freundlich isotherm log X/M=log kc.

Where X is the mass of gold adsorbed on the carbon (Mg0)
M is the mass of carbon taken (g)
C is the equilibrium gold concentration in the filtrate (mg/ℓ).
k and n are constants

The X/M value at 1 mg/ℓ equilibrium concentration, determined from the isotherm plot is the loading capacity.

C. DETERMINATION OF GOLD ADSORPTION RATE OF ACTIVATED CARBON MESH SIZE 8 x 16

PRINCIPLE

The activated carbon is contacted with a known low gold concentration solution at 25°C, in the presence of sodium hydroxide and sodium cyanide.

At periodic intervals, small aliquot portions of the solution are taken, and gold content is determined by the solvent extraction procedure.

The rate of gold adsorption is obtained from a calibration graph of per cent gold adsorbed on the carbon plotted against time.

PROCEDURE

2.1 Transfer 5 g of the wet carbon to the top of a 1,68 mm sieve. Screen through and collect on a 1,41 mm sieve.

2.2 Transfer the carbon, collected on top of the 1,41 mm sieve, to a 100 mℓ beaker and cover with water (50 mℓ is sufficient).

2.3 Position the beaker with carbon in an ultrasonic bath.

2.4 After 2 minutes, decant the water and wash the carbon twice. Repeat the procedure four times.

2.5 Transfer the carbon to an oven and dry at 105°C to constant mass. Cool in a desiccator.

2.6 Measure, accurately, 1,000 g of the dried carbon sample. Moisten the sample with 1 mℓ of the buffer solution (see B 2.2, above), and allow to stand for 30 minutes.

2.7 Allow the 10 mg/ℓ standard gold solution to stand for 30 minutes in an incubator at 25°C, to adjust the temperature of the solution.

2.8 Measure 1 ℓ of the gold solution into an oblong plastic container (150 x 150 x 140 mm), and add the prepared carbon sample.

2.9 Position the container on a linear shaker-incubator, the temperature (25°C) and speed (130 oscillations/min) of which should have been adjusted previously. Commence shaking immediately.

2.10 Withdraw 5 mℓ aliquots after 0, 15, 30, 45, 60, 75 and 90 minutes.

2.11 Proceed with the determination of gold using a solvent extraction procedure.

CALCULATION

3.1 The results from the solvent extraction will give the residual gold in the solution (mg/ℓ Au) at the time of the aliquot withdrawal: Value B.

3.2 Subtract the residual gold concentration from the original gold concentration (10 mg/ℓ Au). Calculated, in per cent, the quantity of gold adsorbed on the carbon at any given time:

$$\text{R value (\%)} = \frac{(A-B) \times 100}{A}$$

Where A is the Au concentration (mg/ℓ) of the standard solution
and B is the residual Au concentration (mg/ℓ) at the time of aliquot withdrawal.

3.3 Plot a calibration curve of percentage gold adsorbed, against time, to obtain the mean percentage of gold adsorbed onto the carbon after 60 minutes.

Technical Sessions

Session No. 2

Plant Practice 1

GOLD RECOVERY AT DETOUR LAKE MINE

D. Rollwagen*, P. Kresin** and C. Lam***

* Project Metallurgist, Lakefield Research, Lakefield, Ontario, Canada KOL 2HO, (formerly Sr. Metallurgist, Detour Lake Mine)

** Mill Superintendent, Bell Creek Mine, Canamax Resources Inc., Porcupine, Ontario, PON 1CO (formerly Mill Superintendent, Detour Lake Mine)

*** Senior Metallurgist, Detour Lake Mine, Timmins, Ontario, Canada P4N 2S9

ABSTRACT

Detour Lake Mine is located in northeastern Ontario (50°00'N, 79°40'W)not far from the Quebec border. Campbell Red Lake Mines Limited has, under a joint venture arrangement, a 50 percent working interest in, and is operator of the Detour Lake Mine. The remaining 50 percent interest is held by Amoco Canada Petroleum Company Ltd. The paper will discuss the start-up, modification and operation of the gold recovery circuit. This includes crushing, grinding, leaching, CIP, carbon stripping, electrolytic gold deposition and refining to dore'.

KEYWORDS

Optimization, C.I.P., Carbon Stripping, Electrolytic gold deposition, Refining, Gold Recovery

INTRODUCTION

Detour Lake Mine is located 200 km northeast of Timmins, Ontario and, 13 km west of the Quebec border. Employees are flown to and from work from Timmins and Cochrane. Non supervisory and first line supervisors work a seven day in - seven day out schedule while other staff work a four and three day schedule alternate weeks. The history of the mine now encompasses 13 years of exploration, development, construction, start-up and operation. Amoco Canada Petroleum Company Ltd. first discovered copper and gold in early 1974. Exploration of the property by Amoco Canada Petroleum Company Ltd. continued until October 1979 when Campbell Red Lake Mines and Dome Mines Limited entered into a joint venture agreement with Amoco Canada Petroleum Company Ltd. Campbell Red Lake and Dome Mines were to earn 25% each with the remaining 50% going to Amoco. By early 1981, Campbell Red Lake and Dome Mines had completed extensive underground exploration, metallurgical testing and feasibility studies. At this time, the decision was made to bring the property into production at a rate of 2000 (later increased to 2,500) tonnes per day by August 1, 1983. On December 29, 1982, Campbell Red Lake Mines acquired Dome Ltd's 25% working interest in Detour Lake Mines giving Campbell Red Lake, as operator, a 50% working interest in the mine with 50% still remaining with Amoco Canada Petroleum Company Ltd. Both companies equally shared the capital expenditure of 136.088 million dollars required to bring the property into production. Crushing of waste rock commenced July 8, 1983 with the grinding circuit turning over on July 17, 1983. Commercial production started on November 1, 1983 milling open pit ore, making Detour the largest Carbon-in-Pulp gold mill operating in Canada at that time. (Wright and Ward, 1981)

GEOLOGY AND PROCESS SELECTION

Geology. Gold at Detour is primarily associated with quartz and sulphide mineralization. Significant quantities of pyrrhotite, pyrite and chalcopyrite are present.

The Main Zone where most of the ore reserves are located, is hosted in a cherty iron sulphide formation at the interface between basaltic and ultramafic rocks. Gold is also found associated with sulphide mineralization in the ultramafic rocks and with quartz veining in the basaltic rocks.

Process selection. The dense, competent nature of the gold bearing ore was found to be compatible to semi-autogenous grinding. Due to the location of the minesite the most economical method of gold recovery was investigated and found to be gold cyanidation followed by Carbon-in-Pulp technology. A mill flowsheet consisting of a single grinding unit followed by leaching and C.I.P. circuit presented a compact concentrator layout at favourable capital cost.

PROCESS DESCRIPTION

The current flowsheet used for gold recovery illustrated in Figure No. 1 has undergone significant change since start-up. Modifications to the unit processes and equipment are discussed below.

Crushing

Description. Crusher feed is trucked from the pit directly or from the stockpile in 50 tonne capacity Euclid trucks. A 1.524 x 4.266 m (60" x 14') Hydrostroke reciprocating feeder feeds a 1.220 x 1.520 m (48" x 60") Kue Ken crusher set at 165 mm (6.5") on the closed side setting. The crusher is capable of a maximum throughput of 500 tonnes per hour at a 7" closed side setting. The crushed ore is transferred by a 1.067 m (42") transfer belt to the primary haulage conveyor which carries the mill feed over 200 meters at an incline of 13.4% to feed the coarse ore storage bin. (Kilborn Ltd. May 1983)

Modifications. Improvements in the crusher area include an enclosed heated control room, television monitored jaw opening and improved dust control measures. An interior steel lining was added to the coarse ore bin to reduce wear on the concrete sides and the number of feeders located in the base was reduced from four to two. The two feeders were also positioned to maintain a "centre" drawdown on the bin and minimize wall wear. The feeder chutes were also enlarged to handle oversize slabby material.

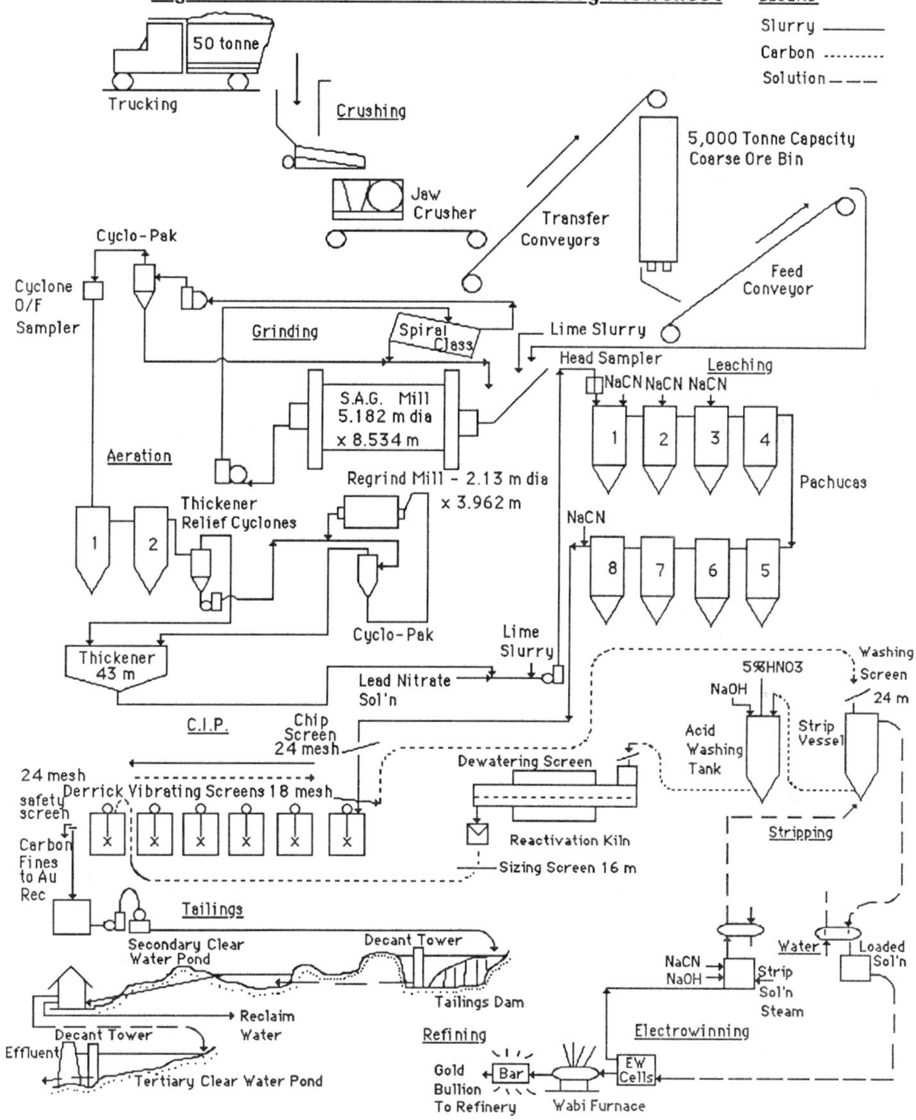

Fig. 1 - Detour Lake Mine Gold Recovery Flowsheet

Grinding

Ore is withdrawn from the coarse ore bin by two centrally located, variable speed, 1.067 x 3.048 m (42" x 10') Forano vibrating feeders. It is then conveyed by means of a 1.067 m (42") wide belt over a weighscale to allow automatic regulation of the feedrate by the Rosemount-Diogeres Controller.

The conveyor discharges to the feed chute of a 5.182 m diameter x 8.534 m long (17' x 28' long) Allis Chalmers semi autogenous grinding mill driven through an air clutch by a 3500 Hp 200 rpm synchronous motor. The mill rotates at 14.19 rpm or approximately 75% of critical speed. The mechanism of grinding is a combination of crushing and attrition grinding. Grinding throughput is highly dependent on the size and competency of the mill feed. While fines (-1") increase mill tonnage as expected, intermediate size material (2-4") has proved to be hard to grind. A change of feed from basaltic to talcose material is accompanied by dramatic increase in tonnage. The steel load is normally set at 12-13 percent mill volume with 15 percent being the upper operational limit for optimum throughput.

Mill discharge slurry is classified by a 1.22 m dia. x 8.128 m (4' dia. x 26'8") spiral classifier which typically runs with a circulating load of between 25-50%. The classifier overflow is cycloned using 3 or 4 -280 mm dia (15") cyclones. Cyclone overflow with a K80 of 80 microns is pumped to aeration, while the cyclone underflow returns by gravity to the SAG mill at a circulating load ratio of 100-200%.

Grinding throughput has been below the expected 2500 t/d level, but has adequately met the original 2000 t/d criteria. Optimization of existing equipment is ongoing. A summary of the grinding data since start-up has been tabulated in Table No. 1.

TABLE 1 - Grinding Data Summary

Period	Dry Tonnes Milled	Milling Rate Tonnes/Hour	% Utilization	Operating Work Index (Wio at Pinion)	Steel Cons. Kg/t
July-Dec '83	265,693	95.2	76.0	-	1.019
1984	794,561	104.2	86.8	28.09*	1.066
1985	813,039	102.2	90.8	29.32	0.869
1986	783,357	100.4	89.0	28.90	0.979
Jan - April/87	249,324	90.4	95.8	31.41	0.843

* April-Dec. 1984

Modifications. Improvement to the semi-autogenous mill has focussed on the wear part design to increase utilization and reduce costs. The use of higher lifters, rubber liners with built-up wear bar on the shell and changes to the discharge grate slot design and geometry have increased the life of all wear parts to give a 90% utilization for 1986. The use of ceramic wear parts for cyclones helped reduce maintenance requirements and produce consistent fine grinds.

Aeration, Regrinding and Thickening

Aeration. Cyclone overflow product from grinding which s sampled on the upstream from the aeration pachucas is compared to future leach feed samples representative of the same stream. Lime slurry is added to the grinding circuit to control the pH in aeration to 10.6.

Two aeration pachucas each 10.1 m dia. x 15.7 m (33' x 51.5') high with a 60° cone receive the cyclone overflow slurry at approximately 45 - 48% solids. A total retention time of approximately 12 hours raises the dissolved oxygen content from 1.0 ppm or less to between 3-4 ppm at the overflow. Aeration overflow slurry feeds two 500 mm (20") thickener relief cyclones.

Regrinding. The original purpose of the relief cyclones was to reduce the load on the thickener to avoid sliming. These cyclones allowed 25 tonnes of solids per hour to pass directly to leaching and sent the overflow of fines to the thickener for dewatering. Thickener underflow was pumped back to the leach feed pump box.

Since early 1985, the relief cyclone underflow was re-routed to be reground in a 2.133 m x 3.962 m (7' x 13') regrind mill (previously installed as the Cu Rougher Concentrate Regrind mill) for further gold liberation. The regrind mill is driven at 23 rpm (77% critical speed) via a gear reducer by a 300 Hp, 1800 rpm induction motor. Four 250 mm (10") cyclones classify the combined regrind circuit feed and mill discharge. The regrind cyclone overflow and relief cyclone overflow are dewatered in the 43 m (143') thickener. The thickener underflow (leach feed) is typically 89% -75 μm.

Thickening. The pre-leach thickener, receiving the relief and regrind cyclone overflows functions as a surge tank to buffer the leach circuit from surges in grinding tonnage. The thickener being 44 meters in diameter (143') requires only a pH 10.5 with lime to assist settling. Thickener underflow discharges by gravity to either of two variable speed 125 mm x 100 mm (5" x 4") SRL pumps. The pump speed is adjusted automatically to maintain a leach density of 48% solids. Thickener overflow flows by gravity to a 50 cubic meter process water tank for re-use in the mill circuit.

Cyanidation

Lime slurry is added to the thickener underflow to maintain a pH of 10.6 during cyanidation. Lead nitrate solution (10% w/w) is added to leach feed at a predetermined dose. Metallurgical lab tests confirmed that the use of lead nitrate at 0.1-0.2 kg/tonne reduced sodium cyanide consumption and increased gold extraction for higher grade ores with corresponding higher sulphide content. Lead nitrate at a dose of 0.05 kg/tonne is maintained on all other ore to minimize cyanide consumption.

The leach feed slurry is pumped via a 200 mm x 150 mm (8" x 6") slurry pump through the head sampler to the first leach pachuca. Sodium Cyanide solution (15% w/w) is fed by gravity to as many as five addition points throughout the leach train in order to maintain pre-set free cyanide levels.

Cyanidation takes place in eight air agitated pachucas 10.1 m dia. x 15.7 m (33" x 51.5') high with a 60 ^{o}Cone. At a feed rate of 100 tonnes per hour retention time is approximately 6 hours per pachuca. During the initial start-up period cyanidation was carried out with and without preaeration. Various cyanide and free lime levels were also tested at various tonnage rates and grinds to optimize gold recovery. As a result the leaching circuit retention time was extended to 48 hours by converting two of the four aeration pachucas to leach vessels. As leaching progresses the free cyanide level is allowed to fall to a minimum of O.15 g NaCN in the CIP circuit feed.

The residue from the second leach pachuca was adopted as an indication of leach performance. A set of leach parameters was developed to provide optimum leaching under a variety of ore conditions based solely on leach performance. Table No. 2 below gives a current example.

TABLE 2 - Leach Control Parameters

#2 Pachuca Residue Assay Au g/t	Lead Nitrate Add'n Rate Kg/t	Minimum Cyanide Levels (g/L) #1	#2	#3	#4	#8
Less than 0.5	0.05	0.40	0.40	0.30	0.30	0.15
0.5 - 1.0	0.075	0.50	0.50	0.40	0.40	0.15
Above 1.0	0.150	0.55	0.55	0.55	0.50	0.15

Data for 1984, 85 and 86 indicate that consistent low leach residues can be achieved at satisfactory cyanide consumptions. Table No. 3 gives a summary of the extraction data.

TABLE 3 - Leach Extraction Data

Period	Head Assay Au, g/t	NaCN Consumption Kg/t	Leach Tail Residue Assay Au, g/t	Leach Extraction %
1984	3.31	1.83	0.203	93.87
1985	3.35	1.95	0.199	94.06
1986	3.65	1.70	0.200	94.53
Jan - Apr'87	2.65	1.41	0.129	95.13

Reagent consumption. Typical reagent consumption for cyanide, lime and lead nitrate are 1.60, 1.30 and 0.08 kg/tonne respectively.

Carbon-In-Pulp

Process description. Leach slurry from cyanidation flows by gravity to feed a 1.22 m x 2.44 m long (4' x 8') Derrick vibrating screen for removal of chips and plastic prior to entering the C.I.P. adsorption contactors. The chip screen has three panels of 24 mesh screen cloth and originally discharged the chips by gravity to the tailings agitators. The adsorption circuit is comprised of six contactors, airlifts and corresponding Derrick vibrating screens. Each contactor is 5 m dia. x 10 m (16.4' x 32.8') high providing a slurry retention time of about an hour. Continuous transfer of slurry and carbon to the screen is achieved by external air lifts. Since the slurry height in the tanks must be kept constant for optimum operation of the draught tubes, the airlifts are fed by overflow from the contactors. Each contactor has a 1.22 m x 2.44 m long (4' x 8') Derrick screen, angled at 15° from horizontal, and fitted with three panels of 18 mesh stainless steel cloth located above it to separate slurry from carbon. Counter-current flow of carbon is attained by intermittent diversion of the carbon flow to the immediate upstream tank. Slurry from the last contactor flows by gravity over the safety screen (24 mesh) for fine carbon recovery. The fine carbon is stored in the spare strip vessel and stripped when full. Slurry from the safety screen proceeds to the tailings pump box for pumping to the tailings pond. Loaded carbon is educted at a rate averaging 1.1 tonnes per day over a washing screen into the strip vessel.

Soluble gold loss. Dissolved gold losses in the C.I.P were considerable during the start-up phase. Most of these losses were attributable to initial operating procedures and problems encountered with plant equipment. Frequent shut-downs and overflowing contactors created a metallurgical nightmare. Figure 2 displays the soluble loss data from August 1983 to the end of 1986. Improved procedures brought the average soluble loss to 1.1 percent in 1984, 0.47 percent in 1985 and 0.53 percent in 1986.

Fig. 2 Soluble Loss Data

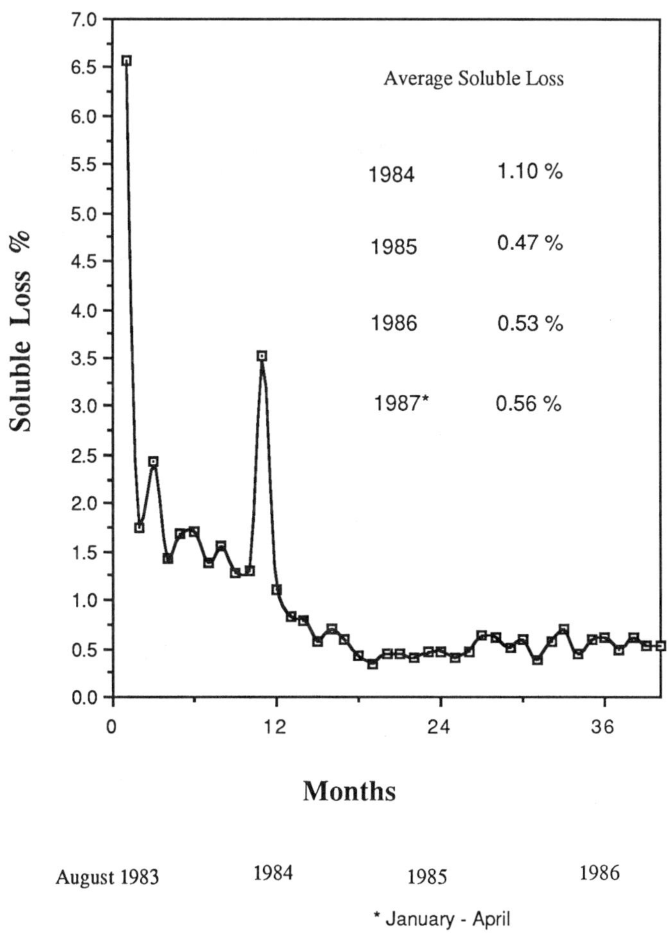

Effect of dissolved copper on C.I.P. The effect of dissolved copper on gold adsorption by carbon is well recognized (Nicol and Fleming, 1984). With adequate free cyanide present in the CIP slurry (0.15 g/L solution), copper adsorption on carbon is inhibited and gold loading remains satisfactory. The importance of maintaining adequate free cyanide in the CIP circuit with high solution copper assay was illustrated by the CIP circuit upset in June 1984. Copper dissolution in the leach circuit was significant (20-30%) with CIP feed solution assaying upwards to 500 mg/L Cu. Without adequate free cyanide content in the CIP circuit and high copper content in CIP feed solution (300-500 mg/L), the entire CIP circuit carbon inventory was loaded with copper. Gold adsorption was impaired resulting in soluble loss ten times the normal levels. When the copper contaminated carbon was stripped, the pregnant strip solution copper level contained up to 2,000 mg/L, five times the present normal amount. Free cyanide level was subsequently increased to 0.25 g/L in the CIP circuit. To remove the copper from the CIP carbon inventory, stripping sequence was accelerated and electrowinning barren solution was extensively bled off. Ten stripping cycles were required to restore the CIP circuit performance. Subsequently, loaded carbon samples are now assayed for copper to flag potential copper interference on CIP operation. Typical copper assay is <0.1% on loaded carbon.

C.I.P. screen wear. The separation of carbon and slurry using external vibrating screen decks causes the screens cloth to wear significantly. An understanding of the factors affecting that wear must be considered in the light that changes may affect the gold recovery capability of the circuit. Factors affecting the screen wear included carbon concentration, milling tonnage, type of screen cloth, deck maintenance and operator care. Emphasis was shifted to operator training and increasing the screen cloth wire thickness in order to improve screen life without affecting the process flow. Plant trials on alternative screen cloths were conducted involving box weave screens similar to that of the 20 mesh screen but with heavier gauge wire. The screens tested were the 16 mesh and 18 mesh screens. Table 4 summarizes the plant trial data and operational data on screen consumption.

TABLE 4 - Screen Consumption Data

Screen Mesh	Period	Wire Diameter mm	Opening Size mm	% Open Area	Consumption screens/1000 tonnes milled	Remarks
20	Aug 1-Dec 1 1983	0.406	0.86	46.2	0.495	Start-up
20	Oct. 1984	0.406	0.86	46.2	1.35	Operating data
16	April 1984	0.711	0.89	30.5	-	Test cancelled due to plugged screen
20	Jan. 1985	0.406	0.86	46.2	1.27	Plant Trial
18	Jan. 1985	0.508	0.90	41.1	0.46	Plant Trial
20	Jan-Mar 1985	0.406	0.86	46.2	1.35	Operating data
18	Apr - Dec 1985	0.505	0.90	41.1	0.682	Operating data

As indicated in Table 4 testwork involving the 16 mesh box weave was stopped as it tended to plug with carbon fines and then allow carbon and slurry to flow over the end of the screen. The plant trial involving the comparison of 18 mesh to the initially installed 20 mesh compared favourable with the available operating data. The plant trial with 18 mesh screens indicated a consumption that was 63.8 percent less than the 20 mesh consumption. After the 18 mesh screens were phased in the actual consumption was 49.5 percent less than the previous 20 mesh consumption.

Carbon levels and consumption. The concentration of carbon in the adsorption circuit followed initially the design parameter of 15 g/L (slurry) and was then modified to meet the needs of the circuit. Table 5 summarizes the adjustments made in light of operating problems and influences from outside the C.I.P. circuit.

TABLE 5 - Operating Carbon Levels in the C.I.P.

Period	Carbon Concentration in CIP g/L slurry 1	2	3	4	5	6	Reason for new Levels
Aug 1 - Dec 1 '83	15	15	15	15	15	15	- Design parameters
Dec 1 '83 - June '84	20	20	20	20	20	20	- To reduce soluble gold loss
June 1'84 - Jan1'85	25	23	23	23	23	23	- To reduce soluble gold loss
Jan 1'85 - June 1'85	15	15	20	20	20	30	- To maintain good soluble loss and reduce carbon consumption
June 1'85 - Dec 1'85	10	10	20	20	20	20	- To reduce gold inventory in CIP and hold soluble loss stable
Dec 1'85 - June '86	15	15	23	23	23	23	- To adjust for increasing soluble loss.
June '86 - April 87	15	15	23	23	30	30	- To control soluble loss.
Apr. 1987	10	10	15	15	20	20	- Lower tonnage with lower grade ore

Carbon consumption was higher than anticipated due both to the method of carbon-slurry separation and the higher carbon concentrations. Improvements in carbon consumption were primarily due to improved operating conditions in the area of air lifts and losses from the last contactor to the safety screen. Table 6 displays a summary of carbon consumption since start-up.

TABLE 6 - Carbon Consumption

Period	Carbon Consumption grams/tonne
Start-up to Dec. 31, 1983	118
1984	75
1985	53
1986	65
1987 Jan-April	62

Modifications. Extensive modifications were completed in the C.I.P. area during and after the initial start-up period. The original chip screen discharge to the tailing's box allowed gold solution carryover to tailings along with the chips. In order to correct this situation the discharge line was redirected to wooden collection boxes with a drainage grid. From these any gold bearing solution was allowed to flow back to the leaching circuit via a floor sump. The slope of the chip screen was also reduced from 25 degrees to 15 degrees and an additional spray bar was added to increase gold solution washing from the chips.

The airlifts used to move the overflow slurry from each contactor to the Derrick vibrating screen are U-shaped with two 90° turns. The original design injected air through a 5.1 cm (2") pipe positioned vertically upwards at the second 90° turn. The design was inefficient and did not lend itself to handling surges. Modifications to the airlifts simulated the current practice at Homestake Mines, Nevada. A removable section of 20.3 cm (8") pipe with air injection holes located radially around the pipe is enclosed as an air manifold which receives a high flow of low pressure air (150 Kpa). A pressure regulator allows for air volume adjustments to handle surges. The airlifts discharge CIP slurry into deaeration boxes prior to flowing over the Derrick screens. The original deaeration box design utilized a small closed top, V-shaped box with a 10 cm diameter vent. The box and vent were too small to adequately handle surges while allowing for proper slurry deaeration. The resulting slurry-air mixture was blown onto the screen cloth and out the vent causing splatter and spillage. New deaeration boxes were built, placing a larger box (1.2 m x 1.2 m x 1.4 m high) with a 36 cm diameter vent pipe on top of a V-shaped base to allow for proper deaeration. Subsequently a 36 cm elbow was mounted over the vent to direct slurry or mist downward to the screen surface. With the carbon flow returning to contactors 5 and 6 close to the overflow, up-comer baffles were installed at the overflow to minimize short circuiting.
Slurry leaving the last C.I.P. Derrick screen flows by gravity to the 24 mesh safety screen. The original feed box discharging onto the screen produced a high velocity slurry flow over the screen. To improve slurry drainage, the slope of the screen was reduced to 15° and an original CIP tank Derrick screen feed box installed.

In-tank launder Screens. A 12 foot in-tank screen launder was installed in the last CIP tank in February 1987 to replace the Derrick vibrating screen. The launder was EPAC design, complete with six 18 mesh screen panels each at 38 cm x 104 cm (15" x 41"). Slurry agitation at the screen surface was provided with eighteen bubbler tubes with 3.2 mm (1/8") diameter ceramic inserts. The screen launder has worked well, handling tonnages up to 115 tonnes per hour. Screen wear has been minimal after two months of operation.

Carbon advance was made with an in-tank air lift made of 15 cm (6") steel pipe with a 2.54 cm (1") airline insert. Experience to date has been good.

Conversion of all CIP tank to screen launder practice is being considered and will likely be phased in during the next 2 years.

Carbon Stripping and Reactivation

Description of carbon stripping. Loaded carbon is educted and washed in 2 advances to fill a strip vessel. Two strip vessels are available, each being 1.5 m dia. x 4.0 m (5' x 13') high with a 50° coned bottom. Each vessel is capable of stripping 2.5 tonnes of loaded carbon. Strip solution, composed of recycled electrowinning barren solution and fresh water make-up, is stroed in a 4.0 m dia. x 5.5 m high (13' x 18') stainless steel tank heated by an external steam jacket. Strip solution is maintained at 90°C. Caustic and sodium cyanide solutions added to adjust the solution concentration to 10 and 4 gpL respectively. Steam heat is provided for stripping by a Clayton, oil-fired steam boiler capable of generating steam at 2270 kg/hr. The strip solution is pumped via a 38 cm x 25 cm stainless steel pump through an in-line plate and frame heat exchanger to raise the solution temperature to 145°C prior to entering the strip vessel. The vessel is fed up-flow at a rate of 8.5 Cu m/hr (2.2 B.V./hour) at an operating pressure of 450 Kpa. Hot strip solution discharging from the vessel is cooled by an in-line heat exchanger employing water to reduce the temperature to 90°C to prevent flashing. The loaded or pregnant solution is stored in a tank identical in size to the barren storage tank. Stripping continues for 7 hours at which time steam supply to the in-line heat exchanger is stopped to allow cooling of the vessel prior to washing. The cool down cycle continues for one hour after which the pressure can be released from the vessel. Washing water is sprayed over the carbon bed at a rate of 10 cu m/hr and allowed to drain by gravity to the loaded solution tank. After an hour of washing the cycle is complete and the carbon bed is allowed to drain. Table 7 illustrates a typical strip sequence.

TABLE 7 - Strip Cycle

Procedure	Flowrate cu m/hr	Flowrate B.V./hour	Time hrs	Total Flow	Temperature ºC	Pressure Kpa
Pressurizing	8.5	2.2	0.5	4.25	90	0-450
Stripping	8.5	2.2	6.5	55.25	140-150	450
Cool down	8.5	2.2	1	8.5	90	450
Washing	10	2.6	1	10.0	Ambient	0
			Total	78		
Electrowinning Flow	2.5	-	20	50	70-80	-
Electrowinning Bleed to Leach	2.5	-	11.2	28*	50-60	-
			Total	78		

* Equivalent to fresh water make-up required for next strip cycle, plus washing water

Results. Desorption of gold and silver peaks within the first 5 (19.5 cu m) bed volumes of flow and is almost complete by 10 bed volumes. Copper desorption peaks within 5 bed volumes and levels off to about 150 ppm until the strip temperature is reduced at the end of the strip cycle. Figure 3 displays a typical elution curve.

Fig. 3 Typical Elution Curves From Carbon Stripping

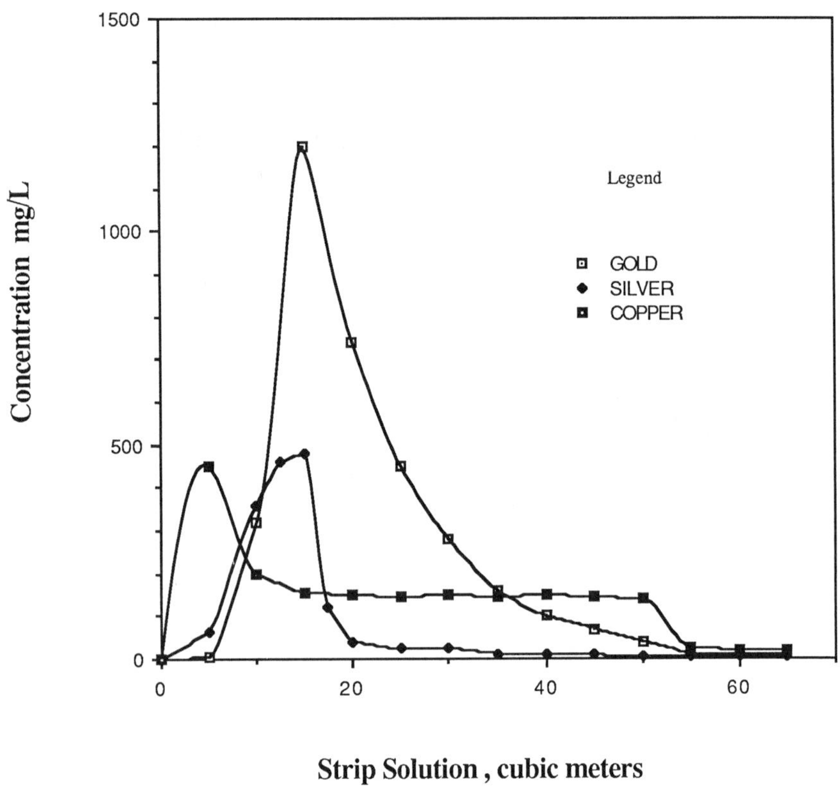

Samples of the loaded and stripped carbon are recovered from each cycle for gold and copper assay. Table 8 displays the strip data recovered for gold since start-up 1983.

TABLE 8 - Strip Cycle Data

Year	No. of Strip Cycles	Average Assay Au g/t Loaded Carbon	Average Assay Au g/t Stripped Carbon	Desorption Efficiency %
1983*	47	N/A	N/A	N/A
1984	164	6244	126	98.0
1985	185	5564	106	98.1
1986	187	5806	182	96.8
1987**	49	5002	111	97.8

* August - December ** January - April

Modifications to stripping. The stripping vessel and associated equipment performed very well once the operators had time to familiarize themselves with the procedures. The internal screens had to be checked every strip cycle due to fine carbon particles and melted plastic. The automatic pressure control worked well. Most modifications were procedural or instrumentation related.

Reactivation. Stripped carbon is educted to the acid wash tank which is flat topped and almost identical in dimensions to the strip vessel. Acid washing with 5% HNO_3 was chosen as a pretreatment to reactivation to remove all acid soluble compounds and in order to maximize the reactivation effect of the kiln. After a thorough acid wash the carbon is water rinsed and rinsed again with a 1.0% sodium hydroxide solution. Fresh acid and caustic solutions are made up as required. When the pretreatment is complete the carbon is educted over a Derrick dewatering screen, fitted with 18 mesh panels, and discharged into the kiln hopper. The kiln hopper holds one tonne of carbon which must be completely drained prior to feeding the reactivation kiln. The kiln, a rotating horizontal tube type, is externally fired with propane and has two heating zones. The first zone raises the temperature to 600°C and creates the stream atmosphere while the second zone (red zone) increases the temperature to 650°C completing the reactivation. The retention time in the red zone is approximately 20 minutes. Extra steam is introduced into the discharge end counter current to the carbon flow displacing the contaminant ladden steam to the feed end exhaust pipe.

Reactivation Results. During 1983-85 activity testwork was conducted using a standard carbon sample with mill pregnant solution to comparing various process carbons. This comparison method indicated a fairly good activity which fluctuated between 80-100% of that of fresh carbon. With optimization of the reactivation system in mind the manufacturer of the process carbon, Calgon, was asked to assist in double checking the carbon activity. Calgon conducted testwork to determine the "R-value" using the "Gold Adsorption Rate Test" (Calgon Corporation, August 1983). This "R value" test was adopted and used as the standard activity test for studying process carbons. This allowed Detour and Calgon to discuss carbon performance in terms that both parties understood. Table 9 displays "R value" data on fresh and various process carbons. The values for reactivated carbons from 1984 and 1985 were significantly lower than that of fresh carbon at 0.1 R value and indicated that the original test procedure was not producing sufficient data to justify the confidence that was being placed upon that activity number.

TABLE 9 - Typical R-Value Data

Carbon Sample	R Value
Fresh GRC-22 (minimum spec)	0.100
Loaded Carbon #1 CIP 1984	0.022
Loaded Carbon # CIP 1985	0.015
Loaded Carbon #1 CIP 1986	0.023
Stripped Carbon + Acid Washed 1984	0.029
Plant reactivated carbon 1984	0.045
Plant reactivated carbon 1985	0.042
Lab tests on stripped and acid washed carbon using steam injection	
- 600°C 40 minutes duration	0.102
- 700°C 40 minutes duration	0.109
- 700°C 25 minutes duration	0.108
Plant results after steam lance installed and operating at 5-10 p.s.i.	0.100

Reactivation results since the steam lance installation indicate the carbon is at, or very close to, fresh carbon levels.

Modifications to the reactivation kiln. Since startup the kiln tube has suffered from blistering and bulging due to insufficiently drained carbon. Tube repairs, modifications to the kiln exhaust, and improvements to the dewatering screen were required to rectify this problem. The steam lance previously discussed was installed and kiln ventilation altered to draw the steam off at the feed end.

Carbon sizing. Reactivated carbon is discharged from the kiln and quenched in a water-filled cooling cone. This carbon is educted over a 0.61 m x 1.8 m long Derrick vibrating screen fitted with 16 mesh screen panels. Sized carbon is recycled to the last C.I.P. contactor.

Fine Carbon Handling. Fine carbon from sizing screens and the stripped fine carbon from the strip vessel is settled in a small single tray 5 meter diameter thickener. Thickener underflow is manually dumped to a screen box for drainage and then shovelled into empty wooden cyanide containers for shipment to an off site refinery. Overflow water is discharged directly to tailings.

Refining and Overall Gold Recovery

Electrolytic gold deposition and refining. Loaded solution is pumped through a series of two electrowinning cells. Each enclosed cell (1.74 m x .93 m x 1.14 m high) contains 7 cathodes and 8 stainless steel screen anodes. Electric current for each cell is set by individual rectifiers at 300 amperes with a voltage of around 2-2.5. Each cell is subdivided by two baffles creating three compartments to inhibit short circuiting and sludge carry-over. Each cathode basket has 2.3 kg of steel wool layered inside. Electrowinning solution is fed at a rate of 2.5 cu m/hr giving a retention time of 60 minutes. From a loaded solution assaying 200-500 grams Au/t electrowinning produces a barren solution between 2 and 5 grams Au/t. Loaded cathodes are regularly pulled from the front end and all other cathodes are advanced counter current to the solution flow. Cathode weights are used to determine flux weights with Borax, Niter and Silica on a 45, 30 and 25 percent basis respectively. Oil fired Wabi furnaces are used for cathode smelting and is usually performed twice per month. Bullion grades vary but are typically 72-77% Au, 20-27% Ag and 1-3% Cu. Regular bleeding of the electrowinning barren prevents copper build-up in the stripping and CIP circuits. Overall plant recovery is based on Bullion Produced, Tailings Losses and the mill circuit inventory change. Gold recoveries are summarized in Table 10.

TABLE 10 - Gold Recovery Data

Period	Recovery %
Start-up - Dec. 1983	86.69
1984	92.77
1985	93.59
1986	94.00
Jan - April 1987	94.57

Plant Services and Tailings Disposal

Plant services for the concentrator. Process air for leach slurry agitation and CIP air lift operation is provided by three Ingersoll-Rand CENTAC compressors. Each compressor is rated at 8753 Nm^3/hr at Kpa (55 psi) of free air required for pachuca and airlift operation. Oil and interstage air cooling for each compressor is provided by a closed circuit cooling water system which operates year round. Plant air is provided by a rotary screw type compressor rated at 833 Nm^3/hr at 760 KPa (110 psi). Instrument air is provided by a reciprocating compressor, complete with a 0.5 m^3 air receiver and dryer. It is rated at 170 Nm^3/hr at 760 kPa (110 psi). Electric power is provided by a 115 KV line from the Ontario Hydro generating station at Island Falls, a distance of 140 km southwest of the minesite. Heating of the concentrator is achieved by a central hot water system which also services all on-site complexes.

Fresh water for process and potable use is withdrawn from Little Hopper Lake and pumped a distance of 6.5 km to the concentrator site storage tank for distribution. Reclaim water is pumped 4 km from the secondary clearwater pond to the concentrator for distribution.

Tailings disposal and cyanide destruction. CIP tailings, at 47-49% solids, are pumped via one of two 250 mm (10") pipelines (Sclairpipe) to a shallow tailings impoundment area about 4 km north-east of the mill site. The impoundment of approximately 140 ha is bounded by an engineered structure on the east and west sides and by naturally rising ground on the north and south sides. The embankments were built sufficient in size to provide 5 years of storage at 2500 tonnes/day. A small polishing pond of 2.4 ha was installed as an add-on to the east embankment and is connected to the main tailings pond via a concrete decant tower and 350 mm (14") Series 160 Sclair pipe. A similar decant structure and piping in the polishing pond allows effluent to flow through the reclaim water pump house (sump) to a small and shallow lake of 112 ha in size.

The mill was designed using the alkaline chlorination process to destroy cyanide from CIP tailings slurry. This circuit consisted of three 4.900 m by 5.500 m high (16 ft. x 18 ft. high) chloriantion tanks; each providing approximately 25 minutes of retention time. Chlorine was added using a liquid transfer system from the evaporators located in a separate chlorine building. This transfer method provided massive problems and was within two months (October 2, 1983) changed to a vacuum chlorine gas transfer method.

With almost every month since start up, chlorine usage in the alkaline chloriantion circuit was consistently increased from 22.7 tonnes in August 1983 to 165.9 tonnes in May 1984. During the same period, monitoring of total cyanide (CN_T) content in the final effluent (polishing pond overflow) showed an increase from 0.4 mg/L to approximately 100 mg/L. Weighted averages (to actual operating hours) for the period November 1983 to July 1984 indicated that the CN_T and copper levels in untreated CIP tailings were 336 mg/L and 199 mg/L respectively. For the same period treated tailings averaged 265 mg/L in CN_T and 151 mg/L copper; thus showing an alkaline chlorination efficiency of 21.1% for CN_T and 24.1% for copper. During the same time period 868.9 tonnes of chlorine and 1,574 tonnes of lime were consumed with a total treatment cost of $572,894 or $1.07/tonne milled.

Because of high cost and treatment inefficiency alkaline chloriantion was abandoned in favour of natural degradation on July 11, 1984. To study the natural degradation capabilities of the tailings pond, zero discharge conditions were maintained from July 11 to November 22, 1984. During this time period CN_T and copper levels were reduced from an average of 350 mg/L and 196 mg/L respectively to 21.1 mg/L and 29.5 mg/L. Allowing for dilution by runoff during this same period, CN_T and copper reduction by natural degradation were in the order of 86% and 78% respectively within the main tailings pond.

The small, shallow lake adjacent to the tailings pond had become rapidly impaired during the winter and spring of 1983/84 because of the poor performance of the alkaline chlorination treatment method. Cyanide and copper concentrations in this lake were also monitored and fell to well within the Ontario Ministry of Environment's Mining Effluent Guidelines during the course of the natural degradiation study.

During the winter 1984/85 the natural degradation study was summarized and compared with sodium metabisulphite and hydrogen peroxide treatment alternatives. As a result of this work, natural degradation was proposed as the treatment method along with the use of the small lake as a tertiary clearwater pond. Ministry of Environment approval was obtained with condition that further chemical treatment must be provided if their guidelines cannot be met at the discharge of this tertiary clearwater pond.

With the spring melt in 1986 the guidelines were again exceeded. A pilot plant was set up at the tailings dam in June 1986 to test and demonstrate the hydrogen peroxide treatment method. Personnel from Degussa Canada Inc. operated the pilot plant. As a result of the good test results a decision was made in late 1986 to construct a hydrogen peroxide treatment plant adjacent to the tailings pond.

Engineering of the treatment facility was given to Kilborn. Degussa provided the detoxification expertise. The facility was designed to treat an average flow of 220 cubic metres/hour effluent with flowrate peaks to 450 cubic metres/hour. At time of writing, site preparation work for this facility had just begun with start-up expectations of mid July 1987.

From pilot plant demonstration test results it was predicted that effluent from the polishing pond after treatment with hydrogen peroxide would contain CN_T in the range of 0.4 to 1.0 mg/L and copper concentrations from 1 to 2 mg/L. Their concentration in effluent from the tertiary clearwater pond was expected to be <0.1 mg/L and <0.3 mg/L respectively.

ACKNOWLEDGEMENT

The authors would like to thank the joint venture partners; Campbell Red Lake Mines and Amoco Canada Petroleum Co. Ltd. for permission given to proceed with this paper. The assistance of all personnel at Lakefield Research who helped in the preparation of this paper is also appreciated.

REFERENCES

Kilborn Ltd. (May 1983) Operating and Training Manual

Nicol, M.J. and C.A. Fleming (1984) The absorption of gold cyanide onto activated carbon. III. Factors influencing the rate of loading and the equilibrium capacity. J.S. Afr. Inst. Min. Metall. Vol 84, pp 85-93.

Wright, K.P. and I.R. Ward (1981) Detour Lake - A new gold Mine Presented at a fall C.M.P. Meeting, Timmins, Ontario

GOLD ISOTHERMS

J.A. King

Placer Development Limited
P.O. Box 49330, Bentall Postal Station,
Canada V7X 1P1

ABSTRACT

Operating isotherms from a number of full-scale plants and pilot-scale plants are presented. A spreadsheet analysis for simulating CIP carbon and solution profiles from isotherm equations is described. Using profile data and unit cost data, a method is presented for optimizing the number of CIP stages.

KEYWORDS

Carbon-in-pulp, operating isotherms, carbon profiles, solution profiles, stage optimization.

INTRODUCTION

The adsorption of gold onto carbon in a CIP or CIL operating plant is best described by the 'operating isotherm' as opposed to the equilibrium isotherm. The operating isotherm gives the relationship between gold loaded onto carbon in a particular CIP or CIL stage and the gold in solution in the pulp in contact with the carbon. The equilibrium isotherm gives the relationship between gold loaded on the carbon and the gold value of solution in contact with the carbon after an infinite time. Many days are usually required to approach this equilibrium isotherm. In a CIP circuit where there is typically only one or two hours per stage for absorption to take place, carbon loadings are significantly lower than those predicted by the equilibrium isotherm.

To predict the CIP circuit performance of a new operation, an operating isotherm that describes gold adsorption onto the carbon may be obtained from a small-scale continuously-operated pilot plant. The equation of the operating isotherm may be used to predict CIP carbon loadings and CIP carbon and solution profiles for a specified number of CIP stages and barren solution assay. An economic analysis of the CIP circuit may then be undertaken to determine the number of CIP stages to be installed and the tonnage of carbon to be loaded and stripped per day.

To enable the construction of operating isotherms, plant data and samples of carbon and solution were obtained from Cortez Gold Mines in Nevada, U.S.A., Golden Sunlight Mines Inc. in Montana, U.S.A., Kidston Gold Mines Limited in

Queensland, Australia and Pine Creek Goldfields Limited in the Northern Territory, Australia. Additionally, isotherms derived from recent continuous pilot-plant testing of two ores from Papua New Guinea, from the Porgera and Misima deposits, are included for comparison. In the full-scale plants, samples of pulp were removed from each tank down the CIP/CIL bank and pulp samples were immediately screened and filtered to obtain the carbon and solution samples for assay. In the pilot-scale plants, operating conditions were similar to those in a full scale plant, including the use of carbon sourced from Cortez and Golden Sunlight Mines that had been subjected to many plant cycles of adsorption, desorption, regeneration and acid washing.

A summary of the wide range of ore types and operating conditions in the full-scale plant operations and the pilot-plant operations is given in Table 1.

TABLE 1 Operating Parameters in Full-Scale Plants and Pilot-Scale Plants

Plant	Type	Au/Ag Ratio In Feed	Solution Grade, g Au/t	Stage Time, h	Temp. °C	Carbon Conc. g/L	Residual NaCN At End of CIP Bank, ppm NaCN
Cortez	CIL	25:1	0.10	4	10-20	8	100
Golden Sunlight	CIP	1:1	0.45	2	10-20	18	640
Kidston	CIP	3:2	0.6-0.8	2	20-30	20	200
Pine Creek	CIP	2:1	1.0-1.75	1	20-30	20	200
Porgera							
Run 1	CIP	1:1 to 1:4	0.5-4.0	2	35	24	200
Run 2	CIP	1:1 to 1:3	2.0-8.0	2	25	24	125
Misima	CIP	1:20	0.6-1.4	2	30	24	450

The operating isotherms from Cortez and Golden Sunlight are shown in Fig. 1. Despite significant operational differences between the two plants, the operating isotherms are seen to be particularly close. The equilibrium isotherm, given for comparison, is a plot of data obtained by Hussey, Salisbury and Potter (1979). The operating isotherms for Kidston and Pine Creek are shown in Fig. 2. The data from Pine Creek was obtained shortly after start up when significant difficulties were experienced with the carbon regeneration kiln. The low amount of carbon regenerated per cycle resulted in lower loadings for a given solution concentration when compared to the isotherms for the other operations. Kiln problems have since been resolved and the loadings at Pine Creek have improved significantly over those shown.

Operating isotherms obtained from continuous pilot-scale testing of CIP circuits for the Porgera and Misima deposits are given in Fig. 3 and 4. The isotherm for the Misima pilot plant diverges most from all other isotherms presented. Gold loadings were lower for the lower solution values, most likely due to the high silver/gold ratio in the Misima ore and silver competing for available adsorption sites on the carbon. At higher solution values and carbon loadings, gold will displace silver from carbon, so that at the higher solution values, the Misima isotherm shows carbon loadings closer to those expected from a gold ore with a lower silver content.

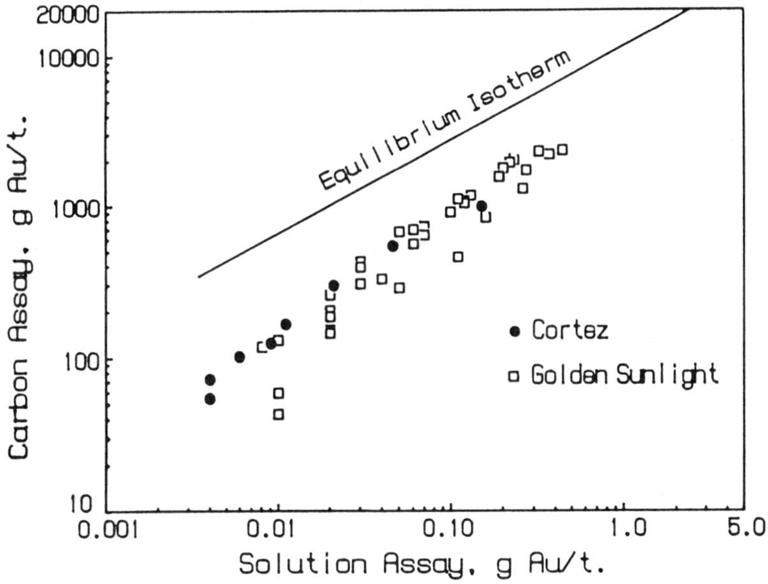

Fig. 1 Cortez and Golden Sunlight Operating Plant Isotherms

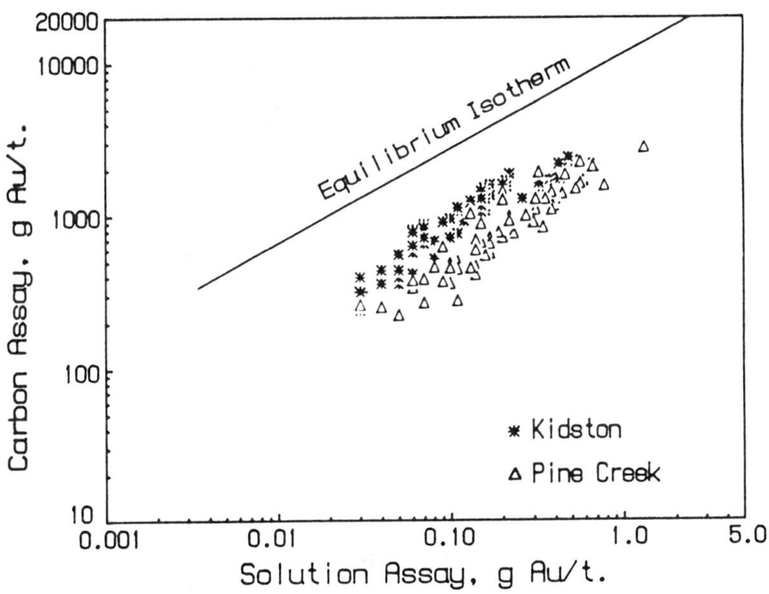

Fig. 2 Kidston and Pine Creek Operating Plant Isotherms

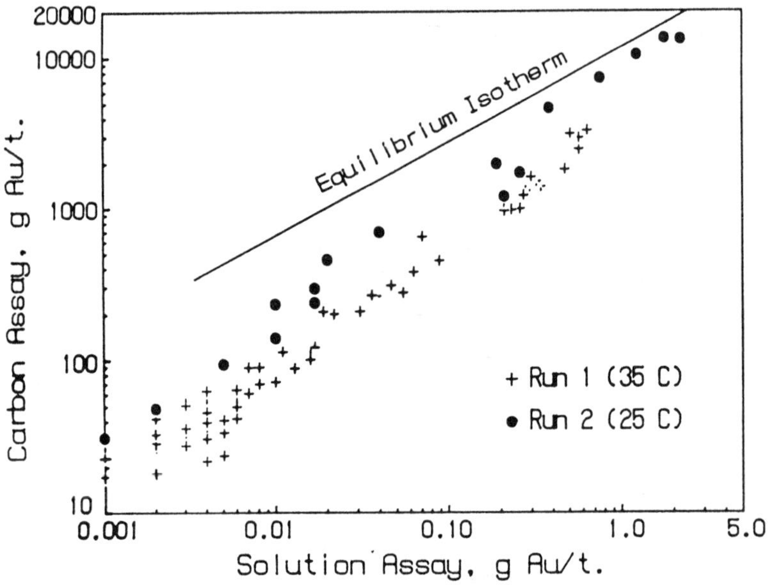

Fig. 3 Porgera Pilot Plant Isotherm

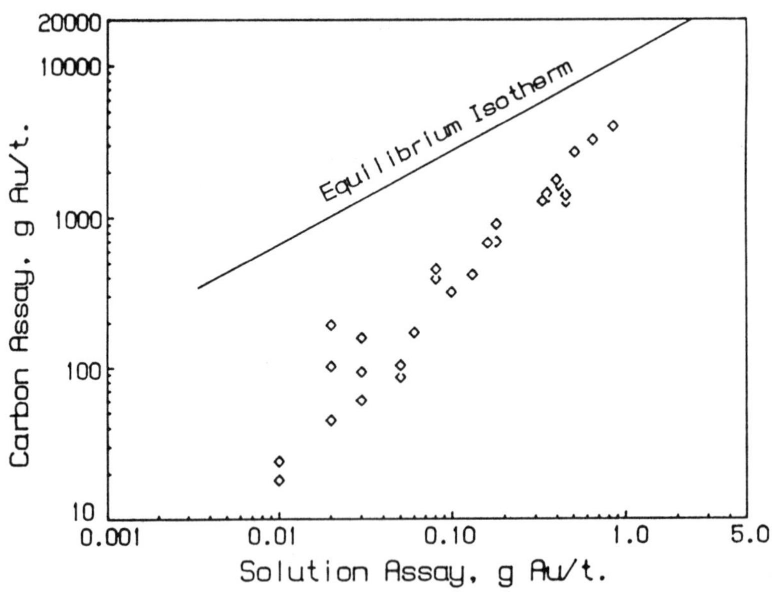

Fig. 4 Misima Pilot Plant Isotherm

The equations that describe the isotherms from the full-scale and pilot-scale plants may be expressed in the form:

$$\text{Carbon Assay} = m \times (\text{Solution Assay})^{n}$$

where m is the gold loading in contact with a solution assaying 1.0 g Au/t and n is the slope of the log-log plot of carbon assay vs. solution assay.

Plant	Equation of Gold Isotherm
Cortez	$\text{Carbon Gold Assay} = 5000 \times (\text{Solution Gold Assay})^{0.78}$
Golden Sunlight	$\text{Carbon Gold Assay} = 5400 \times (\text{Solution Gold Assay})^{0.85}$
Kidston	$\text{Carbon Gold Assay} = 4020 \times (\text{Solution Gold Assay})^{0.70}$
Pine Creek	$\text{Carbon Gold Assay} = 2800 \times (\text{Solution Gold Assay})^{0.75}$
Porgera (Run 1)	$\text{Carbon Gold Assay} = 4200 \times (\text{Solution Gold Assay})^{0.83}$
Porgera (Run 2)	$\text{Carbon Gold Assay} = 7900 \times (\text{Solution Gold Assay})^{0.81}$
Misima	$\text{Carbon Gold Assay} = 4800 \times (\text{Solution Gold Assay})^{1.06}$

It is apparent that for a 'typical' and well operated CIP circuit treating a low-grade gold feed, the operating isotherm will approximate to the equation:

$$\text{Carbon Assay} = 4800 \times (\text{Solution Gold Assay})^{0.8}$$

Calculation of Predicted CIP Carbon and Solution Profiles

Once the operating isotherm has been determined for a particular CIP feedstock, it may be used to calculate carbon and solution assay profiles down the CIP bank for a hypothetical number of CIP stages and barren solution assay.

The calculation uses the operating isotherm to predict the assay of the carbon leaving each stage. The assay of the solution entering a stage is calculated by performing a gold mass balance on the stage. After assuming a desired tail solution value and after assuming a gold value on the recycled barren carbon from the strip circuit, the balance is calculated by stages from the tail end of the circuit proceeding to the head of the circuit. It is conveniently carried out on any computer spreadsheet. Initially, an arbitary carbon advance rate is assumed. Once all known data has been entered, the balance is calculated. After the first calculation, the correct carbon advance rate is determined by iteration to achieve a perfect gold balance over the entire circuit. The balanced is achieved when the gold in the solution feed and the barren carbon precisely equals the gold in the loaded carbon and the barren solution leaving the circuit. The progam does not take into account any additional gold leaching that may occur in the CIP circuit.

A typical spreadsheet printout is shown in Fig. 5. The depicted CIP balance is for treating 10 000 t/d of low grade solution assaying 1.0 g Au/t and producing a barren tail solution assaying 0.005 g/t. It is based on an operating isotherm of:

$$Y = 4800\ X^{0.8} \qquad \text{where } Y = \text{carbon gold assay, g Au/t}$$
$$X = \text{solution gold assay, g Au/t}$$

The CIP balance shows that 4.502 t/d of carbon will be loaded to 2231 g Au/t.

```
SIMULATED CARBON IN PULP BALANCE, 4 STAGES

Inputs:
Solution Feed     -          10000 t/d
                  -            1.0 g Au/t
Barren Carbon     -          4.502 t/d   @       20 g Au/t
Soln Tail Assay   -         0.0050 g Au/t

Gold into circuit:10090.04 g Au/d
Gold ex circuit:   10092.25 g Au/d

Input: Soln feed tonnage and assay, barren carbon assay, and barren
       soln assay.  Then iterate loaded carbon tonnage until
       gold balance is achieved.

                                           :  Solution Feed
                                           :        10000 t/d
Loaded Carbon                              :        1.0000 g Au/t
   4.502 t/d                               :
 2230.62 g Au/t                            V
                      <------------ Stage 1
                                           :
                                           :  Solution ex Stage  1
                                           :        10000 t/d
Carbon ex Stage 2                          :        0.3837 g Au/t
   4.502 t/d                               :
  861.16 g Au/t                            V
                      . . . . . .   Stage 2
                                           :
                                           :  Solution ex Stage  2
                                           :        10000 t/d
Carbon ex Stage 3                          :        0.1168 g Au/t
   4.502 t/d                               :
  268.25 g Au/t                            V
                      . . . . . .   Stage 3
                                           :
                                           :  Solution ex Stage  3
                                           :        10000 t/d
Carbon ex Stage 4                          :        0.0272 g Au/t
   4.502 t/d                               :
   69.25 g Au/t                            V
                      . . . . . .   Stage 4
                                           :
                                           :  Solution ex Stage  4
                                           :        10000 t/d
Barren Carbon                              :        0.0050 g Au/t
   4.502 t/d                               :
   20.00 g Au/t                            V
```

Fig. 5 CIP Balance for Treating 10000 t/d of 1 g Au/t Solution in Four Stages

Simulated CIP balances have been calculated using computer spreadsheet programs similar to those in Fig. 5. The balances assumed a feed solution of 10 000 t/d assaying 1.0 g Au/t, and were calculated for a CIP circuit having between four and ten stages operating with barren solution assaying 0.005, 0.010 and 0.015 g Au/t. The recycled carbon from the strip circuit was assumed to contain 20 g Au/t. The results of the simulated balances are presented in matrix form in Table 2. The solution and carbon profiles for a CIP circuit operating with a CIP barren solution of 0.005 g Au/t are shown graphically in Fig. 6 and Fig. 7. The relationship between the amount of carbon loaded per day and the gold content of the carbon is presented in Fig. 8.

The results of simulated CIP balances indicate that by increasing the number of CIP stages, either the carbon loading may be increased for a fixed CIP barren solution assay, or the CIP barren solution assay may be reduced for the same carbon loading. By increasing the number of CIP stages to increase the carbon loading, carbon strip costs are reduced and carbon losses due to fragmentation after regeneration are also reduced. However the cost of agitator power increases, the physical losses of carbon from the CIP circuit due to abrasion will increase, as will losses of gold contained in the abraded carbon assuming no change in the circuit carbon concentration. The benefits of increased carbon loading also have to be weighed against the increased returns from a reduction in the CIP barren solution assay.

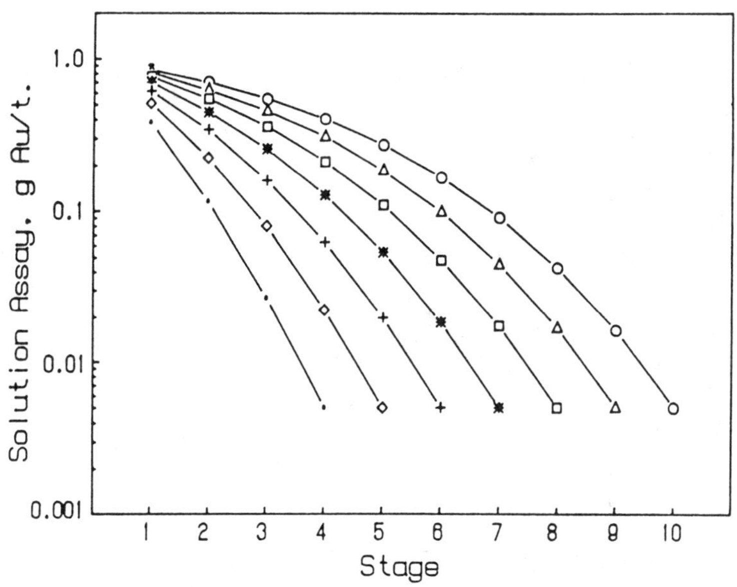

Fig. 6 Solution Profiles for 0.005 g Au/t CIP Barren Solution Based on 10000 t/d Feed Solution Assaying 1.0 g Au/t

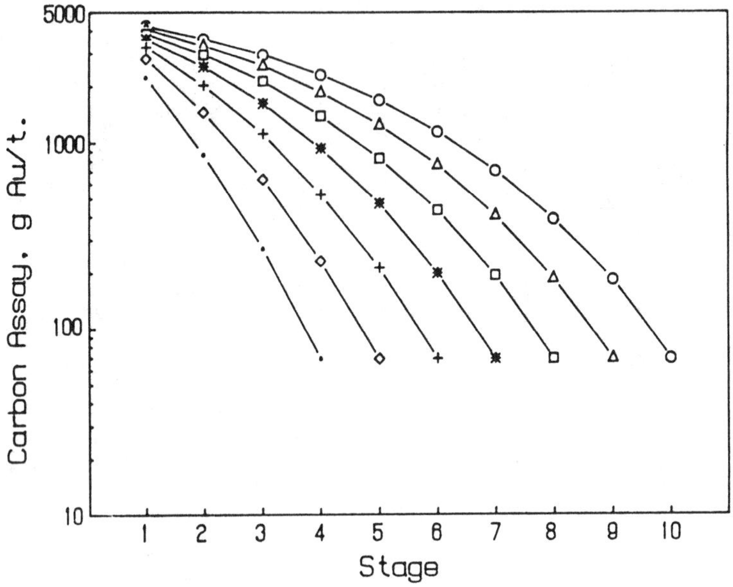

Fig. 7 Carbon Profiles for 0.005 g Au/t CIP Barren Solution
Based on 10000 t/d Feed Solution Assaying 1.0 g Au/t

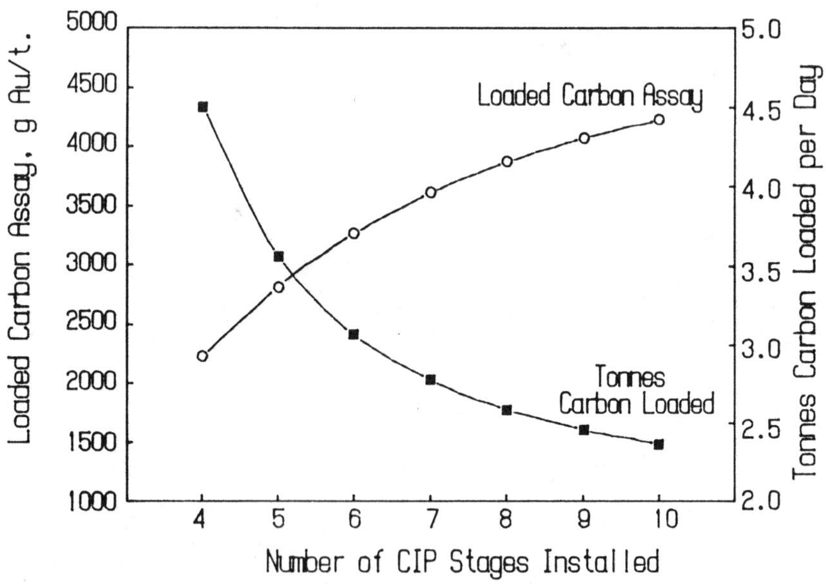

Fig. 8 Effect of Number of CIP Stages on the Assay and Tonnage of Loaded Carbon

Table 2 Simulated Carbon and Solution Profile Matrix
(4 to 10 Stages of CIP with Barren Solution Assays of 0.005 to 0.015 g Au/t)

Solution Treatment Rate: 10000 t/d
Solution Grade: 1 g Au/t
Barren Carbon Assay: 20 g Au/t

CIP Stages	Barren Solution g Au/t	Carbon Treated t/d	Solution Assay Leaving Each CIP Stage, g Au/t									
			1	2	3	4	5	6	7	8	9	10
4	0.0050	4.503	0.3839	0.1168	0.0272	0.0050						
4	0.0100	3.897	0.4560	0.1703	0.0492	0.0100						
4	0.0150	3.598	0.5003	0.2084	0.0678	0.0150						
5	0.0050	3.556	0.5137	0.2241	0.0800	0.0225	0.0050					
5	0.0100	3.198	0.5823	0.2954	0.1255	0.0422	0.0100					
5	0.0150	3.014	0.6230	0.3426	0.1598	0.0592	0.0150					
6	0.0050	3.063	0.6183	0.3394	0.1607	0.0634	0.0201	0.0050				
6	0.0100	2.826	0.6798	0.4183	0.2268	0.1044	0.0384	0.0100				
6	0.0150	2.699	0.7146	0.4674	0.2724	0.1362	0.0546	0.0150				
7	0.0050	2.771	0.7010	0.4495	0.2580	0.1291	0.0545	0.0186	0.0050			
7	0.0100	2.601	0.7527	0.5271	0.3364	0.1907	0.0925	0.0362	0.0100			
7	0.0150	2.509	0.7831	0.5747	0.3880	0.2349	0.1228	0.0518	0.0150			
8	0.0050	2.582	0.7646	0.5469	0.3598	0.2132	0.1109	0.0490	0.0177	0.0050		
8	0.0100	2.456	0.8090	0.6197	0.4430	0.2900	0.1695	0.0852	0.0347	0.0100		
8	0.0150	2.385	0.8338	0.6625	0.4951	0.3416	0.2124	0.1144	0.0500	0.0150		
9	0.0050	2.454	0.8148	0.6307	0.4580	0.3069	0.1861	0.0995	0.0455	0.0171	0.0050	
9	0.0100	2.357	0.8520	0.6961	0.5397	0.3916	0.2611	0.1559	0.0804	0.0337	0.0100	
9	0.0150	2.301	0.8728	0.7340	0.5894	0.4460	0.3126	0.1979	0.1090	0.0488	0.0150	
10	0.0050	2.363	0.8533	0.7003	0.5470	0.4017	0.2730	0.1683	0.0919	0.0431	0.0166	0.0050
10	0.0100	2.286	0.8835	0.7568	0.6229	0.4874	0.3576	0.2416	0.1466	0.0770	0.0330	0.0100
10	0.0150	2.261	0.9014	0.7900	0.6686	0.5411	0.4133	0.2929	0.1879	0.1051	0.0479	0.0150

CIP Stages	Barren Solution g Au/t	Carbon Treated t/d	Carbon Assay Leaving Each CIP Stage, g Au/t									
			1	2	3	4	5	6	7	8	9	10
4	0.0050	4.503	2232	861	268	69						
4	0.0100	3.897	2561	1165	431	121						
4	0.0150	3.598	2758	1369	558	167						
5	0.0050	3.556	2817	1451	636	231	69					
5	0.0100	3.198	3114	1810	912	381	121					
5	0.0150	3.014	3287	2037	1107	500	167					
6	0.0050	3.063	3267	2022	1112	528	211	69				
6	0.0100	2.826	3525	2390	1465	787	354	121				
6	0.0150	2.699	3668	2612	1696	974	469	167				
7	0.0050	2.771	3613	2532	1624	933	468	198	69			
7	0.0100	2.601	3824	2876	2008	1275	715	337	121			
7	0.0150	2.509	3947	3082	2251	1507	897	450	167			
8	0.0050	2.582	3872	2962	2119	1394	826	430	191	69		
8	0.0100	2.456	4051	3273	2503	1783	1160	669	326	121		
8	0.0150	2.385	4150	3453	2735	2033	1390	847	437	167		
9	0.0050	2.454	4074	3320	2570	1866	1250	758	405	185	69	
9	0.0100	2.357	4223	3592	2931	2268	1639	1085	639	319	121	
9	0.0150	2.301	4305	3748	3145	2516	1893	1313	815	428	167	
10	0.0050	2.363	4228	3610	2963	2314	1699	1154	711	388	181	69
10	0.0100	2.286	4347	3841	3287	2701	2109	1541	1033	618	313	121
10	0.0150	2.261	4417	3975	3478	2937	2367	1798	1260	792	422	167

CIP Plant Design

When designing a CIP circuit for a new operation, the decision has to be made as to how many CIP stages will be installed. It is suggested that the optimum number of CIP stages is that number which minimizes the sum value of all operating costs and precious metal losses associated with the CIP circuit, namely:

Cost of stripping carbon.

Physical losses of carbon.

Cost of agitator power.

Precious metal losses in:
- Carbon escaping circuit.
- Carbon fragmented after regeneration.
- Solution leaving the circuit.

Cost of Stripping Carbon Using the Zadra Process

The cost of stripping carbon is the sum of a reagent/power component related directly to the tonnage of carbon stripped, plus a fixed labour component. Although no two plants are operated in a like manner, the reagent requirements and costs for the Zadra strip process are generally as follows:

Reagent	Consumption (kg/t Carbon)	Reagent Cost	
		(US $/kg)	(US $/t Carbon)
Sodium Hydroxide	40 kg/t carbon eluted 10 kg/t carbon washed	0.58 0.58	23.20 5.80
Sodium Cyanide	5 kg/t carbon eluted	1.30	6.50
Hydrochloric Acid (35%)	60 kg/t carbon washed	0.65	39.00

The reagent costs total $74.50 U.S. per tonne of carbon stripped.

Power requirements for stripping and regenerating carbon are substantial. If electrically operated heat exchangers and regeneration kilns are employed, power requirements will approximate to 500 kW.h/t carbon for stripping and 2000 kW.h/t carbon for regeneration. Assuming a unit power cost of 4¢/kW.h, power costs for stripping and regeneration will total:

$$(500 + 2000)\ \text{kW.h} \times 0.04 = \$100\ \text{U.S./t carbon}$$

Labour to operate the carbon plant is estimated at 20 manhours/day, including scheduled time off, paid for at $17/hour including overhead costs.

Total carbon plant operating costs:

$174.5 U.S./t carbon stripped, washed and regenerated
+ $340 U.S./day for labour

Physical Losses of Carbon

Carbon is lost due to abrasion in the CIP circuit and due to fragmentation during regeneration.

Losses in the CIP circuit are best measured in terms of percent carbon lost per day contact in the circuit. A survey of several plants indicates the loss per day is in the order of 0.05% of all the carbon in the circuit.

If the CIP plant is designed for:

1 hour/stage
10 000 cubic metres of solution per day
30 percent solids pulp density
2.7 solids SG
20 grams carbon per litre of pulp

and if 0.05 percent of all carbon in the circuit is lost per day contact, the daily loss of carbon per stage will be 4.828 kg.

The loss of carbon during regeneration usually represents the largest loss of carbon and is in the order of 3% of the weight of carbon regenerated.

If the current cost of carbon is $3600 U.S. per tonne, the value of carbon lost from the overall circuit will be:

$17.38 U.S. per day per stage

plus $108.00 U.S. per tonne carbon stripped/regenerated

Cost of Agitator Power

Sufficient agitation is generated in a CIP tank by providing 0.033 kW of drawn power per cubic metre of agitated pulp. Knowing the pulp volume per stage for the required residence time per stage, the drawn power per stage is calculated. The cost of agitation is the product of the power draw per stage, the number of stages and the power unit cost.

Gold Losses from the Circuit

Gold is lost from the CIP overall circuit in three streams, namely gold in the CIP barren solution, gold in carbon fragmented after regeneration/acid washing and gold in carbon escaping the CIP circuit.

The value of gold in the carbon fragmented after regeneration/acid washing is easily quantified. Knowing that approximately 3% of the total carbon stripped is fragmented during regeneration/acid washing, and knowing the assay of the barren carbon after stripping, the gold value of the fragmented carbon may be calculated. Although this gold is lost from the overall CIP circuit, the refinery screen washings and undersize may be sent to a carbon settling pond and typically 80% of this carbon may be recovered for sale to a smelter.

Gold lost in carbon escaping the CIP tankage is difficult to quantify precisely. If gold is abraded equally from each stage and if there were no desorption of gold as abraded carbon travelled with the pulp flow to exit the CIP circuit, the daily gold loss in abraded carbon would be the product of the number of CIP stages, the tonnes of carbon per stage, the percent carbon abrasion loss per day contact and the arithmetic mean of the carbon loading down the CIP circuit.

It has, however, been established that gold does desorb under these conditions although it appears to do so at a very slow rate, (Fleming and Nichol [1984]). This fact was confirmed in testwork recently carried out at the Placer Research Centre. Approximately equal weights of loaded and stripped carbon were contacted by bottle rolling in a CIP barren pulp containing 150 ppm free NaCN for various contact times. The results are detailed in Table 3 and in Fig. 9. After 20 hours, 16% of the gold had been transferred from a -16 +20 mesh loaded carbon that initially assayed 13 000 g Au/t to a +12 mesh stripped carbon that initially assayed 25 g Au/t. Other carbons with lower gold loadings were tested and it was found that the rate of desorption was slower, in terms of both percent and weight of gold transferred.

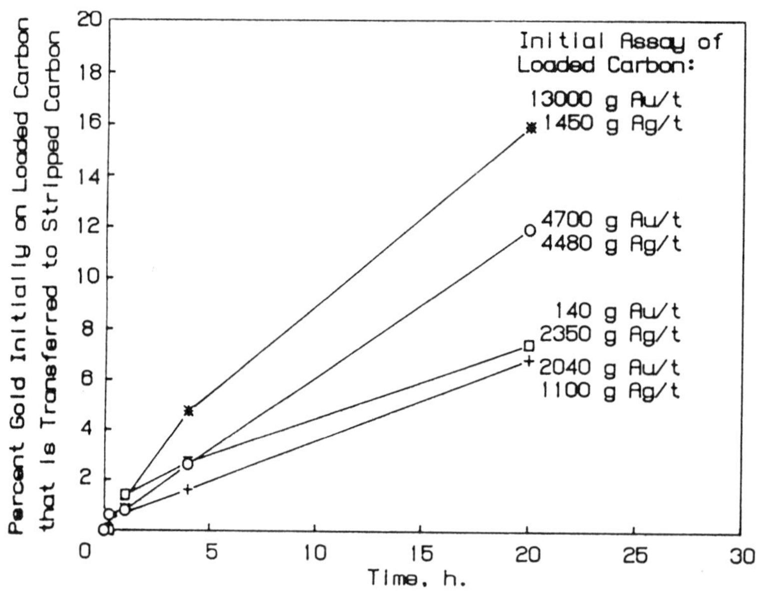

Fig. 9 Percent Gold Transferred from Loaded to Stripped Carbon

To evaluate the effect of particle size on the rate of desorption, 13 000 g Au/t loaded carbon was pulverized to 100% passing 70 mesh and contacted with +12 mesh stripped carbon assaying 23 g Au/t. As might be expected, the rate of desorption was faster from the -70 mesh loaded carbon than from the -16 +20 mesh loaded carbon but after 20 hours contact still only 22% of the gold had transferred.

It will be noticed from Table 3 that the gold value of the solution in the CIP pulp increases with the addition of the loaded carbon. In plant practice, where the ratio of the escaping loaded carbon to the lower loaded carbon normally resident in a particular tank is extremely low, this solution build up will not be so great and the desorption will likely be faster.

In view of the slow desorption rates measured experimentally and the typical CIP residence time of one hour per stage, it is conservatively estimated that the assay of the escaping abraded carbon is at least 80% of the arithmetic mean assay of carbon in the CIP circuit. This mean assay may be determined from the simulated carbon profile data given in Table 2.

Table 3 Precious Metal Transfer from Loaded Carbon to Stripped Carbon

Loaded Sample Sized at -16 + 20 mesh. Adsorbing Carbon Sized at +12 mesh, assaying 25 g Au/t, and 10 g Ag/t.

Loaded Carbon Sample	Contact Time	Adsorbing Carbon Data			Desorbing Carbon Data			Precious Metal Transferred				Backcalculated Loaded Carbon		Assay of Solution Phase of Slurry	
		Weight	Assay		Weight	Assay		ug		%					
	h	g	g Au/t	g Ag/t	g	g Au/t	g Ag/t	Au	Ag	Au	Ag	g Au/t	g Ag/t	g Au/t	g Ag/t
1	0.25	5.9	59	26	5.18	13058	1350	201	94	0.3	1.3	13097	1368	0.203	0.280
	1.00	5.72	166	69	5.28	12930	1380	807	337	1.2	4.4	13083	1444	0.213	0.240
	4.00	5.77	633	358	5.79	12366	1210	3508	2008	4.7	22.3	12972	1557	0.230	0.180
	20.00	5.71	2169	698	5.78	11215	746	12242	3928	15.9	47.7	13333	1426	0.190	0.220
2	0.25	5.8	49	55	4.84	4582	4350	139	261	0.6	1.2	4611	4404	0.055	0.580
	1.00	5.93	57	95	5.08	4794	4380	190	504	0.8	2.2	4831	4479	0.053	0.480
	4.00	5.84	138	478	5.57	4482	4020	660	2733	2.6	10.9	4600	4511	0.049	0.570
	20.00	5.82	573	1690	5.66	4158	2800	3189	9778	11.9	38.2	4721	4527	0.048	0.530
3	0.25	6.04	25	36	5.24	140	2420	0	157	0.0	1.2	140	2450	0.005	0.180
	1.00	5.93	27	123	5.57	148	2400	12	670	1.4	4.8	150	2520	<0.003	0.160
	4.00	5.91	29	95	5.86	144	2100	24	502	2.7	3.9	148	2186	<0.003	0.170
	20.00	5.76	35	448	6.04	119	1870	58	2523	7.4	18.3	129	2288	<0.003	0.200
4	0.25	5.56	33	22	5.47	2128	1050	44	67	0.4	1.1	2136	1062	0.040	0.130
	1.00	5.88	38	19	5.68	1949	1110	76	53	0.7	0.8	1962	1119	0.040	0.110
	4.00	5.79	57	97	5.81	2009	1080	185	504	1.6	7.4	2041	1167	0.035	0.140
	20.00	5.43	167	286	5.56	1888	843	771	1499	6.8	24.2	2027	1113	0.035	0.130

Loaded Sample Sized at -70 mesh. Adsorbing Carbon Sized at +12 mesh, assaying 25 g Au/t, and 10 g Ag/t.

Loaded Carbon Sample	Contact Time	Adsorbing Carbon Data			Desorbing Carbon Data			Precious Metal Transferred				Backcalculated Loaded Carbon		Assay of Solution Phase of Slurry	
		Weight	Assay		Weight	Assay		ug		%					
	h	g	g Au/t	g Ag/t	g	g Au/t	g Ag/t	Au	Ag	Au	Ag	g Au/t	g Ag/t	g Au/t	g Ag/t
1	0.08	5.96	83	39	-	-	-	346	173	0.5	2.2	-	-	-	-
	0.25	5.7	104	59	-	-	-	450	279	0.6	3.5	-	-	-	-
	1.00	5.9	434	182	-	-	-	2413	1015	3.3	12.7	-	-	-	-
	4.00	5.96	1227	452	-	-	-	7164	2634	9.9	33.1	-	-	-	-
	20.00	5.92	2740	697	-	-	-	16073	4067	22.3	51.0	-	-	0.190	0.205

Metal transfer based on loaded carbon weight of 5.50 g, assaying 13121 g Au/t, and 1449 g Ag/t.

Optimum Number of CIP Stages

To demonstrate the method for determining the number of CIP stages for a new project, the calculation has been made for a hypothetical feed of 10 000 t/d of solution grading 1.0 g Au/t and an operating isotherm of:

$$Y = 4800\ X^{0.8}$$

where Y is the carbon loading and X is the solution assay, both numbers expressed in g Au/t. All other input data for operating conditions and costs is given in Table 4.

Table 4 CIP Optimization Operating Parameters and Unit Cost Input Data

1. Cost of Stripping Carbon

Caustic: 40 kg/t carbon eluted
10 kg/t carbon washed
58 US cents/kg

Cyanide: 5 kg/t carbon eluted
130 US cents/kg

Hydrochloric Acid 35%:

60 kg/t carbon washed
65 US cents/kg

Power: 500 kW.h/t carbon for stripping
2000 kW.h/t carbon for regeneration
4 US cents/kW.h

Labour: 20 manhours/day, including scheduled time off
17 US \$/hour including overheads

Total Strip Operating Costs:

174.5 US \$/ton carbon stripped PLUS
340 US \$/day for labour

3. Cost of Agitation Power

Volume of Pulp per one hour stage: 482.8 m3

Drawn power per cubic meter: 0.033 kW/m3

Drawn power per stage: 382.3 kW

Cost of agitation per stage: 15.295 US \$/stage

2. Value of Carbon Lost

Carbon lost in CIP tailings pulp:

Assume: 1 hour/stage
10000 cubic meters of solution per day
30 percent solids pulp density
2.7 Solids SG
20 grams carbon per liter of pulp
0.05 percent of all carbon in circuit is lost per day contact

Carbon lost per stage: 4.828 kg/day

Carbon lost after regeneration:

3 % of carbon eluted

Cost of carbon: 3.6 US \$ per kg

Value of Carbon Lost: 17.38 US \$/day per stage PLUS
108.00 US \$/ton of carbon eluted

4. Gold Losses

Gold in carbon escaping in CIP tail =
Carbon lost/stage x No. of stages
x 80% of the Mean Au assay of carbon in circuit

Gold in carbon lost after regeneration =
0.03 x 20% x carbon eluted/day x barren carbon assay

Gold lost in barren solution
= 10000 x barren solution assay

Gold price = 12 US \$/gram

The operating costs and gold losses have been calculated in matrix form for four to ten CIP stages and for barren solution values of 0.005 to 0.015 g Au/t. The results are given in Table 5 and are shown graphically in Fig. 10 and Fig. 11.

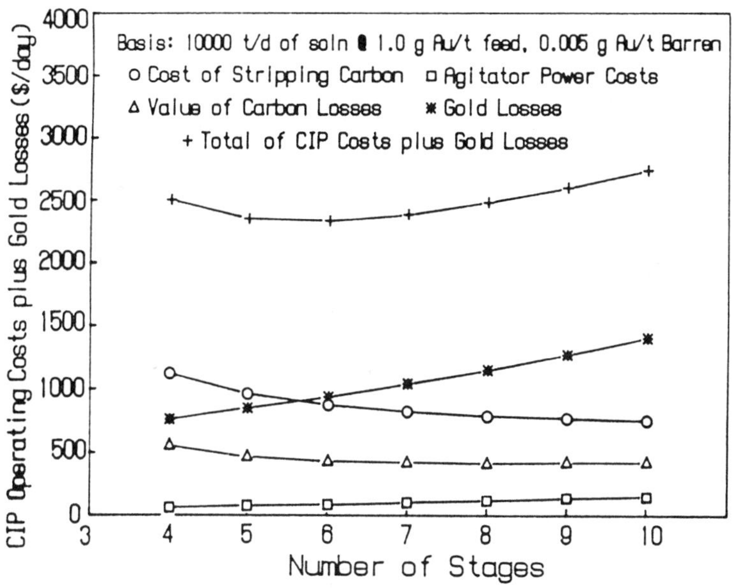

Fig. 10 CIP Operating Costs and Gold Losses vs Number of CIP Stages

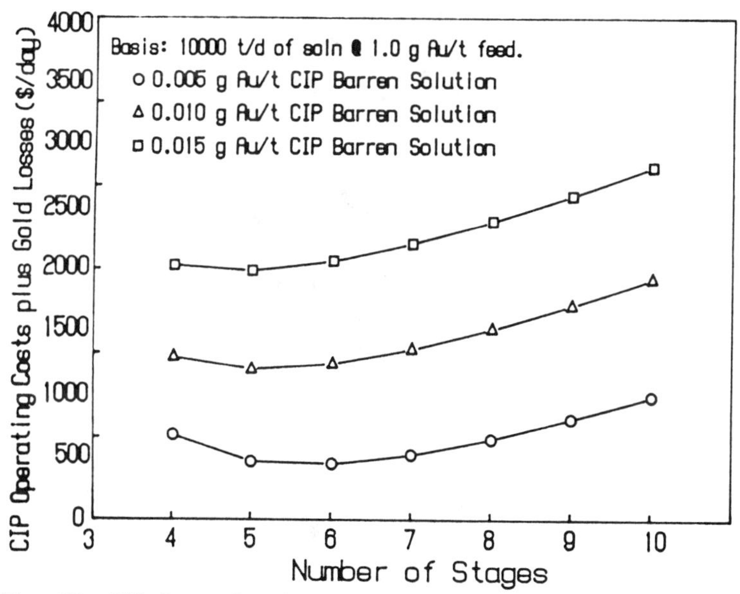

Fig. 11 CIP Operating Costs and Gold Losses vs Number of CIP Stages

Table 5 CIP Operating Costs and Gold Losses as Function of Number of CIP Stages and Barren Solution Assay

CIP Stages	Barren Solution g Au/t	Carbon Treated t/d	Cost of Stripping Carbon $/day	Cost of Stripping Carbon $/t	Value of Carbon Lost $/day CIP Tail	Value of Carbon Lost $/day Regen	Value of Carbon Lost $/day Total	Cost of Agitation $/day	CIP Circuit Gold Losses, $/day Carbon In CIP Tails	CIP Circuit Gold Losses, $/day Carbon After Regen	CIP Circuit Gold Losses, $/day CIP Tail Soln	CIP Circuit Gold Losses, $/day Total	CIP ASSOCIATED OPERATING COSTS + VALUE OF GOLD LOSSES, $/DAY
4	0.0050	4.503	1126	250	70	486	556	61	159	6	600	765	2508
4	0.0100	3.897	1020	262	70	421	490	61	198	6	1200	1404	2976
4	0.0150	3.598	968	269	70	389	458	61	225	5	1800	2030	3517
5	0.0050	3.556	961	270	87	384	471	76	241	5	600	846	2354
5	0.0100	3.198	898	281	87	345	432	76	294	5	1200	1498	2905
5	0.0150	3.014	866	287	87	326	412	76	329	4	1800	2133	3488
6	0.0050	3.063	874	286	104	331	435	92	334	4	600	939	2340
6	0.0100	2.826	833	295	104	305	409	92	401	4	1200	1605	2939
6	0.0150	2.699	811	300	104	291	396	92	444	4	1800	2248	3547
7	0.0050	2.771	824	297	122	299	421	107	437	4	600	1041	2393
7	0.0100	2.601	794	305	122	281	403	107	517	4	1200	1721	3024
7	0.0150	2.509	778	310	122	271	393	107	570	4	1800	2374	3651
8	0.0050	2.582	791	306	139	279	418	122	550	4	600	1154	2484
8	0.0100	2.456	769	313	139	265	404	122	644	4	1200	1847	3142
8	0.0150	2.385	756	317	139	258	397	122	705	3	1800	2508	3784
9	0.0050	2.454	768	313	156	265	421	138	672	4	600	1275	2603
9	0.0100	2.357	751	319	156	255	411	138	779	3	1200	1983	3283
9	0.0150	2.301	742	322	156	249	405	138	850	3	1800	2653	3937
10	0.0050	2.363	752	318	174	255	429	153	803	3	600	1406	2740
10	0.0100	2.286	739	323	174	247	421	153	923	3	1200	2126	3439
10	0.0150	2.261	735	325	174	244	418	153	1002	3	1800	2805	4110

The results indicate that for the assumed conditions, the sum of CIP operating costs and gold losses will be minimized by installing six CIP stages, operating with a CIP barren solution assay of 0.005 g Au/t. Operating with a barren solution assay higher than 0.005 g Au/t, the mean loading of carbon in the CIP circuit is increased (refer to Table 2), as are the gold losses in carbon abraded from the circuit. For this reason, the economic analysis is indicating the use of five stages when the barren solution assay is 0.015 g Au/t.

The result of the above analysis will vary from ore to ore, depending on the equation of the operating isotherm, the values of the cost operating parameters selected, and the unit costs for reagents and power that prevail at a given property. Although areas of the analysis require better definition, especially the gold losses in carbon escaping in the CIP tailing pulp, it does provide a useful indication of the required number of CIP stages.

SUMMARY

The equations of operating isotherms from full-scale and pilot-scale plants were found to be relatively close despite significant differences in operating parameters and conditions. A CIP plant treating a low grade gold feed is expected to have an operating isotherm approximating to the equation:

$$Y = 4800\ X^{0.8}$$

where Y = gold loading on carbon in a CIP stage
X = gold assay of solution in the same CIP stage

Knowing the equation of the operating isotherm, simulated carbon and solution profiles may be readily calculated for a range of barren solution assay and CIP stages, using a computer spreadsheet.

An economic analysis in matrix form, for a low grade gold CIP feed, indicated the optimum number of CIP stages to be five or six, depending on the achievable gold content of the CIP barren solution.

ACKNOWLEDGEMENT

The author would like to express his gratitude to Placer Development Limited for permission to publish this paper and to the operating plants who kindly supplied data and samples for construction of the plant operating isotherms. Thanks are also due to Larry O'Connell and Henry Wong for their assistance with analytical services.

REFERENCES

Hussey, S.J., H.B. Salisbury, and G.M. Potter (1979). U.S. Bureau of Mines Report of Investigations 8368

Fleming, C.A., and M.J. Nichol (1984). J.S. Afr. Inst. Min. Metall., 84, 85-93

AGGLOMERATION PRACTICE IN THE TREATMENT OF PRECIOUS METAL ORES

Rory L. Tibbals

ABSTRACT

The use of agglomeration in the heap leach treatment of precious metal ores has been gaining rapid acceptance. Methods in use for agglomerating precious metal ores are varied and innovative. The difficultly of experimenting with operating heaps necessitates that laboratory practice provide good operating parameters. Current plant practice is reviewed in light of laboratory experience and theory.

KEY WORDS

Heap leaching; gold; silver; precious metals; agglomeration.

INTRODUCTION

The use of the heap leaching for the recovery of gold and silver has caused a major metamorphosis of the precious metals mining industry. In the United States, heap leaching has grown from zero production in the mid-seventies to one of the largest metal producing techniques for gold and silver. Interest in the technique is currently spreading worldwide.

Perhaps one of the most significant technological advances stimulating such rapid growth was the development of methods for agglomerating precious metal ores. Since the application of agglomeration technology, the failure rate of new heap leach projects has decreased dramatically. Agglomeration has not only provided great technological advantage, but it has also allowed heap leaching to gain an atmosphere of respectability.

Agglomeration is used to insure good solution percolation in operating heaps. Good percolation has been shown to increase overall gold and silver recoveries and to lower the leach time required to achieve maximum recovery. Agglomeration aids percolation by preventing the segregation and migration of fines and by creating void spaces open to leach solution flow.

There are currently dozens of heap leach operations in production in the western United States. The majority of these operations are using some form of agglomeration. The agglomeration techniques employed are highly varied and innovative, but they all have certain factors in common. The agglomeration process can be reduced to five unit processes: preparation of feed; mixing of ore and binder; addition of agglomerating solution; mixing of ore and solution with agglomerate formation; and curing. Each of these aspects is discussed along with design considerations and methods of control.

AGGLOMERATION DEFINED

Webster defines agglomerates as, "a mass of things clustered together". As used in the mineral processing field, agglomerates are collections of mineral particles clustered together. They can take one of two basic forms:

-Fine particles adhering to coarse particles.

-Fine particles adhering to each other.

In green agglomerates, prior to wetting, the adhesion is a result of the surface tension of interstitial water. In cured agglomerates, the adhesion is a result of chemical bridging between particles as the result of the addition of a chemical binding agent. The chemical bridging survives the wetting of agglomerates, whereas the adhesion caused by surface tension fails upon the wetting of agglomerates during leaching.

Two critical factors affecting agglomeration are therefore the addition of water for green agglomerate strength and the addition of a binder for cured agglomerate strength. Since the formation of chemical bridging is due to the growth of hydrated crystals, water addition is also critical to the curing process.

AGGLOMERATION DERIVED BENEFITS TO LEACHING

The adhesion of particles in agglomeration benefits the percolation in operating heaps in three ways.

-By the prevention of fines segregation.

-By the prevention of fines migration.

-By the creation of greater void spaces.

There are many good discussions of these effects in the literature (McClelland, 1986; Chamberlin, 1986). They will be reviewed very briefly.

All present methods of stacking heaps create segregation of fines in unagglomerated ore. Segregation of fines causes disparity in percolation rates. In operating heaps, the disparity can result in much reduced leach rates. Severe disparity of percolation rates can result in areas of a heap not being leached at all. Often the result is significantly lower leach recoveries, extended leach times and lower pregnant solution grades. Segregation of fines in heap construction can be eliminated by agglomeration with solution alone. In ores agglomerated with water or cyanide solution the adhesion of fines to the coarse rock is sufficiently strong to withstand the heap building process and the segregation of fines is prevented.

The addition of a binder to agglomeration causes the growth of crystals which form bridges between particles in the resulting agglomerates. Although the adhesion caused by the binder is relatively weak, it can hold the fines to the larger particles and to each other. Fines can migrate through operating heaps and move into areas of lower permeability. If sufficient fines are allowed to migrate into these areas, the fines can effectively seal large areas of the heap from further solution flow. The net effect is the same for fines segregation. Leach solution cannot percolate into these areas resulting in lower recoveries and longer leach times.

The formation of agglomerates also causes a large decrease in the bulk density of the agglomerated material. The decrease in bulk density is due to the increase in average particle size which causes an increase in void spaces between particles. Since percolation rates depend on the channels formed as a result of the interaction of the void spaces, an increase in the void spaces improves percolation rates. For ores requiring agglomeration, agglomeration always results in a great increase in the percolation rate which can be sustained by the ore.

The binder must then create bonds which are sufficiently strong to maintain the agglomerates as discrete particles, or the void spaces will collapse.

Figure 1 is a sketch of the increase in void space as a result of agglomeration.

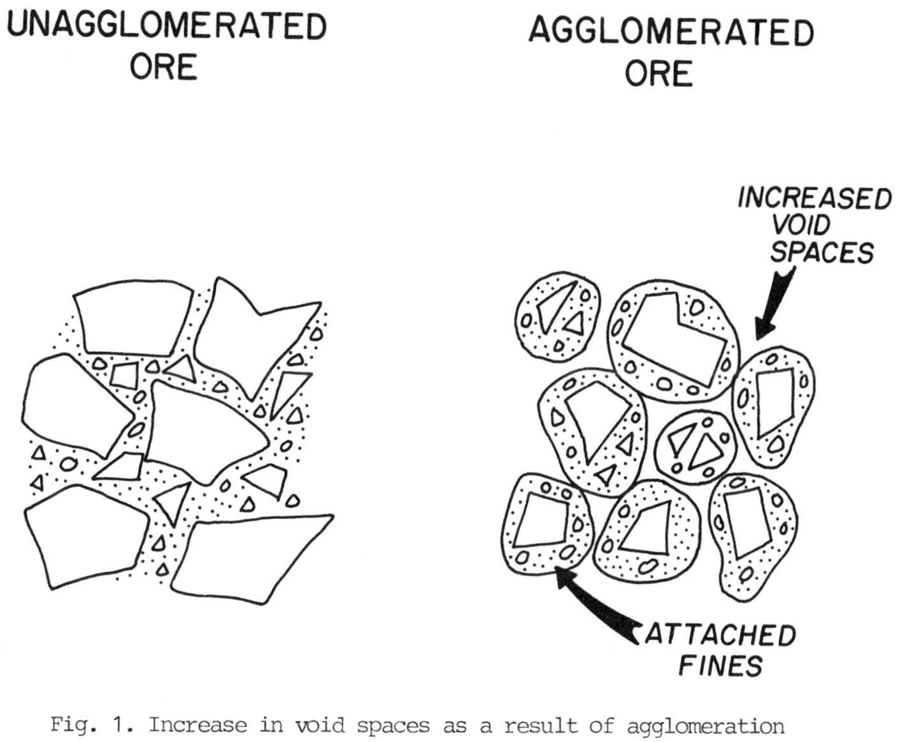

Fig. 1. Increase in void spaces as a result of agglomeration

The increase in void spaces increases the permeability of the heaps and consequently the flow rates of solution. It also reduces hold-up of solution in the heap and thereby decreases the wash time required once dissolution has occurred.

Example:

At the Little Bald Mountain Mine belonging the Northern Dynasty, two heaps were built in successive years. The first heap was stacked and leached in the summer of 1985. The ore placed on the heap was agglomerated with water. A laboratory study performed on core samples had shown that the need for agglomeration was marginal for the first year's production and higher for the remaining reserves. The decision was made not to add cement to agglomeration the first year.

There were several indications that the first year's heap would have benefited from agglomeration with cement. The heap was observed to subside by about 24 percent during the life of the heap. Pregnant solution produced by the heap contained large quantities of suspended solids, a sure sign that fines migration is occurring. In addition, the heap had a high hold-up of solution. Saturation of the ore was evident and there was ponding of solution on top of the heap. Slumping of the heap was a problem along the perimeter. In spite of the problems with percolation and fines migration, the heap was a success and produced the projected recovery.

The ore for the second year's heap was higher in clay content. The consulting laboratory had recommended using 10 lbs.of cement per ton of ore for agglomeration. The ore was agglomerated on a transfer belt between the crushing plant and the radial stacker. Cement was added to the ore at the crusher and mixed through transfer points before the addition of water. The water was added to the ore on the transfer belt and mixing was accomplished by plows which turned the ore on the belt. The ore then traveled down the conveyor train. Additional mixing occurred at the transfer points of the conveyor and as the material was discharged from the radial stacker onto the heap.

The ore agglomerated with cement subsided much less than the first year's heap. The reduction in heap height after subsidence was about 8 percent as compared to 24 percent for the heap without cement agglomeration. The pregnant solution produced by the heap was much lower in suspended solids and the heap was found to drain much freer.

CURRENT AGGLOMERATION PRACTICE

The purpose of agglomeration is to bond coarse and fine particles into stable agglomerates. There are several steps involved in the production of agglomerates for heap leaching. Each of these steps requires some measure of control to be effective.

Preparation of Feed

The steps involved in the preparation of agglomerates for heap leaching depend upon the material being agglomerated. For operations treating rock, the first step is crushing and screening to obtain the desired crush size. For the majority of ores, crushing and screening to size are the only pretreatment necessary.

Ores containing very high percentages of clay and finely ground material such as tailings may require drying prior to agglomeration. For a material to be properly agglomerated it is necessary to ensure that the binder is completely mixed with the contained fines. In materials containing high percentages of fines and a high moisture content, it is not possible to achieve sufficient blending of the fines and the binder. These materials must first be dried to an acceptable level prior to the addition of binder.

Examples:

Solar drying in which the material to be leached is spread out in thin layers and turned periodically until an acceptable moisture content is obtained has been employed successfully at several operations. The technique was used at Macacona in Costa Rica to treat a primary ore containing mostly kaolinite clay. The mine was in a high rainfall area and because of the very high clay content, agglomeration was critical to the success of the project.

A similar technique is in use at the Mount Gaines gold mine in California. The ore, a stamp mill tailings, is collected from an abandoned dump and agglomerated prior to heap leaching. Areas of the tailings have high clay contents and high moisture contents. To achieve proper blending of the binder prior to agglomeration these high clay materials must be dried and broken-up. The action is accomplished by spreading the material over a large area and mixing with a backhoe until the material can be agglomerated.

The problem with wet, high clay content ore has not been dealt with at a high tonnage mine. New techniques will have to be adopted to handle these material on a large scale.

Mixing of Ore and Binder

Prepared feed is next mixed with the binder and pH modifier. Portland cement and lime are being used in practice as both binders and pH modifiers. Because of its high alkalinity, the addition of portland cement during agglomeration tends to raise the pH of the resulting agglomerate. Lime has also been used effectively as a binding agent. As a general rule, however, cement is the most effective binding agent and lime is the most effective pH modifier. The proper mix must be determined from laboratory testing of agglomerate strength and stability and from the consideration of cost and availability of the two products.

Laboratory testing of agglomerates for strength and stability have been done by a number of methods. Bateman employs a jigging procedure in which the cured agglomerates are placed in a ten mesh screen and jigged several times under water. The quantity of fines passing the ten mesh sieve is determined and plotted versus binder addition. An optimal binder addition is then selected as the minimum amount required to prevent the production of fines during jigging.

Other tests have been employed for the selection of agglomeration binder requirement, and they all produce comparable results.

Years of experience with a multitude of ores has shown that portland cement is a more effective binder than lime, producing a more stable agglomerate in the vast majority of ores tested. The selection of a binder and pH modifier will depend on the availability and cost of the two products. In practice, there is no set rule. Some ores are being agglomerated with just cement, some with cement and lime, and some with lime alone.

As mentioned earlier, it is essential that the lime and cement be thoroughly mixed with the ore prior to the addition of water. The growth of chemical bridging requires that particles of cement and lime be intimately mixed with the particles of fines.

The mixing of ore and binder is accomplished in many ways. Some operators add the cement to the crushing and screening equipment. Others add the binder to the belt and allow the mixing to occur at transfer points. Operators using drum agglomerators and pug mills often add the binder to the agglomerator feed and allow dry mixing to occur in the first portion of the equipment, prior to the addition of solution.

The importance of the thorough mixing of binder to the quality of the final agglomerate cannot be over emphasized.

Addition of Agglomerating Solution

Agglomeration can be done using water, strong cyanide solution or weak cyanide solution. All of these will yield good quality agglomerates if added properly. Cyanide solutions can decrease the leach time considerably, and for this reason most operators prefer to use a dilute solution of cyanide.

For ores with high cyanide consumptions, there is occasional justification for adding a strong cyanide solution to agglomeration. The addition of a strong cyanide solution provides cyanide to react with the cyanide-consuming species and leaves added free cyanide to get the leaching started. However, some operators have found that the use of strong cyanide solution in agglomeration increases the cyanide consumption. The use of strong cyanide can also be a safety hazard because of the danger of over-spray contacting personnel during windy conditions and because of the liberation of toxic gases during the cyanide reactions.

Bateman Metallurgical Laboratories routinely uses 2 pounds per ton sodium cyanide solution for agglomeration of clients' ore samples. This is the practice at many of the other reputable consulting laboratories performing gold and silver agglomerated column leach tests.

Common practice at operating heaps is to use barren leach solution for agglomeration. This practice provides a flexible amount of dilute cyanide solution which is relatively safe to handle. No additional mixing equipment for the cyanide is required. For the majority of ores the cyanide content of barrens leach solution is of sufficient strength to get the leaching started during the agglomerate curing process. The initial pregnant solutions from these heaps will contain some free cyanide. The only allowance when using cyanide during agglomeration is that the mixing and agglomerating equipment has to be contained in a lined area. This would marginally add to the capital cost of such a plant.

The addition of solution during agglomeration has been performed in a wide variety of ways. Current theory requires that the solution be sprayed in coarse droplets which become nuclei around which the agglomerates form. While the theory has some validity for tailings and other finely crushed or ground ores, the method of solution addition becomes less critical for coarser ores. The amount of energy input to the material during agglomeration is more important than the method of water addition. For the majority of mines where crushed ores are being treated, the method of addition is not a major factor affecting agglomerate quality.

Example:

The Hog Ranch Mine has been treating crushed ore using only conveyor transfer points for agglomeration. The ore is stage crushed to a nominal 1-1/2 inches. Cement is added at a rate of about 10 pounds per ton of ore and mixed into the ore at two belt transfer points prior to the addition of barren leach solution. Leach solution is pumped from the barren pond and discharged directly onto the ore through open pipes. No attempt is made to spray the solution onto the ore. The volume of solution being added is controlled by valves on the feed pipes. From the transfer conveyor the ore is discharged onto a line of grasshopper conveyors which carry the ore to a radial stacker. The ore is discharged by the radial stacker directly onto the heaps for curing and leaching. The ore agglomeration takes place at the belt transfer points and as the material cascades down the sides of the heap.

The ore on the heaps appears to be well agglomerated and the heaps are leaching well. The operators of the mine are quite happy with the results they are obtaining.

The critical factors concerning the addition of solution to agglomeration are control of the solution addition rate and sufficient mixing of the ore and solution. The strength of green agglomerates is dependent upon the quantity of water added during agglomeration. The agglomerates gain strength with increasing water content to a point and beyond that point their strength declines rapidly. Fortunately the optimal moisture content can be easily determined visually by a trained operator. Control becomes a matter of operator diligence.

For the agglomeration of fine materials such as tailings, the method of solution addition becomes more critical. The energy input during agglomeration must be less during the formation of the agglomerates or they will disintegrate as fast as they form. For this reason the amount of mixing that can be done is less than for crushed ore and the distribution of solution becomes more dependent upon the spray technique.

In the agglomeration of crushed ore, the large ore particles become the nuclei around which the agglomerates build. For fine ore material, no such nuclei exist and the agglomerates must build upon water droplets. The droplet size and distribution of the solution is therefore critical for the formation of agglomerates of fine materials. The rolling action imparted by drum and disk agglomerators then packs the agglomerates and gives them their final green strength.

Agglomeration

Agglomeration occurs as solution is added to the prepared feed and particle adhesion begins to cause the growth of agglomerates. The energy input of the agglomerating equipment mixes the ore and solution and packs the agglomerates in a similar manner to the building of a snowball. Many types of equipment are being used for agglomeration, but they all perform the same basic functions of mixing and compaction.

The following is a list of equipment being used for agglomeration in the precious metals heap leaching industry. Also included are examples of operations employing the equipment:

- -Drums: Alligator Ridge, Mount Gaines, Goldfields
- -Disks: None operating
- -Pug Mills: Haywood-Santiago, Florida Canyon
- -Reverse Belts: Tombstone
- -Belt Agglomeration: Hog Ranch, Little Bald Mountain, Borealis
- -Stockpile Agglomeration: Maggie Creek, Carlin, Carson Hill

There are several good discussions of agglomerating equipment in the literature (McClelland, 1986; Chamberlin, 1986).

Drum and disk agglomerators are the two items of equipment which can be used effectively for agglomerating fine materials. All of the above equipment is effective for agglomerating crushed ore. The selection of a method for agglomerating crushed ore often depends upon the likes and dislikes of the mine management, rather than upon sound engineering criteria.

Perhaps the equipment with the most potential is the pug mill. The capital cost of a pug mill is from 25 percent to 50 percent of that of an agglomerating drum. The power consumption is less than a drum and the mixing action of the pug mill is as good as or better. The pug mill does not impart the rolling action to the ore which compacts the agglomerates, but if the put mill discharges onto transfer belts, the compaction will occur at the belt transfer points. New operations considering the use of agglomeration should give full consideration to pug milling as an agglomeration technique.

Curing

The curing of agglomerates is time dependent in much the same way as the curing of cement. At ambient temperatures the agglomerates will begin to show noticeable wet strength in about 6 hours. They will continue to get stronger with curing for weeks, but will have gained a working strength within 72 hours. A curing period of 72 hours is considered standard in the industry.

Curing of the agglomerates can be done on the heaps or in separate curing piles. Both methods are employed, but separate curing piles tend to be live storage for operations tramming ore from agglomeration to the heaps via trucks or loaders. The green agglomerate strength produced by the surface tension of the agglomerating solution is sufficient to keep the heaps from subsiding until curing takes place. If the heaps are sprayed with leach solution before the curing process is complete, the bonds formed by solution bridging will fail, resulting in the failure of the agglomerates.

CONTROL OF AGGLOMERATION

The factors which have been shown to be critical to the formation of quality agglomerates are:

- -pH modifier addition
- -Binder addition
- -Water addition
- -Moisture content of feed
- -Degree of mixing of ore and binder
- -Degree of mixing of ore and solution
- -Curing time

Of the seven factors which control the quality of agglomerates produced, the degrees of mixing of ore and binder and ore and solution are fixed once the plant equipment has been erected and are not subject to operator control. The curing time is taken to be 72 hours by most operators and leach solution is not introduced to the ore until the curing period has elapsed. The moisture content of the feed is only a consideration for finely ground ore or very clayey ores. The control of the moisture content of material entering the agglomerating circuit will depend upon visual inspection of the material and the operator's judgement.

Control of the pH Modifier Addition

The control of the pH modifier is important, particularly in highly acidic ores.

If insufficient lime is added during agglomeration the acid contained in the ore will react with the cement binder and reduce its effectiveness. When agglomeration is being done with cyanide solution, it is also important to maintain a protective alkalinity to prevent the loss of cyanide into the air.

On the other hand, if excess lime is added to agglomeration, high alkalinities will result in the leach solutions. High alkalinities can cause reduced leach kinetics and scaling problems with pumps, pipes and carbon.

The usual method for controlling the addition of pH modifier is to take grab samples of the finished agglomerates and pulp them with water. The pH of the solution is then measured and the addition rate modified as required. The frequency of sampling is a function of the variability of the feed.

Control of the Binder Addition

The control of the binder addition is dependent upon the quality of the agglomerates formed. Unfortunately, the normal curing time for agglomerates is 72 hours. Laboratory testing by the author has shown that there is no correlation between the green strength of agglomerates and the final cured strength. No way has been found to tell the quality of the agglomerates being produced until they have completed curing.

The usual method for controlling the addition rate of binder is either to set the addition rate at some safe level, usually determined by laboratory testing, or to vary the addition rate depending upon the operations experience with ore types. At Hog Ranch, the rate of binder addition is controlled by management depending on the amount of clay present and the pit area from which the ore is being mined. There is a certain risk inherent in the lack of control. It is not unusual for areas of operating heaps to have received too little binder during agglomeration and to suffer with poor percolation as a result.

One method reported in the literature for increasing the control of the binder addition (Milligan, 1983) is the use of an accelerated cure. A sample of agglomerates is taken and heated to about 200 degrees Fahrenheit for six hours. The time required for curing is thereby reduced to 6 hours and the agglomerate can be tested for strength and stability. The method does not answer the question of instantaneous control, but it could reduce the chances of producing a large tonnage of inadequately agglomerated ore.

The actual feed rate of binder can be controlled by a variable speed feeder, linked to a weightometer. The rate of binder added per ton of feed can therefore be controlled automatically.

Control of the Solution Addition

As mentioned previously, the control of the addition of solution to agglomeration is relatively simple. A diligent operator can control the addition of agglomerating solution quite effectively simply by visual observation of the product.

CONCLUSIONS

The use of agglomeration has revolutionized the precious metals heap leaching industry. As the process becomes better understood, it will contribute more and more substantially to the gold mining industry.

REFERENCES

McClelland, G.E. (1986). Heap Leaching and agglomeration: heap leaching practice in the United States. Gold 100 Proceedings of the International Conference on Gold, Vol. 2, Extractive Metallurgy of Gold.

Chamberlin, P.D. (1986). Agglomeration: cheap insurance for good recovery when heap leaching gold and silver ores. Mining Engineering, December, 1105-1109.

Milligan, D.A., and Engelhart, P.R. (1983). Agglomerated heap leaching at Anaconda's Darwin silver recovery project. Au and Ag Heap and Dump Leaching Practice, Proceedings of SME Fall Meeting, 29-39.

THIOUREA LEACHING OF GOLD IN A CONTINUOUS PILOT PLANT

Rein Raudsepp* and Robert Allgood**

* Bacon Donaldson and Associates Ltd., 2036 Columbia Street, Vancouver, B.C. V5Y 3E1

** Sonora Mining Corporation, Box 1418, Highschool Road, Highway 108, Jamestown, California 95327

ABSTRACT

Sonora pyritic gold concentrate was continuously leached in a two-stage thiourea pilot plant which was operated at 750 g/hr. The pilot plant incorporated automatic solution potential control and solution recycles. Gold recovery by cementation onto atomized aluminum was done on a batch basis. The paper presents leaching results, operating experience and reagent consumptions. A summary of the capital and operating costs for a full scale plant is also presented.

KEYWORDS

Thiourea leaching; gold extraction; pilot plant; reagent consumptions; capital and operating costs.

INTRODUCTION

In the development of the Jamestown Mine in Jamestown, California, the Sonora Mining Corporation considered the use of thiourea leaching as an option for the extraction of gold from their concentrate. Bacon, Donaldson and Associates was engaged to develop the concentrator metallurgy and a thiourea leach circuit. Over 1984 and 1985, a mill flowsheet was evolved, concentrate was produced, and thiourea leach parameters were delineated. In late 1985, a bulk sample was mined and processed to concentrate. Using this feed, Bacon Donaldson operated a 750 g/hr thiourea pilot plant for a three week period in early 1986. A cost estimate for gold extraction by thiourea leaching and aluminum cementation was prepared by Wright Engineers Ltd. based on data generated during this pilot plant test program.

The Chemical Fundamentals of Thiourea Leaching

Thiourea, $(NH_2)_2CS$, strongly complexes gold and silver in solution and so reduces the oxidizing potential required to dissolve these metals. The oxidizing agent in solution is an oxidized form of thiourea: formamidine disulphide. Thiourea and formamidine disulphide form a reversible electrochemical couple as shown in the equation below.

$$\underset{\text{Thiourea}}{2\ (NH_2)_2CS} \underset{SO_2}{\overset{H_2O_2,\ Fe^{3+}}{\rightleftharpoons}} \underset{\text{Formamidine disulphide}}{((NH_2)_2\ CSSC(NH_2)_2)^{2+}} + 2e^-$$

$$E^o = 420 \text{ mV vs SHE}$$

Unfortunately formamidine disulphide is not completely stable in solution and decomposes irreversibly. The first decomposition products are an undetermined sulphinic compound and thiourea; the sulphinic compound then decomposes irreversibly to elemental sulphur and cyanamide. Decomposition occurs rapidly above pH 4.

Initially the rate of gold leaching is mass transfer controlled and so, fairly rapid. In the later stages a thin layer of elemental sulphur forms from formamidine disulphide decomposition which slows the leach process. High concentrations of thiourea which produce relatively thick layers of sulphur do not always give better gold extractions than low concentrations.

To successfully employ thiourea leaching the solution redox potential, pH and dimer decomposition must be controlled.

The Advantages of Thiourea Leaching

Thiourea leaching of gold and silver is a viable alternative to cyanidation in specific instances. Its advantages over cyanidation are outlined below:

1. Thiourea leaching is carried out under acid conditions: pH 1.5 to 2.0, and so, gold and silver can be recovered from materials which are unstable in alkaline solutions and/or react directly with cyanide. Examples of such materials are elemental sulphur and unstable arsenic and antimony minerals, such as stibnite and tetrahedrite.

2. Thiourea leaching is feasible for acidic feeds, such as acid pressure oxidation slurries, as these must be neutralized prior to cyanidation. Thiourea is compatible with small quantities of dissolved metals, such as copper and iron.

3. Thiourea is only slightly toxic. Its ultimate degradation products are elemental sulphur and cyanamide, a material which can be used as a fertilizer.

4. Thiourea leaching is rapid. High gold extractions can be achieved in less than four hours. Gold can be readily recovered from solution.

Thiourea leaching is not without its limitations.

1. Thiourea leaching is not suitable for materials which
   i) react with acid, eg. carbonate minerals
   ii) are strongly oxidizing eg. ferric hydroxide
   iii) strongly adsorb the gold-thiourea complex from solution eg. active carbon or certain clay minerals.

2. Thiourea is a suspected secondary carcinogen.

3. Very few thiourea leach plants are in operation and the process is not widely understood.

CONCENTRATE DESCRIPTION

The Sonora concentrate is produced from Jamestown ore with an average assay of 0.071 oz/ton gold and 0.062 oz/ton silver. The concentration ratio is approximately 28:1 with a 93% recovery of gold and a 73% recovery of silver. The concentrate is essentially pure pyrite with 1% copper as chalcopyrite and a trace of talc. The locked cycle cyanidation test recoveries from concentrate reground to 77% -400 mesh were 97.5% for gold and 75% for silver. The cyanide consumption was 3 to 4 lb/ton; the lime consumption was 8 to 10 lb/ton.

BATCH THIOUREA TEST RESULTS

In batch testwork the optimum thiourea leach conditions were found to be:

i) 5 g/l thiourea, 20 to 50% oxidized to formamidine disulphide
ii) 15 g/l sulphuric acid
iii) 40°C
iv) 40% solids

After two leaching stages of two hours each, gold extractions of up to 96% were achieved. The solution potential was controlled with addition of hydrogen peroxide to oxidize the thiourea and SO_2 gas to reduce the formamidine disulphide.

PILOT PLANT DESCRIPTION

The thiourea pilot plant was sized to match the relatively small quantity of concentrate available. The plant consisted of 2 stages of 6 mixers each in cascade. Each mixer was fabricated from a 2 l polyethylene beaker with a mean active volume of 930 ml. The total residence time per stage was 560 min. Agitation was supplied by 7 cm marine style propellers driven by 1/8 hp variable speed DC motors. Each mixer was heated to 40°C by warm water pumped through a stainless steel coil.

Solids were fed to each stage as a 75% solids by weight slurry by a variable speed progressing cavity pump. Acidic thiourea leach solutions were fed by variable speed perisaltic pumps. The target solids density in each stage was 40% solids. As the liquid and solid flows were quite small, the pumps were controlled by a variable cycle timer which was on 10% of the time. After passing through each stage the slurry overflowed to a vacuum filter which was periodically drained. The solids from the first stage were filtered and washed with thiourea solution and then water before they were added to the second stage feed. The second stage solids were repulped with thiourea solution and then filtered and washed with thiourea solution and water.

Control of the solution redox potential was maintained with 4 potential controllers using platinum/saturated calomel electrode pairs. The first mixer in each stage was kept sufficiently oxidized (20%) by the addition of 5% hydrogen peroxide, the third and fifth mixers were kept sufficiently reduced by the addition of SO_2 gas. Manual potential control was also applied as necessary.

Gold was recovered from solution prior to solution recycle by cementation onto atomized aluminum dust. The solutions were reduced by SO_2 gas to convert all the formamidine disulphide to thiourea and then 600 mg/l of aluminum dust was added and reacted for 30 min. The slurries were then filtered. The mean gold recovery was 99.5%.

The solids and solutions were fire assayed to determine their gold content. Thiourea in solution was determined by titration. Two types of thiourea were distinguished: free thiourea that is present as only thiourea, and total thiourea which is the sum of the free thiourea and thiourea as formamidine disulphide.

PILOT PLANT RESULTS

The results presented are from the third week of the pilot operation. In this period the plant operated well and data for the costing of a full scale plant was developed.

Information from the pilot plant was recorded with respect to the operating hours. Throughout the entire period concentrate was added to the first stage, and the first stage product solids were advanced to the second stage. The pilot plant was started with fresh thiourea solution added to each stage. At 21 hr second stage product solution, treated for gold recovery, was used for the first stage solution feed. At 52 hr, the treatment was stopped and the second stage solution was advanced as produced.

At 76 hr, the second stage feed solution was made up of 50% fresh solution and 50% treated first stage solution (a 50% total solution recycle). The test was ended at 106 hr.

Figures 1 and 2 (on the following page) show the gold assays of the solids in each mixer as a function of time. Although there appears to be some trends in the data, they do not correlate with the changes in pilot plant operation and have been assigned to minor variations in the pilot plant feed. The trends seem to "die down" in the later mixers of each stage.

Table 1 gives the mean gold assays from each mixer over the week's operation.

Table 1 Mean Gold Assays

Mixer	Gold (oz/ton) 1st Stage	2nd Stage
Feed	2.093	0.209
1	1.013	0.119
2	0.546	0.110
3	0.354	0.112
4	0.288	0.104
5	0.255	0.104
6	0.220	0.105
Product	0.205	0.107
Gold Recovery (%)	90.2	94.9
Product % solids	42.9	37.9

It is clear that the majority of the gold extraction occurred in the first stage with a steady decrease in the gold content of the solids through all six mixers. In the second stage most of the extraction occurred in the first mixer and there was very little change in the subsequent mixers. Batch second stage leach tests that were done concurrently with the pilot program showed that the second stage extraction was very rapid. The evidence indicates that this rapid extraction is due to desorption of gold from the solids, adsorbed during the first stage where the gold concentration in solution is much higher.

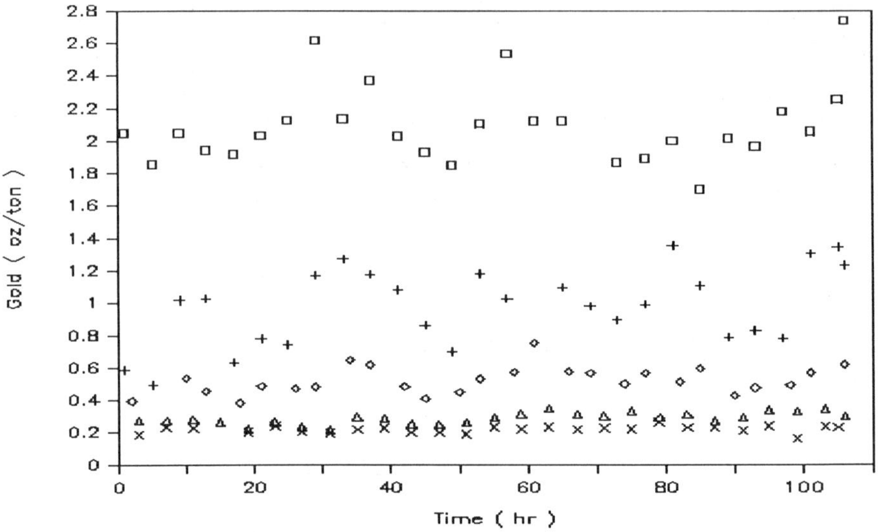

Fig. 1. First stage solids gold assays.
(□ = Feed, + = Mixer 1, ◇ = Mixer 2, △ = Mixer 4, X = Mixer 6)

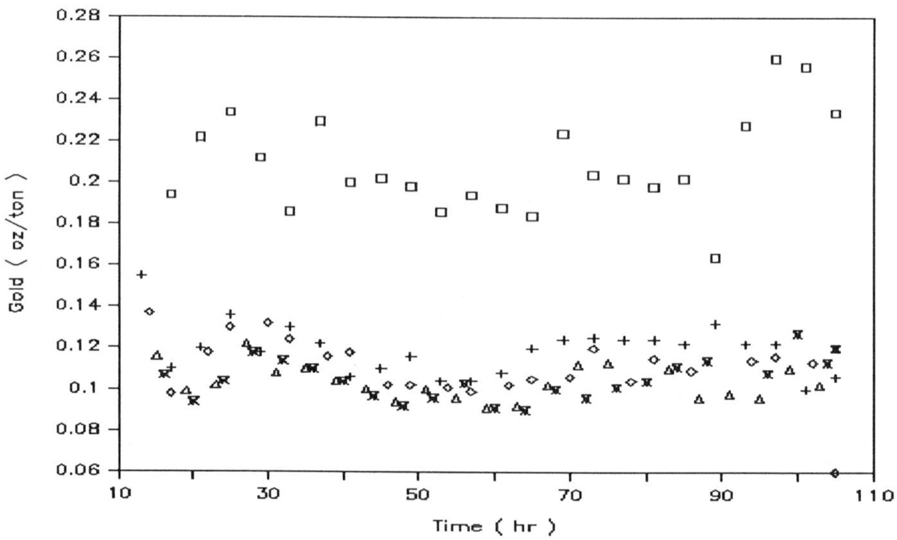

Fig. 2. Second stage solids gold assays.
(□ = Feed, + = Mixer 1, ◇ = Mixer 2, △ = Mixer 4, X = Mixer 6)

The mean gold concentration in the product solution was 45.2 ppm in the first stage and 2.1 ppm in the second. The gold concentrations were affected by the manner that the solutions were advanced and so, varied over the test period.

Table 2 shows the mean total thiourea assays. The free thiourea assays were dependent on the extent of thiourea oxidation which was controlled between 20 and 50%.

Table 2 Mean Total Thiourea Assays

Mixer	Total Thiourea (g/l) First Stage	Total Thiourea (g/l) 2nd Stage
Feed	5.00	5.00
1	4.78	4.97
2	4.72	4.91
4	4.68	4.82
6	4.60	4.75

The total thiourea consumption over the two stages averaged to 0.65 g/l. The first mixer of the first stage was the major thiourea consumer with the loss probably due to adsorption of thiourea on the feed solids.

Reagent Consumptions

Thiourea. The true consumption of thiourea was 0.65 g/l in both stages which is 0.95 kg/ton at 40% solids. With a 50% recycle of first stage product solution to the second stage and makeup, the consumption would be 4.1 kg/t. If the recycle was increased to 80%, the consumption would be 1.9 kg/t of thiourea.

Sulphuric Acid. No sulphuric acid is consumed by the reaction with concentrate. The consumption due to solution losses at 50% recycle was 11 kg H_2SO_4/t.

Hydrogen Peroxide. Sufficient hydrogen peroxide is required to oxidize 20% of the thiourea in the second stage solution: 1.7 kg H_2O_2/t.

Sulphur Dioxide. Sulphur dioxide use was not monitored in the pilot plant. In a full scale plant, SO_2 is likely only to be used to reduce the first stage solution prior to gold recovery. This would consume 3.2 kg SO_2/t assuming 40% solids and 50% thiourea oxidation.

Aluminum Dust. It was estimated that in excess of 99.5% of the gold can be recovered with 500 ppm of aluminum dust. This consumption is 0.75 kg Al/t.

CAPITAL AND OPERATING COSTS

Wright Engineers of Vancouver, B.C., prepared a cost study for a 200 ton per day thiourea leach circuit based on the data generated in the pilot study. Figure 3 (on the following page) shows a preliminary flowsheet. Wright's noted that "The circuit is relatively simple and uses easily maintainable equipment commonly found in the hydrometallurgical industry".

The total capital cost including contingency was estimated to be $2,403,900 US.

The total annual operating cost was estimated to be $1,504,000 US for an annual production of 130,000 oz of gold. The table below shows the breakdown.

Table 3 Components of Operating Cost

Item	$US/oz Au Produced
Operating labour	$ 2.59
Maintenance	0.28
Power	1.35
Reagents	5.56
Fuel Oil	1.14
Spares	0.36
Operating consumables	0.29
Total	$11.57

CONCLUSION

Over three weeks of pilot plant operation it was shown that thiourea leaching of Sonora concentrate could be carrried out on a continuous basis achieving gold extraction results comparable to batch testing, approximately 95%. A 50% solution recycle was found to produce results comparable to the use of fresh leach solution and it is anticipated that a higher solution recycle is feasible. Gold recovery was done by cementation onto aluminum powder on a batch basis and gave removals averaging 99.5%.

In 1987, Sonora Mining plans to operate a larger thiourea pilot plant as their mill comes into operation.

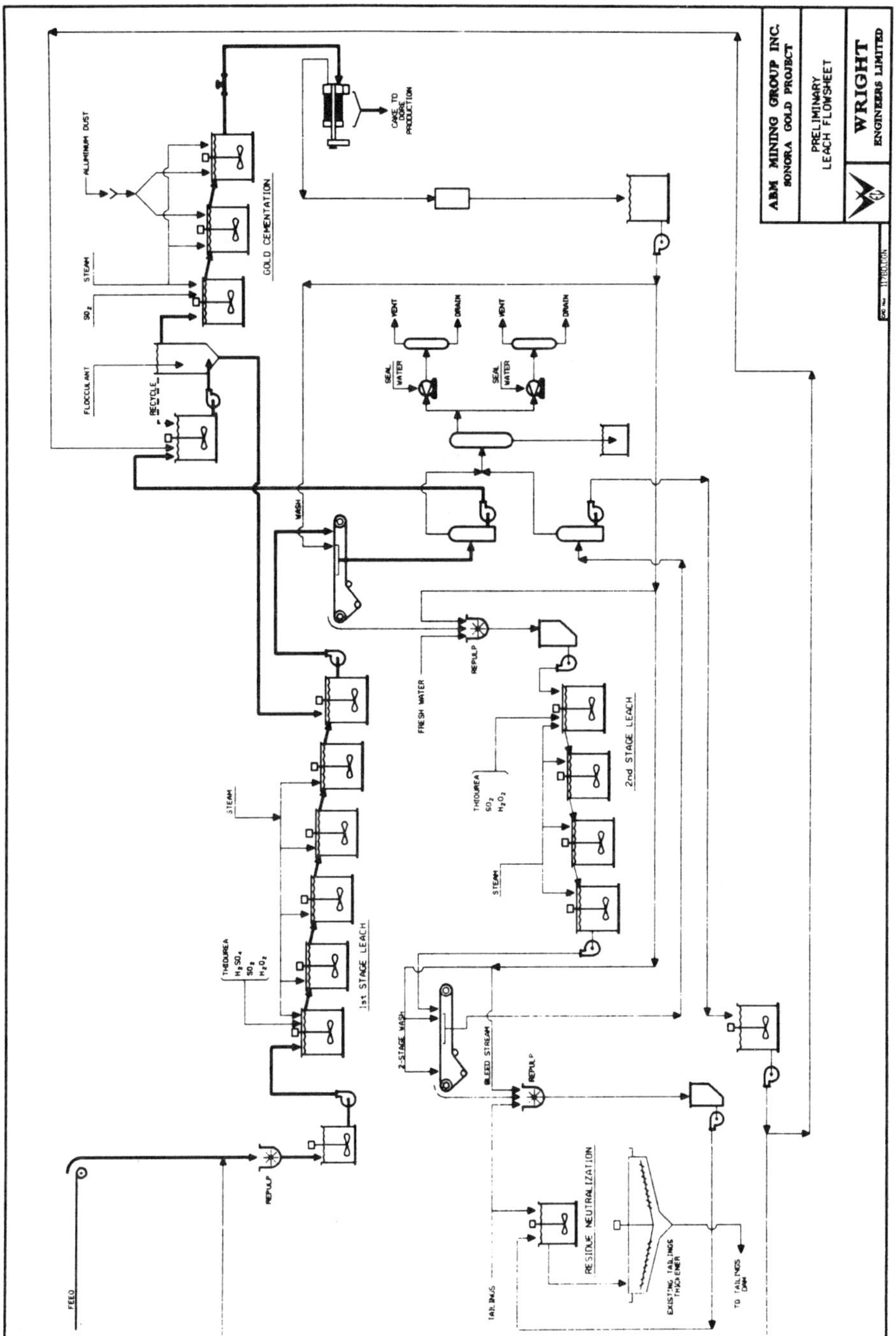

FIGURE 3. Preliminary Sonora Thiourea Leach Plant Flowsheet.

THE LINEAR SCREEN AS A CLASSIFIER IN A MILLING CIRCUIT

J.A. Tumilty, M.J. Wilkinson, M.P. Collins

Anglo American Research Laboratories, Johannesburg,
Republic of South Africa

ABSTRACT

The Linear Screen* has overcome the problems normally associated with fine screening in milling circuits by moving the screen continuously and at the same time using upward washing to remove entrained fines and prevent screen blinding. A small unit of 0,5 m^2 screening area has been tested on a bleed stream from a cyclone feed.

The results of this testwork have indicated that replacing a hydrocyclone with a linear screen in a milling circuit will result in less overgrinding of minerals, reduce their hold up within the circuit and also lead to significant increases in mill capacities. Screening installations will be large in comparison to cyclones and a compromise will have to be reached between screen open area which is directly related to capacity and filament thickness which is related to screen life.

Small scale closed circuit tests are in progress and the next step in the programme will be to totally replace a cyclone by a linear screen in one of the Anglo American Corporation's milling circuits.

*Patented by AAC and sold under licence by Delkor Tecknik (Pty) Ltd.

KEYWORDS

Linear screen; milling circuit; hydrocyclones; separation efficiency; increased capacity; screen wear; screen blinding.

INTRODUCTION

Screens have an inherent advantage over hydrocyclones when they are used as classifiers in milling circuits in that they differentiate on size alone rather than a combination of size and density. A hydrocyclone underflow also contains an inherent proportion of fine particles associated with the water. These particles are returned to the mill, resulting in inefficiency and overgrinding, particularly of the heavy minerals and this leads to losses in subsequent processes such as flotation and gravity separation. In the past, problems associated with high cloth wear rate and severe blinding have militated against the use of fine screens in milling circuits.

The Linear Screen has overcome these two problems by continuously moving the screen cloth, as in a belt filter, and at the same time using upward washing to remove entrained fines and prevent blinding. This also obviates the need to vibrate the screen, which is probably another major factor in screen wear.

THE LINEAR SCREEN

A diagram of the linear screen is shown in Fig. 1.

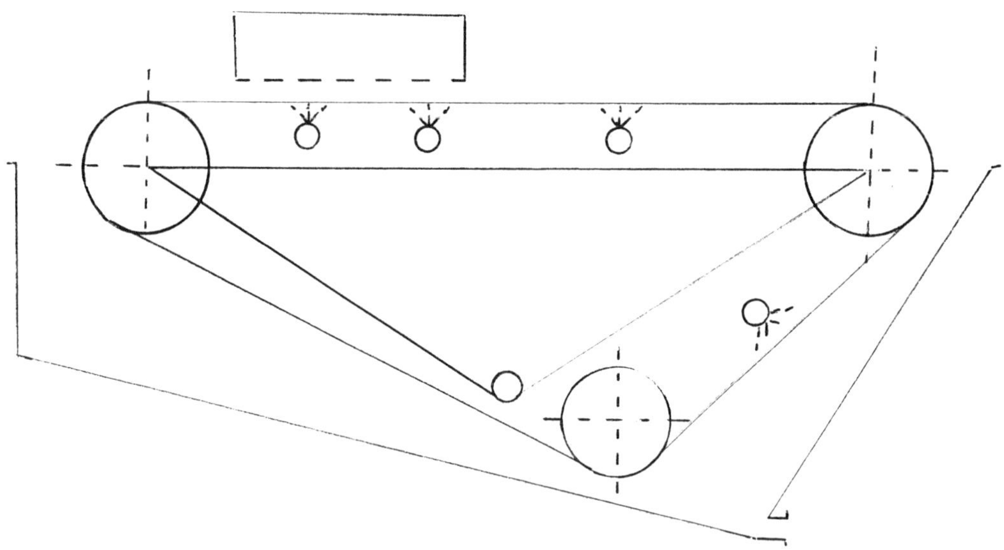

Fig. 1 The linear screen

It consists essentially of an endless screen cloth which is moved by a drive roller and supported by smaller rollers to prevent pooling of material on the screen cloth. Pulp is fed into the feed box which distributes it across the width of the belt. In the prototype unit, the feed box covers approximately a quarter of the screen area but it is believed that an overflow type feed arrangement producing a curtain of pulp will prove to be a better design. As the pulp falls onto the screen cloth, the majority of the undersize material will pass through the screen cloth and be collected in the undersize collection box below the cloth. The oversize and the remainder of the undersize will form a bed on the cloth. A series of wash water jets across the width of the screen disturb this bed, allowing more fines to pass through the screen cloth in the adjacent quiescent zone. In this test unit, a further two rows of sprays were used to rewash the bed on the belt. The number of spray bars used and the length of the belt will clearly be function of the material being screened, the amount of oversize material and the amount of misplaced material that can be tolerated in the oversize.

The oversize material is carried over the end of the drive roller and is washed off the screen into the oversize collection box. Washing is again effected from behind the cloth. The washed screencloth then continues over the tracking mechanism and a clean screening area is presented to the feed.

The long screen life and lack of blinding can be attributed to various factors. Firstly, blinding is prevented by washing the bed from underneath the cloth. Thus any near size material is lifted away from the cloth aperture rather than being forced into it as with a conventional screen that is washed from the top. It is also minimised by again using wash water from behind the screen cloth to remove the oversize. This effectively presents what is essentially a new screen cloth to the feed.

Screen wear is also minimised by this backwashing mode of operation. In addition, instead of particles flowing across the screen cloth, as in a conventional inclined screen, they only move in a direction perpendicular to the screen, thus minimising abrasion. In addition, because particles are not forced through the screen by water washing from above, gouging of the screen cloth fibres is prevented. Indeed, particles passing through the screencloth are carried in a lubricating medium (water) and this further reduces wear. The final factor is that because the screen cloth is not vibrated, abrasion between the warp and weft threads is avoided.

The linear screen has been successfully used for woodchip removal ahead of CIP/CIL plants and in this application screen cloth lives of 3 months or longer have been achieved.

TESTWORK

General

It was decided to test the concept of using a linear screen to replace a cyclone by installing a small test unit of 0,5 m^2 screening area at milling plant on one of the gold mines administered by the Anglo American Corporation of S.A. This milling circuit utilises a single stage cyclone. A bleed stream was taken from the cyclone feed and passed over a 5 mm scalping screen to remove tramp oversize. The tramp oversize represented around 6 per cent of the feed to the unit and its removal was necessary to prevent the feed distribution box from choking.

The initial testwork was aimed at establishing the effect of the following parameters on screen performance using a 180 micrometre screen cloth

belt speed
wash water flowrate
wash water nozzle type
wash water nozzle positioning

The capacity of the screen under a given set of conditions was determined by increasing the pulp feed rate until the screen flooded. The feedrate was then reduced by around 5 per cent and this system worked well.

The results were assessed by calculating the proportion of each size fraction that was recovered to screen oversize (mill return) and screen undersize (product). A plot of the results allowed the cut size (d50) to be determined as well as the separation efficiency. The latter is defined as the ratio of 80 per cent recovery size to the mill return to the 20 per cent recovery size and is an indication of the slope of the Plitt curve.

Initial Testwork

The screen worked most efficiently at full belt speed and the wash water nozzle positioning had only a minor effect on screen performance. Not surprisingly, the separation efficiency of the screen was determined by the amount of wash water and the nozzle type. The use of finer nozzles reduced the required amount of wash water from 4 to 1,4 m^3/hr to produce the same split size and separation efficiency. This clearly demonstrates that the function of the wash water is one of disturbing the bed to allow further drainage rather than washing of the fines from the coarse material. Reproducible results were being achieved at this stage, the cut size being between 175 and 180 micrometres and the separation efficiency 2,1 at a feedrate of 3,46 tph/m^2. Attempts to increase the feedrate even slightly (to 3,6 tph/m^2) resulted in the screen flooding and the separation efficiency decreasing to 2,6. Attempts to amplify the bed disturbance by rapping the screen cloth resulted in a small increase in separation efficiency and capacity but could, in the long term, have an adverse effect on screen cloth life.

Tests with Coarser Screens

Further tests were performed using 200 and 300 micrometre screen cloths. Screening efficiency remained virtually the same but there was a significant increase in capacity. The average maximum feedrates for the three sizes of cloth were:-

180 micrometres - 3,5 tph/m^2

200 micrometres - 5,2 tph/m^2

300 micrometres - 7,2 tph/m^2

Screen analysis results for products using the three screen cloths are given below.

TABLE 1. Screen Analysis of Products from Different Sized Screens

Size micrometres	Cum. % passing		
	180 um	200 um	300 um
417	99,7	98,7	99,4
295	99,1	96,9	96,0
208	95,6	90,9	87,0
150	84,5	76,4	73,3
105	72,7	63,4	60,1
75	59,8	51,4	49,1
45	43,5	37,4	35,6

In all three cases, the d50 size was within ten percent of the rated aperture size and the separation efficiency around 2,0. Measurement of the open areas of the screen cloths showed a good correlation between open area it and screen capacity.

TABLE 2. Screen Open Area and Capacity

Cloth Aperture um	Per Cent Open Area	Capacity tph/m^2
180	16,0	3,5
200	24,5	5,2
300	29,6	7,2

It is believed that this will provide a valuable tool in screen sizing although in practice there will have to be a compromise between having as large an open area as possible i.e. thin filaments and screen cloth life.

MODELLING OF THE LINEAR SCREEN

In order to allow a simulation to be made on the effects of replacing a cyclone by a linear-screen, a model had to be developed. This was done using the following formula, the least squares method for minimising the errors being used to evaluate the constants in the equation (Plitt formula).

$$Oi = \frac{Fi \quad \exp(\psi .x\ i/d50) - 1}{\exp(\psi .x1.\phi d50) + \exp(\psi) -2}$$

Fi	-	flow of size class i in feed
Ai	-	flow of size class i in oversize stream
Xi	-	mean size of size class i (micrometres)
d50	-	cutsize (measured for each test) in micrometres
ψ	-	constant depending on separation efficiency.

The data derived from the Plitt formula was in good agreement with the measured data as the data for the 180 micrometre screen shows.

TABLE 3. Measured and Calculated Data

Size um	% Recovery to oversize	
	Measured	Calculated
417	99,3	98,9
356	96,9	96,7
251	83,9	80,8
179	52,4	52,4
125	30,1	28,9
90	17,5	15,4
60	11,0	7,9
45	6,1	5,2

The measured and calculated values for the 200 and 300 micrometre screens also showed good agreement.

COMPARISON OF THE LINEAR SCREEN AND CYCLONE

Data required for comparison was collected over a period of 6 weeks from the plant primary cyclone. All three streams were sampled once a week and analysed. The required flowrates were obtained from a data logger that had been installed as part of another test programme.

The product sizes produced from the plant cyclone and two sizes of screen cloth fitted to the linear screen are given below. The plant cyclone overflow average was slightly coarser than normal.

TABLE 4. Size Analysis of Products from a Cyclone and Linear Screen

	Cum. % passing		
Size um	Cyclone	180 um Cloth	300 um Cloth
417	100,0	100,0	99,9
295	99,8	99,4	96,9
208	99,1	98,4	87,0
150	91,4	83,9	72,8
105	79,3	69,0	60,4
75	66,4	57,2	49,7
45	49,3	41,1	36,1

The product using the 180 micrometre screen has a similar top size to the cyclone overflow but in all other aspects is coarser. The d80 figure is around 140 micrometres compared to 105 micrometres for the cyclone. The product from the 300 micrometre cloth is significantly coarser in all respects.

With respect to the material being returned to the mill (cyclone underflow or screen oversize), around 25 per cent of the minus 75 micrometre material was returned from the cyclone while the corresponding figure for the screen was less than 10 per cent. The separation efficiency of the cyclone was around 3 while the corresponding figure for the screen was around 2 (the ideal separation gives a figure of 1,0).

Of probably greater importance is the deportment of values into the classifier products. In the case of the hydrocyclone, because it classifies by size and density, the gold and sulphide sulphur values are upgraded in the mill return; from 6,5 to 10,7 g/t gold and 1,8 to 3,1 per cent sulphide sulphur. In the case of the screen, the minerals are removed from the circuit as soon as they are of a given size and the results indicated a downgrading into the mill return; from 11 to 13 down to 2,3 g/t gold and from 1,8 down to 0,6 to 0,9 per cent sulphide sulphur. Using a bleed stream to feed the screen naturally tends to exaggerate this phenomenon since if a screen totally replaces a cyclone, the mill return will be bled of values and the screen feed assay will fall. However, it will prevent overgrinding, particularly of the sulphide sulphur minerals and this will have important consequences for subsequent flotation processes.

The products from the linear screen and the cyclone overflow were subjected to a standard cyanidation test.

The product from the 180 micrometre screen and the cyclone overflow produced the same gold extraction (96 per cent) while the products from the 200 and 300 micrometre screens produced lower extractions (91 per cent). However, when these cyanidation residues were subjected to diagnostic leaching, the final residues were as follows.

TABLE 5. Final Residue Values from Diagnostic Leach

Classifier	Final Residue g/t Au
Hydrocyclone	0,06
180 micrometre screen	0,14
200 micrometre screen	0,11
300 micrometre screen	0,08

Diagnostic leaching is a technique where the ore matrix is sequentially destroyed using progressively stronger oxidants and the gold released at each stage recovered by cyanidation. The final residue represents gold present in silica and the above figures therefore indicate considerable potential for using relatively coarse screens providing a flotation-leach type circuit is used.

EFFECT ON MILLING CIRCUIT CAPACITIES

The reduction of fines returning to the mill resulting from the use of a linear screen rather than a cyclone will allow a greater tonnage to be passed through the mill. Computer simulation using selection and breakage functions were used for the S.A.G. mills using data generated from the single stage circuit and data supplied by plant personnel for the two stage circuit. Standard simulation packages were used for the ball mills and hydrocyclones while the linear screen fits the Plitt model well. In the case of the S.A.G. mill, the material which cannot pass through the discharge grate was used to limit tonnage increases; the flow in this stream at 100 tph with a cyclone was taken as acceptable.

In the case of a single stage S.A.G. mill in closed circuit with either a cyclone or a linear screen fitted with an 180 micrometre cloth, the following predictions were made.

TABLE 6. Predicted Circuit Data Using a Cyclone and Linear Screen

	Cyclone	Screen
feed tph	100	110
percentage circulating load	200	172
kwhr/ton -75 um material produced	33	30
% minus 75 um in product	75	76

These figures assume that the removal of fines from the circulating load will not be detrimental to the milling action in the mill.

In the second circuit considered, the product from 3 S.A.G. mills is cycloned. The primary cyclone overflow is recycloned, the cyclone underflow being routed to a ball mill which is in closed circuit with a cyclone. Using linear screens, the circuit would be modified in that the S.A.G. mills would in a closed circuit with a 3 mm screen and become almost open circuit, the recirculating load dropping from 200 to 5 per cent. The undersize from the 3 mm screen would be rescreened at 300 micrometres, the oversize being ground in closed circuit with the ball mill. The computer simulation predicts that the mill throughput would rise from 300 to 390 tph and the power consumption would drop by 18 per cent.

CONCLUSIONS

The indications from this initial testwork using a bleed stream from a plant cyclone have clearly indicated potential for replacing hydrocyclones in milling circuits by linear screens. The envisaged benefits are reduced overgrinding of sulphide minerals in particular, a major decrease in the hold up of valuable minerals in the mill and significant mill capacity increases.

The next step in the programme is the testing of a small scale closed circuit mill and screen followed by the installation of full sized screen in one of the Group's milling circuits to totally replace a hydrocyclone.

ACKNOWLEDGEMENTS

The work described in this paper was sponsored by the Gold and Uranium Division of AAC and is published by permission of AAC. We also acknowledge the help and assistance provided by Delkor Technik (Pty) Limited.

GOLD ORE DRESSING PLANT OF THE POURA MINE (BURKINA FASO, EX-UPPER VOLTA, WEST AFRICA)

G. QUEDRAOGO*, P. OLLIVIER**, B. QUEDRAOGO*,
J.F. THOMASSIN**

* SOREMIB BP 5662 OUAGADOUGOU BURKINA FASO
** ASSIGNED BY BRGM & COFRAMINES

ABSTRACT

SOREMIB, the Société de Recherches et d-Exploitation Minière du Burkina Faso, was formed in 1973 to resume the operation of the Poura gold mine in Burkina Faso.

The capital of SOREMIB is owned by the State of Burkina Faso (60%), COFRAMINES, a BRGM subsidiary (20%), and the Islamic Development Bank (20%). Since 1980, SOREMIB has demonstrated the technical and economic feasibility, raised the necessary funds, and installed the means to work a gold quartz vein whose reserves are estimated at 1,5 million tonnes of ore assaying 14.9 g/t.

The ore processing plant, with a capacity of 600 t/day, was inaugurated on 18 October 1984 by comrade Thomas Sankara, President of CNR, President of Burkina Faso, and by the end of 1986 had already produced over 4 600 kg of dore metal assaying 95% precious metals.

Poura mill is a flotation plant with jigging to recover coarse gold and cyanidation of the floated concentrate followed by CIP and Zadra Processes for gold recovery.

KEYWORDS

Gold process; flotation; cyanidation; carbon in pulp.

GENERAL

Geographic Situation (Figure 1)

The Poura Mine lies at a distance of 180 km by road southwest of Ouagadougou, near the Ghanaian border. It lies at the centre of an exploration concession called the "Permis Volta Noire" (Black Volta Concession) with an area of about 1200 km^2, held by the Societé de

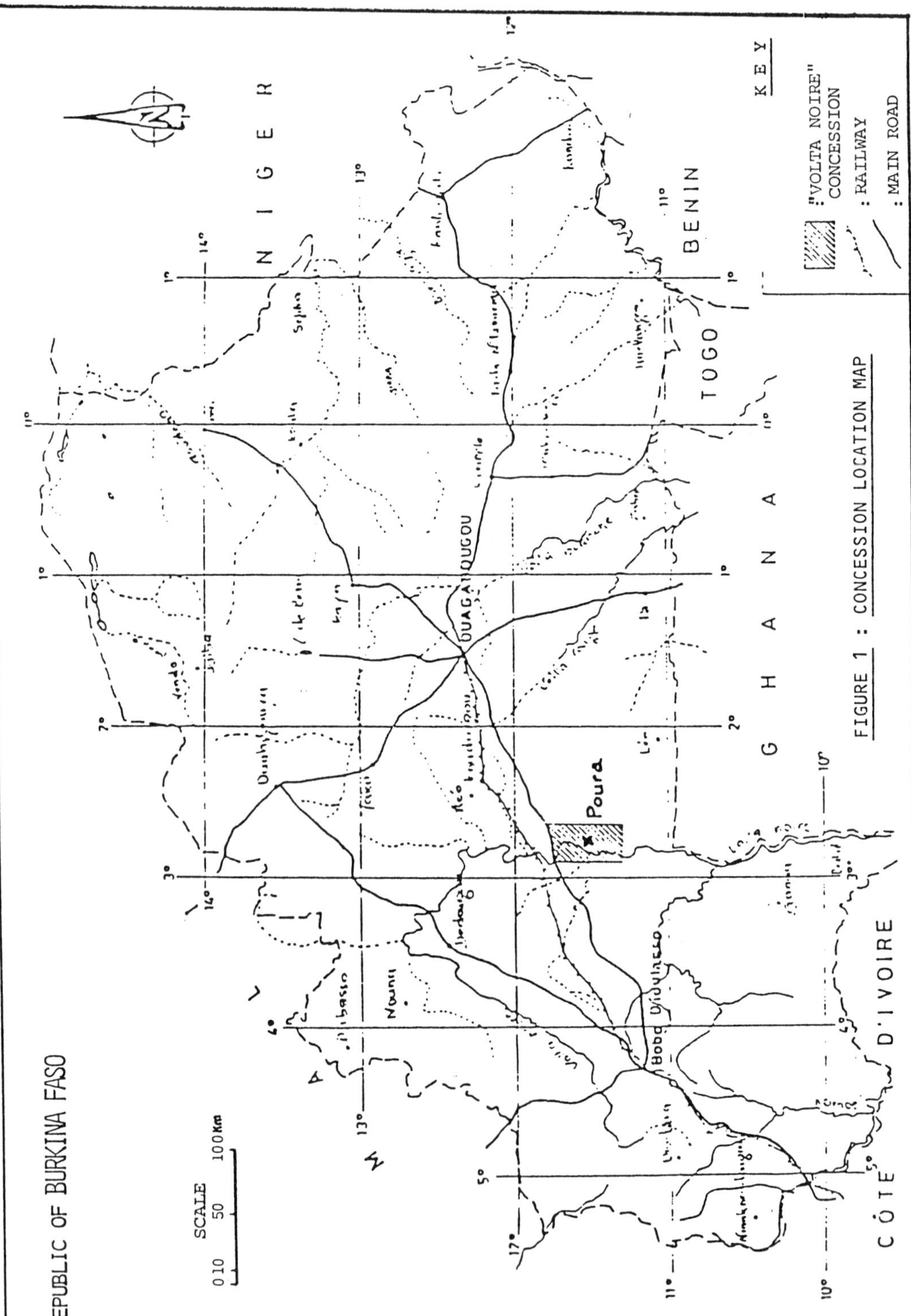

FIGURE 1 : CONCESSION LOCATION MAP

Recherches at d'Exploitation Minière du Burkina (SOREMIB), a semi-public company responsible for mining the Poura vein system and for prospecting the area covered by the exploration concession for both gold-bearing and base metal mineralization.

The climate is of the Sudan/Sahel type with contrasted rainy and dry seasons. The rainy season is from May to October, and the dry season from November to April. The vegetation is in general a shrubby savannah, with rare gallery forests along some streams, and is denser in the rainy season, masking the outcrops.

The region is a broad, gently undulating peneplain with a mean altitude barely exceeding 300 m above sea level. The reliefs consist of three main elements : tabular associated (features) with laterite caps, domes of isolated hillocks consisting of volcanic-sedimentary rocks, and ridges formed by steeply dipping layers (quartz veins in particular).

The Concession area extends on either side of the Black Volta, a permanent river, whose main tributaries in the area are the Bolo, the Grand Balé and the Sanbayorou, which flow with the first rains. More than half of the water consumed by the dressing plant is taken from this river.

General Survey of the Mining District

The Poura region is situated in the western part of the Mossi plateau, consisting of Archean gneissic, migmatitic and granitic rocks. The centre of the Concession, is occupied by Lower Proterozoic Birrimian sedimentary rocks which constitute the mining district of Poura. They form the N-S Boromo belt, extending from the Btié Gaoua area in the south into the north of Burkina Faso. It is characterized by volcanic-sedimentary sequences with basic rocks passing through intermediate to silicic towards the top. The rocks are strongly folded, steeply dipping and affected by major fractures trending mainly N 160° E, N 60° E, and essentially N-S. Lowgrade metamorphism affects the entire belt. The rocks have been strongly affected by later hydrothermal activity and intrusions. The exploration Concession Sector contains many quartz veins, which are contemporaneous with this phase. Among them, the mining district (the Poura vein system) is the most important.

History

The Poura region contains many gold shows long known by the local population. Some sources consider that the old gold washings still visible were carried out for Samoury Toure (1870). The potential mining value of this region has therefore attracted attention and led to many exploration surveys. These resulted in several periods of exploitation the most important being that of the Societé des Mines de Poura, which discontinued its activities in 1966.

SMP mining operations were chiefly on the Plain vein and secondarily on the Montagne vein. Considerable data were obtained on the thicknesses and contents of these quartz veins down to the - 267 m level. The geological data were obtained in the mine workings themselves, and by boreholes from the galleries. SMP has installed a dressing plant in the mining village of Poura at the southern end of the Plain vein. All the authors who have examined the veins and the goldbearing mineralization consider that the gold is essentially free, but rarely visible to the naked eye. Mr. Vincent (1970) employed by SMP to extract the metal : "...the ore was crushed, ground and floated ... the flotation concentrates were amalgamated in an apparatus invented by SMP, part of the free gold was recovered in traps ... The flotation concentrates contained about 1600 g/t of gold. After amalgamation, they still contained about 200 to 500 g/t of gold and were sent to Boliden in Sweden for smelting and recovery of the gold ... The flotation tailings assayed 3 g/t on average".

Between 1960 and 1966, 419 000 tons of ore were extracted, producing 5620 kg of gold and 845 kg of silver. SOREMIB was formed in 1973, and resumed exploration by a drilling programme on the Plaine vein in 1975. The report on this survey in 1976 indicated the existence of 1.5 Mt of ore containing 14.9 g/t above the 327 m level. Following feasibility studies which forecast a mining period of ten years, SOREMIB reopened the mine in 1981.

Description of Raw Ore

The raw ore is a vein of massive to fractured milky quartz in which the gold is deposited essentially in the cracks. In the surface layer down to about 20 m, oxidation has converted the sulphides (pyrite) to iron hydroxide (limonite). Sphalerite, covellite, arsenopyrite and galena are still found. Deeper down, the sulphides commonly fill the fractures. Pyrite and arsenopyrite occur as grains attaining several millimetres. Sphalerite, galena, chalcopyrite, grey copper combined with arsenopyrite are present either included or as independent minerals.

The gold is essentially free, sometimes included in pyrite or arsenopyrite in fine microscopic flecks. The general chemical analysis totals nearly 100% of only three elements expressed as oxides:

. SiO_2	91.7%
. Fe_2O_3	6.3%
. Al_2O_3	1.0%

The sulphur content in the surface ore is 0.013% and the main trace elements have the following average contents in g/t:

. As	200
. Cu	42
. Zn	110
. Co	60
. Pb	38

Preliminary Study

Pilot laboratory. SOREMIB asked BRGM (Mineral Dressing Department) to conduct a laboratory and pilot test campaign to determine an overall ore dressing scheme for the oxidized and sulphide zones. The tests were performed in late 1981 and yielded good results in gold recovery by grinding and flotation, confirmed by a test in the OrLéans La Source pilot plant. These results confirmed the process already planned by the last mining operation. However, the final gold recovery process of amalgamation was replaced by cyanidation, which achieves much better gold extraction.

The carbon-in-pulp process tested in the laboratory was proposed to extract the gold from the cyanide pulp, and combined with carbon desorption by the modified Zadra process, enables the on-site production of gold-bearing ingots with a high precious metal content.

Dressing plant engineering. The basic and detailed engineering studies were entrusted during 1981 and 1982 to the Krebs Company, which gave SOREMIB a complete dossier for ordering the equipment and the erection plans for the plant. The basic and detailed engineering of the mechanical ore dressing shop was carried out by SOFREMINES.

INSTALLATION

Equipment Erection and Time-table

The ore dressing plant was built by two Burkina Faso companies associated with French companies, UDEC with Campenon Bernard for civil engineering, and SOGEFIA with DeLattre-Levivier for the structural work and equipment assembly. The civil engineering began in the last quarter of 1983, and the first units (thickeners) were installed in late December 1983.

Crushing was operational at the end of August 1984 and started up in early September. The main equipment (mills, flotation) and the piping were erected in nine months from January to September 1984. The Hardinge mills were run in at the beginning of October, and the installations were inaugurated on 18 October 1984 by comrade Thomas Sankara, President of Burkina Faso. The ore dressing plant was built and went on stream in less than one year.

Supplementary Works

During construction, SOREMIB, using its own resources, installed the settling basins, the control laboratory, and the tailings treatment system.

Tailings settling basin. The tailings disposal system of the ore dressing plant was planned and constructed in 1984. Three settling basins measuring 200 x 200 m each and 4 m deep were built with excavated material. A bulldozer was leased to spread the earth and to build the walls and access roads.

The water level control sluices for the basins were built by a local contractor. 6000 m of pipes were laid and cemented, by the first class of plant workers trained at the Centre, in three parallel lines of 200 m, two for pulp disposal, and one line for clarified water recycle. A submersible recycle pump was installed on a raft.

Construction of the control laboratory. The analytical laboratory installed since 1981 in a building of the former plant soon became too small, and its enlargement was planned in 1983. An application for funds was filed with the funding institutions, and construction began in September 1983 by a local contractor. This laboratory houses the equipment for analysis by fire assay and for production control. Sanitary blocks were also provided for the staff.

Treatment of cyanided tailings. Special attention was paid to effluent treatment. A separate installation was built to treat the cyanided effluents by destruction of the cyanides with calcium hypochlorite in a agitated tank. After chemical analysis, the detoxified pulp is loaded into a mobile 8 m^3 tank, which is sent to the 5 x 10 m concrete evaporation basins for storage.

MILL OPERATION

The plant processes 600 t or ore per day. The diagram in Fig. 2 shows the operations, from ore loading from stock to the discharge of tailings into the settling basins.

Ore Loading

The ore is brought in by 35-ton Euclid Lorries directly on the crushing platform. Lumps bigger than 500 to 600 mm in size are broken by a hydraulic rock breaker. Steel scrap and wood from the old workings are removed by hand. The crusher is fed by a Caterpillar Loader type 950.

Crushing

Raw material is crushed to grinding size in three stages:

. primary crushing to -100 mm,
. secondary crushing to -30 mm,
. tertiary crushing to -12 mm.

The last two crushing units are sequenced in a single installation.

Primary crushing. The ore picked up by the loader is dumped directly on a Babitles BP21 primary crusher, a 2 m diameter, 0,5 m wide gyratory crusher. The ore crushed to -100 mm falls into a rock box feeding an 800 mm wide conveyor belt. The throughput of the installation is 200 t/h. A stock of 5000 t of ore is built up at the conveyor belt pier.

Secondary and tertiary crushing. The stock of -100 mm ore is withdrawn in a tunnel by two vibrating extractors which feed a 500 mm wide conveyor belt. After automatic detection of steel scrap and removal by hand, the product is sized on a Bergeaud 4 x 1,2 m two-stage vibrating screen, cutting at 30 and 12 mm. The 30 mm oversize is reduced to 15 mm by a 3 foot standard Symons gyratory crusher and the oversize is reduced to 12 mm by a 3 footshort-head Symons gyratory crusher. The products are collected on to the same conveyor belt and returned to the inlet of the Bergeaud screen for rescreening.

The final -12 mm product is sent to the -12 mm stockpile with a capacity of 5000 t.

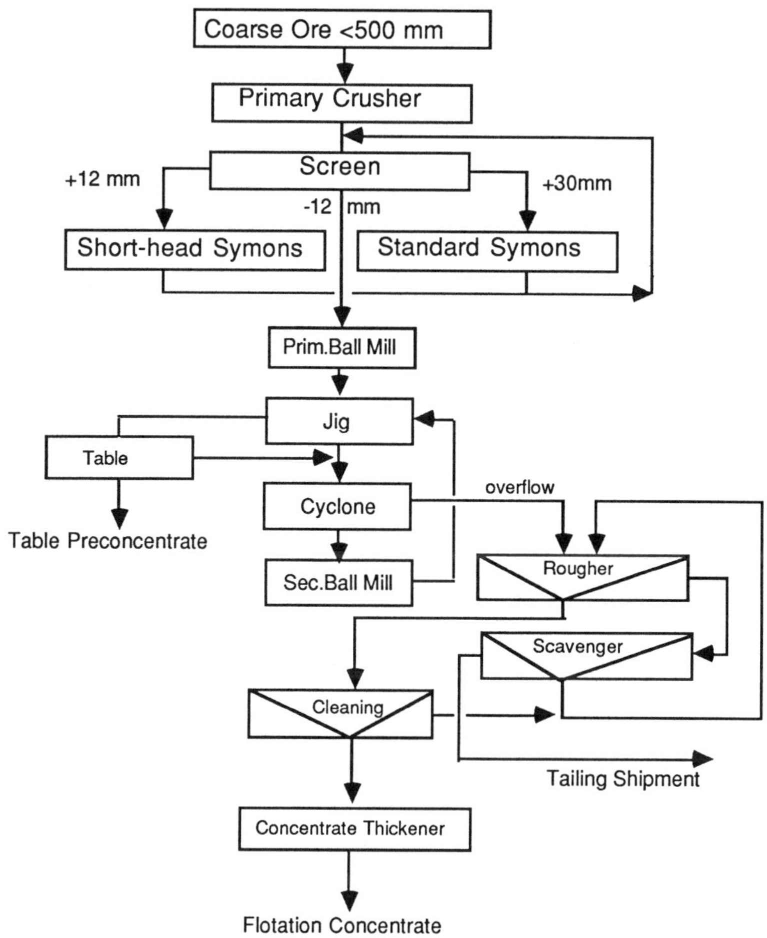

Fig. 2 Flowsheet of physical treatment

Grinding and Sizing

The ore is taken from the -12 mm stockpile by two vibrating extractors and a 500 mm wide conveyor belt and fed to a 10 m^3 surge hopper at the plant flow rate of about 26 t/h.

Sampling. At the conveyor belt fall, a primary rotary bucket sampler takes about 325 kg/h of ore and dumps it into a 50 l hopper which feeds a jaw crusher. The ore reduced to -3 mm is sampled by a rotary type secondary cutter at 4 kg/h. One approximately 30 kg sample is thus prepared per 8 hour shift to determine the gold content on entry to the plant. The sampling rejects are returned to the plant on the feed conveyor belt.

Transport. A belt extractor under a hopper with a speed adjustable by a Texrope mechanical variable-speed drive unit serves to adjust the ore feed rate to the mill. A Hassler integrating balance weighs the ore, and an electronic servo-controlled system correlates the speed of the conveyor belt with a given instantaneous throughput (t/h).

Grinding and classification. Grinding is carried out in two stages by two ball mills separated by classification of the ore pulp by hydrocyclone.

Primary grinding. The Hardinge two-cone, primary mill with a shell length of 1.5 m and 2.7 m diameter is lined with rubber (10 cm thick) and rotates at 19 r/min. It contains 22 t of 80 mm diameter balls of forged steel, and is driven by a 260 kW motor coupled by a Citroen reduction gear.

The entering -12 mm ore is completely ground to -4 mm, exiting in a pulp containing 65% solids.

Secondary grinding and classification. The -4 mm ground pulp is sent to a jig which acts as a trap (see Section Gravity Separation) to recover the coarse gold, and is then dumped into a constant-level tank to feed the cyclone. A rubberized centrifugal pump, with a delivery of 110 m^3/h, raises the pulp containing 50% to 55% solids to a height of about 10 m, and feeds a 500 mm diameter tilting cyclone. The pulp grain size is cut at 200 μm. The fines are sent in a dilute pulp containing 25% to 27% solids to flotation, and the sands (underflow) are dumped into a secondary mill in a thick pulp containing 70% solids. This secondary mill is identical in all respects to the primary mill. It contains 22 t of 40 mm diameter balls. The sands and the cyclone underflow are ground to 1 mm. The circulating load is established at about 140%.

Gravity Separation

The ground ore pulp leaving each of the mills is collected by a rubberized Y-chute, which feeds a Yuba type double-cell jig, each cell measuring 1 x 2 m. The gravity trap placed at the exit of each mill helps to recover the liberated coarse gold larger than 100 µm.

The dense mineral concentrate leaves through the tip of each hutch and is sent by gravity to a Wilfley 76 shaking table measuring 4,7 x 1,8 m, which separates a gold-bearing preconcentrate, stored in a mobile tank, and the table tailings pumped to the cyclone feed tank. Middlings recycle to the secondary mill enables additional liberation of the gold contained in the grains leaving the primary mill.

Preconcentrate treatment is discussed in Gold Room.

Flotation

This section has two lines of cells in parallel.

. Roughing and scavenging cells, two arrays of six cells of 2.8 m^3 each, installed in series, treating about 80 m^3/h of pulp containing 25% solids. An air booster supplies the air required for flotation. The scavenging foams are recycled to the flotation inlet.

. Cleaning cells for the rougher and scavenger preconcentrates, one array of ten cleaning cells of 0.5 m^3 each which are self-exhausting. The cleaning middlings return to the flotation inlet.

The flotation reagents, potassium amylxanthate and phosocresol, are added at the inlet to each array of cells in a dilute solution conveyed by proportioning piston pumps. Skimmer mechanisms extract the mineralized froth on the surface of each cell and rubberized centrifugal pumps convey the pulps. The flotation tailings are pumped continuously to the tailings disposal tank.

The pulp is sampled before and after flotation by a rotating cutter sampler. The pulp sampled is collected, settled and sent to the analytical laboratory for treatment. The volume sampled is 40 l, or about 20 kg of material per 8-hour shift.

Cyanidation of the Flotation Concentrate

Settling. After cleaning, the flotation concentrate is screened to remove any foreign particles (plastic, electric wire, wood fragments). It is sent by centrifugal pump to a 7 m diameter thickener, which allows settling of up to 35% of the solids. The clear supernatant water is returned to the grinding circuit. The addition of flocculant helps to improve the clarification of the recycled water and to increase the settling rate.

Cyanidation. The flotation concentrate, thickened to 35% solids at the bottom of the settling tank is sent in a 13 m^3 pachuca (2 m diameter, 6.2 m deep) for leaching by sodium cyanide. The prepared attack solution containing 300 g/l NaCN is added to the pulp to attain a concentration between 35 and 50 kg of sodium cyanide per ton of concentrate, or 0.1 kg NaCN/t of raw material. Lime (or cement) is added in powder form to keep the pH at 11. A pulp alkalinity test procedure is set up to forestall any problems.

The gold is dissolved during a period of at least 24 h. The gold content of the residues and of the solution is checked during each operation on a pulp sample to monitor the progress of dissolution. The concentrates assay an average of 5000 g/t and the tailings 20 g/t.

Gold recovery by carbon-in-pulp. The gold-bearing cyanided pulp is pumped to a battery of six 5 m^3 capacity pachuca type reactors containing 250 to 300 kg of activated charcoal kept in suspension in the pulp by a compressed air stream. The pulp resides between 8 and 10 h in the reactors, and the dissolved gold is selectively adsorbed on the activated charcoal. The charcoal from the first tanks fixes most of the gold, and the final ones totally exhaust the solution.

The waste pulp is sent to the cyanide destruction unit. The gold-bearing charcoal is extracted by screening on a 700 μm vibrating screen, washed and sent to the treatment unit installed in the gold room for recovery of the gold by desorption and electrolysis.

Gold Room

The gold room contains all the process, allowing both purification of the preconcentrates prepared during treatment, and the fabrication of dore ingots.

Treatment of gravity preconcentrate. The gravity preconcentrate leaving the large table is treated daily in the gold room on a small Wilfley 1313 shaking table measuring 1.27 x 0.61 m, in two passes. The first pass serves to prepare a concentrate containing an average of 70% of precious metal, which is melted directly. The second pass serves to extract a gold-bearing pyrite concentrate containing an average of 30% to 40% of precious metal, which is upgraded by drum amalgamation. The amalgam is then treated in the furnace to distill the mercury around 500°C and to recover a gold sponge which is melted separately into ingots.

Treatment of gold-bearing activated charcoal. The gold-bearing charcoal extracted from the carbon-in-pulp installation is stored in a 1.1 m^3 stainless steel tank and treated in a batch desorption operation by a 1% soda liquor containing 15% alcohol heated to 80°C, with the electrolysis of the gold-bearing liquor.

The complete desorption of a 500 kg batch of charcoal takes less than 36 h. After draining of the column and the circuit, the electrolyser cathodes are recovered, weighed and fluxes added (sodium carbonate, silica, borax, sodium nitrate) for melting in the Wabi furnace at 1100° to 1200° C. The impurities (iron, copper) are slagged and the first-melt metal is cast in the form of buttons. The purity of these buttons is close to 95% of precious metal.

Ingot casting. The first-melt buttons leaving gravity separation, electrolysis or amalgamation are as far as possible combined and melted to cast a dore metal ingot weighing about 6000 g (200 oz).

The second, shaping fusion normally takes place in the presence of borax alone in a silicon carbide and graphite crucible, with casting in a small, cast iron ingot mould. The lighter liquid slag is removed by overflow.

The ingots are marked, weighed, registered and crated for shipment and treatment by specialized refining companies.

Plant Tailings Treatment

The ore dressing rejects are essentially tailings from the flotation shop and tailings from the cyanidation of the flotation concentrate.

Mill tailings and water recycling. The mill tailings are stored in a 14 m^3 duplex tank before transmission by two groups of parallel pumping systems consisting of two centrifugal pumps in series with a total capacity of 75 kW.

The doubled pipeline 1400 m long 150 mm diameter, is made of high-density polyethylene. It dumps the tailings into three settling basins 200 x 200 m each. After the particles settle, the decanted water is recycled to the plant reservoir basin by a raft-mounted 20 kW submersible pump. The return line is 2000 m long and the average delivery of the pump is 45 to 80 m^3/h.

Cyanided wastes. After cyanidation and extraction of the gold by activated charcoal, the concentrate pulp is sent in a pachuca reserved for the destruction of the cyanide. Following analysis of the free cyanide content and adjustment of the pH to 11, calcium hypochlorite is added in a pulp to the agitated pachuca. A ratio of 4 to 5 kg of hypochlorite per kg of cyanide (CN^-) is required to reduce the free cyanide to a content not exceeding 1 mg/l. The pulp is then extracted from the pachuca, conveyed in a tank to evaporation concrete-basins, where residual cyanides are destroyed by the U.V. sunlight.

Electrical Installation

Electric power is supplied to the plant by a thermal power station with four 900 kW generators, in 5000 V MV, directly to the ball mill switchboxes, and in 380 V LV for the electric motors and general equipment.

A mimic panel installed in the control room allows monitoring of the electrical operation of each motor, and lights indicate faults in the LV and MV power supplies. An intercom makes it possible to harmonize the operation of the ore dressing plant with the power station. The installed plant HV capacity is 540 kW and low voltage capacity 688 kW.

Water Circuit

Process water is essentially supplied from the Volta River, pumped to the 1000 m^3 basin of the ore dressing plant, under the responsibility of ONE (Office National des Eeaux). Additional water is obtained from the underground mine (dewatering) at certain times of the day, and from the tailings settling basins.

The amount of water used by the plant to treat 1 t of dry ore averages between 4 and 5 m^3, about half of which consists of recycled water.

Hygiene and Safety

Hygiene. Complete sanitary blocks and dressing rooms are provided for the staff at the plant entrance, to enable them to prepare for their work and to return home clean. The plant bay is ventilated by air extractors installed in the roof. A cold drinking water fountain is also available for the staff.

Safety. A video monitoring system enables movements and operation of the equipment at different points in the plant to be monitored. The personnel is provided with protective equipment suitable for the working conditions, boots, shoes, gloves, dust mask, safety goggles, safety helmet.

The atmosphere is also monitored regularly each day by the shift foreman to measure the cyanide and mercury contents of the atmosphere. The medical department receives the results and is kept informed of the operation of the different workshops and of the specific safety systems available, particularly for the cyaniding shop.

Analytical Laboratory

The analytical and production control laboratory performs all the chemical analyses for the mine and quarry and the ore dressing plant. It performs about 1500 determinations each month, essentially gold assays by atomic absorption.

The essential equipment is housed in two separate buildings for the following:

- mechanical sample preparation,
- chemical attack,
- determination of elements (Au, Ag, Fe, Sn etc),
- determination of gold lead fusion.

Gold and silver are also analyzed in the activated charcoal by x-ray fluorescence (Asoma instrument), and water and effluents are analyzed by colorimetry.

A basic ore treatment unit helps to monitor and, if necessary, improve the treatment process.

STARTUP

Incidents and Specific Problems

Crushing. The crushing installation completed in August 1984 was started up in September and became operational very quickly. The first ores treated consisted essentially of pure quartz with a low gold content. Automatic sequence startup of the main units, conveyor belts, crushers, and screen allowed rapid training of the operators.

The presence of clay fines nevertheless gave rise to two incidents:

. the rain-soaked ore failed to flow normally by gravity intc the hopper of the primary crusher,
. the subsequent drought caused unacceptable atmospheric pollution: this problem was partly solved by spraying the conveyor belts.

Dressing plant. Ore crushing was the only cause of startup problems. The nominal throughput could not be achieved in the first few months because the primary mill sometimes discharged a batch of unground ore, and regularly discharged ore pulp through the mill inlet plate. These two major incidents were solved by the following modifications:

. the balls were increased from 50 to 80 mm in diameter,
. an outlet grate was installed to lower the pulp level in the mill.

The restoration of good mechanical operation helped to achieve the nominal throughput of six months of operation.

Clayey materials. One remaining feature of the ore makes the installation vulnerable to climatic conditions. This is the considerable presence of clay fines that are very adhesive in the presence of moisture. This made operation of the crushing installation very difficult in the rainy season and its output was diminished. The presence of fines is also a problem in flotation, where addition of excess reagents is required to extract the gold.

Training

The 1984 startup team was small:

. 1 Burkina Faso engineer and 2 expatriate engineers,
. 5 expatriate foremen for operation and 3 for maintenance.

The Burkina Faso and expatriate staff proved to be effective and motivated since the erection phase, so that production was built up with a minimum of resources within six months.

Since early 1985, operations have been conducted by three engineers, one local and two expatriates, and seven foremen (four local with IUT training and three expatriates).

PERFORMANCE

Capacity

Continuous operation unit. The plant processes 600 t/day with a stream factor of about 95%. At full operation, it requires four teams of ten workers and a scheduled shutdown of eight hours each week is planned for maintenance. Supply of wet ore ranges between 17,000 and 18,000 t/month.

The crushing shop is capable of higher instantaneous output, 200 t/h by primary crushing and 50 t/h by secondary crushing. It always operates at full capacity, also with four teams of six workers each.

Chemical treatment. The chemical treatment system operates on a continuous day basis with two teams of three workers. It can now treat 40 t of concentrate per month on average, and perform one desorption/electrolysis cycle per week. The initiation of full-time operation with three shifts will help to increase capacity substantially, if necessary.

Gold room. The gold room can produce an average of 300 kg of dore metal in the form of 6 kg ingots. Operation is in a continuous day for all operations, except for table retreatment of the preconcentrate, which takes place with two teams of two workers.

Production Results

The first ingot was cast in November 1984. In 1985, the output of the plant was built up to the maximum in June, and it reached its nominal treatment capacity. However, difficulties due to the rainy season reduced the production of dore metal, which only reached its normal level in October.

The assay of dore metal produced, which was about 85% to 90% of precious metals at startup, gradually rose to over 95%, with the gold content settling at around 84% on the average, with peaks of 88% in early 1986.

At the end of 1985, the plant had been supplied with 100,000 t of mined ore with an average content of 17,5 g/t, and has disposed of rejects with an average content of 1,2 g/t corresponding to a recovery rate of 93 %. The quantity of gold accumulated in the plant is thus about 1630 kg.

By 31 December 1985, since commissioning, 1650 kg of dore metal had been produced and 1595 kg shipped, with an average assay of 82% gold and 12% silver, the remainder consisting essentially of copper. In the first three months of 1986, 1012 kg of dore metal was produced, and 1000 kg shipped.

TECHNICAL DATA SPECIFIC TO THE SECTIONS

Crushing

The raw material particle size is all smaller than 500 mm, with a high proportion of lumps larger than 100 mm. The average size distribution leaving the crushing shop is as follows:

mesh size	% weight
+ 12 mm	0
+ 10 mm	10
+ 8 mm	18
+ 4 mm	38
- 4 mm	34

The average energy consumption of the shop is 8 kWh per ton of raw ore crushed.

The wear parts of the secondary and tertiary mills are replaced every month after having processed about 15 000 t.

Grinding and Flotation

Material balance. The weight and volumetric balance is shown in Fig. 3 for the overall continuous processing flow sheet. The essential data for each stream are combined in one box. The average enrichment factor of flotation is 400/1.

Consumption. Ball consumption is more or less equally distributed between the two ball dimensions of 40 and 80 mm diameter, corresponding to 2.1 kg/t, and the two ball mills consume about 15 kWh/t. In flotation, collector consumption is approximately 250 g/t and that of the frother is 100 g/t.

The low voltage energy consumption for the entire ore dressing plant, except for crushing, ranges between 9 and 10 kWh/t, or a total of 24 kWh/t if the grinding energy is added.

Chemical Treatment

The balance of the cyanidation/adsorption operation is obviously very favourable because of the high concentrate content, which fluctuates between 3000 and 6000 g/t. The tailings content ranges between 20 and 30 g/t, taking account of the additional residence time in the CIP installation, during which cyaniding continues. The cumulative yields of the two operations thus range between 99.5% and 99.8%.

The cyanide consumption involved ranges between 30 and 50 kg/t of concentrate, or 0.08 to 012 kg of raw material. The reagents consumed by the desorption unit during the first year of operation were as follows:

	kg/t concentrate
alcohol	10
charcoal	4
soda	0.5 to 1
iron	0.5
cyanide	30 to 50
lime	5 to 10

Waste Treatment

The treatment of the plant tailings actually only concerns the detoxification of the cyanide concentrate by calcium hypochlorite. This is conducted in batch mode in a pachuca by a pulp of lime and calcium hypochlorite at pH 11. Following a variable reaction time between 4 and 8 h, the effluent is discharged after checking the free cyanide content. Average consumption is 2.5 kg calcium hypochlorite at 70% to 1 to 1.5 kg of lime per free and combined kg of sodium cyanide (NaCN), determined by the Desniges method.

STAFF AND WORKMEN'S TRAINING

Arrival of the Teams

Professional training of the personnel of the ore dressing plant was carried out at the SOREMIB professional training centre for two months for the first period, called the joint session. Six classes graduated from the centre, or about 72 workers. The first two groups participated in erection and startup. The various classes successively joined in to complete the team in June 1985.

Team Operations

Operation with three shifts, four teams, was organized so that the staff works 160 h per month over twenty days. The morning shift is operated for six consecutive days with two days of rest, the afternoon shift for seven consecutive days with two days of rest, and the night shift for seven consecutive days with four days of rest. The Friday morning shift, which is free on this schedule, is allocated to maintenance and repairs with the plant shut down.

Staff Distribution

The Ore Dressing Department employs a total of 139 persons. The department is run by an expatriate departmental head, assisted by two engineers, one local and one expatriate.

It is divided into three units:

- ore preparation, with a local foreman and 30 workmen,
- the dressing plant, with six foremen, including four local, and 83 workmen and other employees,

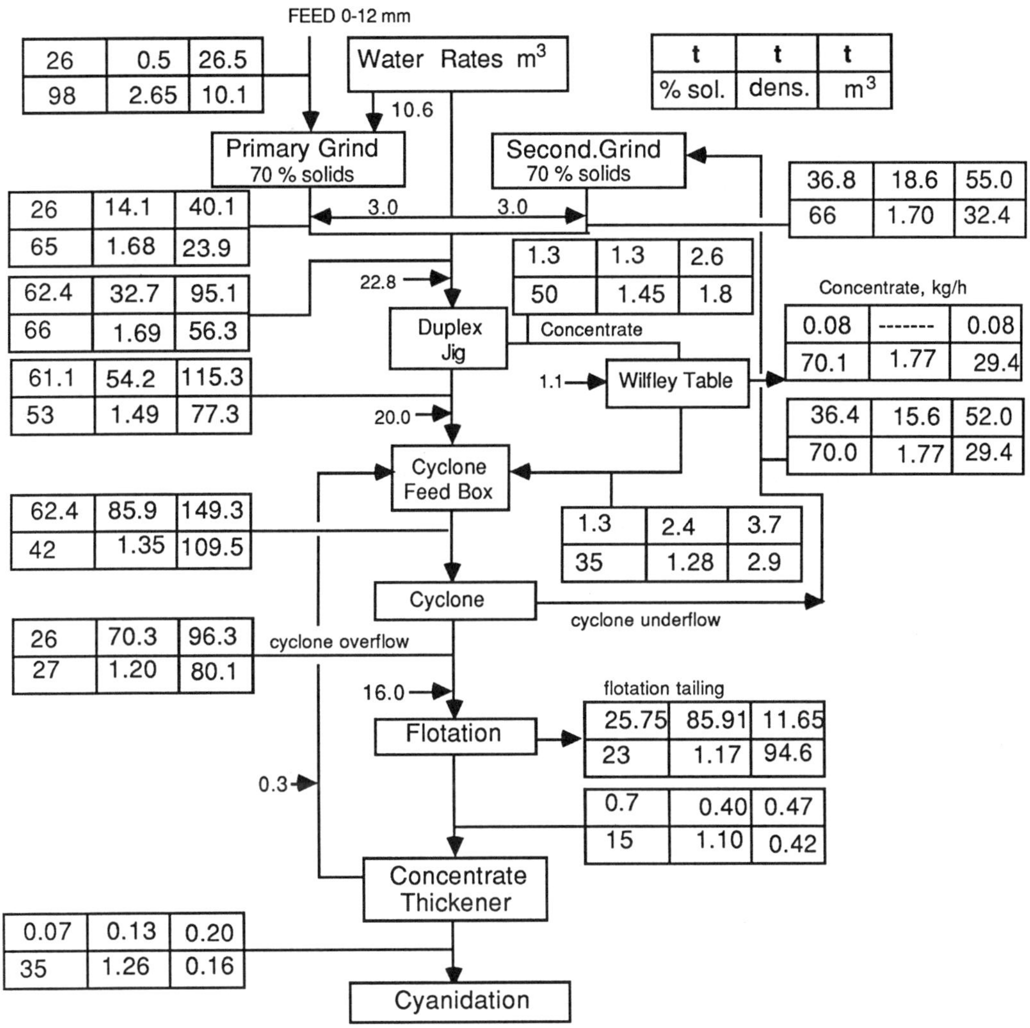

Fig. 3 Material balance-physical treatment

- the laboratory, with a local head and 15 employees, mainly chemists.

CONCLUSIONS

The construction and commissioning of the Poura Gold Mine dressing plant can rightly be considered as a success at the human and technical levels. The mine is in a country with a limited industrial past, and with no modern mining past. It was therefore necessary to import all the equipment, adapt the technique to the environment and quickly train all the operating personnel and most of the supervisory staff.

The absence of national capital ready to invest in an operation of this type also meant relying on loans for nearly 90% of the total funding of 20,000 million CFA francs. This resulted in high financing costs and serious constraints in company management, causing delays in project implementation.

The decision to start up the project was taken at the end of 1980. The first work on the site (professional training centre and living base) began in mid-1981. Mining began in September 1982 after the quarry equipment reached the site. The fixed installations, on which construction began in late 1983, were totally operational by the end of 1984.

This four-year interval to build an industrial installation of this scale, at a location far from any industrial facility and from the production centre of the materials and equipment employed, together with a production buildup over a period of nine months, constitute a performance that could only be achieved thanks to the determination and team spirit of all those who participated, both nationals of Burkina Faso and foreigners.

ACKNOWLEDGEMENT

The authors would like to express their gratitude to the SOREMIB Company which allows publication of flow sheet and general results of the industrial plant.

Many thanks are due to our BURKINABE comrades for their efficiency.

Technical Sessions

Session No.3

Effect of Mineralogy on Flowsheet Development

PROCESS MINERALOGY OF SILVER AND GOLD AT KIDD CREEK, FROM ORE TO ANODE SLIME

J. D. Scott

Kidd Creek Mines Ltd., A Falconbridge Group Company,
Box 2002, Timmins, Ontario, Canada P4N 7K1

ABSTRACT

The Kidd Creek mine orebody contains considerable silver, about 25% of which is in solid solution in the chalcopyrite. This silver must be coprocessed with selenium, also in solid solution in the chalcopyrite, in the smelter and refinery. In the copper anodes, silver occurs as the beta solid solution and the selenium as inclusions of a superstoichiometric $Cu_{2.10}Se$. During electrolytic refining, the silver forms both metallic silver grains and, by replacing the copper selenide, CuAgSe and Ag_2Se; gold forms tiny metallic particles and minor amounts of $Ag_4Cu_4AuSe_5$. There is also an anodic-oxidative electrolytic reaction series for the copper selenide grains in which copper is progressively stripped from the original $Cu_{2+x}Se$ to produce cupric ions in solution and a series of crystallographically distinct copper selenides. Predominance diagrams for copper and silver species, which permit the prediction of the bulk reaction products in the slimes for a wide range of current densities and silver to selenium molar ratios in the anodes, are presented. A detailed description of the silver speciation in the zinc plant circuits is given in which it is demonstrated that only silver sulfide reaction rims on unroasted sphalerite particles are present in the neutral leach residue but silver sulfate impregnations in amorphous silica gel masses (20% Ag) are the dominant form in high acid leach residue.

KEYWORDS

Process mineralogy; solid solution; silver selenide; copper selenide; gold selenide; electrorefining; anode slime; predominance diagram; jarosite.

INTRODUCTION

The Kidd Creek mine orebody contains no recoverable gold. There is only one small area at the extreme top of the bornite zone which grades a few grams per tonne, and it can not be mined separately. The average gold content of the ore is 44 ppb and the level in the copper concentrates is about 200 ppb. Kidd Creek Mines Ltd. nevertheless produced 62,800 ounces of gold in 1986 (Falconbridge 1986 Annual Report), from its Hoyle Pond and Owl Creek mines, and much of that gold passed through the copper smelter and refinery at the metallurgical site (Macfarlane, 1987).

The orebody however contains considerable silver, with some 6.3 million ounces of sales in 1986 (Falconbridge 1986 Annual Report). Current geological ore reserves at the Kidd Creek mine are about 56.8 million tonnes, including a 10% dilution factor, grading 66 ppm Ag, 3.21% Cu, 5.15% Zn and 0.16% Pb. These are split into a zinc rich high silver 'C' ore (229 ppm Ag) and a copper rich low silver 'A' ore (48 ppm Ag). The detailed silver mineralogy is quite complex (Pringle and Thorpe, 1980; Robinson, 1975; Wilshaw, 1982), but, on a practical basis, the orebody can be divided into three parts. The majority of the rare species occur in the bornite zone (Pringle and Thorpe, 1980), the 'A' ore has the bulk of the silver in solid solution in chalcopyrite, and the 'C' ore contains only four principal sources of silver: native silver veinlets (with minor acanthite) in pyrite and sphalerite, argentian tetrahedrite (freibergite, 28 to 32% Ag) in galena, minor dyscrasite and pyrargyrite in galena, and silver in solid solution in chalcopyrite. In bulk terms, about 25% of the mined silver is in solid solution in chalcopyrite, with 150 ppm in 'A' ore and 1400 ppm in 'C' ore (Cabri and co-workers, 1985).

The mineralogy and composition of the bornite zone has, however, a disproportionate effect on the metallurgical processing of the silver in the ore. There are two main reasons for this. The first is physical in that the bornite zone is severely microfractured, to the point that it rapidly breaks up during ore handling into something resembling coarse Demerarra sugar, and thus it contaminates the entire ore-transfer system as fines. The overriding reason is chemical; the bornite zone and its chalcopyrite halo are the main high-selenium areas of the orebody and the silver-selenium association has a profound effect on the smelting and refining of copper (the place where all of the silver recovery occurs).

Relatively little has appeared in the literature on the bulk nature of the bornite zone (Macfarlane and co-workers, 1985), and, given its importance to this discussion, a brief description is appropriate. The zone forms an elongated sigmoidal pod between the 1200' and the 2600' levels, with a pronounced thickening between 1600' and 2400' levels. In that area it is some 200 by 25 feet in plan, and, above the 2000' level, it has a high grade chalcopyrite halo extending outward by an additional 20 to 100 feet. The zone itself consists of massive intergrown bornite and chalcopyrite, with minor tennantite, pyrrhotite and cobaltian pyrite and trace amounts of several rare minerals (Pringle and Thorpe, 1980). The bulk of the silver is in solid solution in the chalcopyrite (500 ppm Ag) but there are local concentrations of several silver selenide minerals (principally eucairite, naumannite and bohdanowiczite); neither the bornite nor the tennantite contain detectable silver in solid solution. The bornite zone halo consists of almost solid chalcopyrite (150 ppm Ag), with minor to trace intergrown sphalerite, pyrite and carrollite. With the exception of some of the Zaire deposits, the chalcopyrite and bornite in this area have the highest average selenium levels ever recorded from a volcanogenetic massive sulfide deposit. Proton microprobe selenium values (Cabri and co-workers, 1985) are: chalcopyrite 3500 ppm [zone] 1000 ppm [halo], bornite 7000 ppm [zone]. Chalcopyrite in the run of mine 'A' ore contains 200 ppm selenium and in the 'C' ore, 135 ppm. Ore reserves (in percent and ppm) in this area are:

	tonnes	Cu %	Zn %	Pb %	Ag	Se	Co	Bi
zone	275,000	20.30	0.31	0.02	721	1200	450	310
halo	250,000	19.97	2.05	0.02	161	700	630	60

PROCESS MINERALOGY OF SILVER IN THE COPPER SMELTER/REFINERY

The overall process mineralogy of the metallurgical site is complicated by the co-presence of silver and selenium in solid solution in the chalcopyrite. In the concentrator the two ore types are kept separate, with three grinding and flotation divisions processing 'A' ore and one processing 'C' ore (Macfarlane and co-workers, 1985). The combine 'A' copper concentrates are smelted in the Kidd smelter and the 'C' copper concentrate is toll smelted or sold. During late April, early May 1987, processing normal ores with minor bornite-zone contamination due to spalling in one stope, the partitioning of silver and selenium (in ppm) to the concentrates was:

	'A'			'C'			
	head	copper	zinc	head	copper	silver	zinc
silver	50	350	35	275	10000	22000	190
selenium	170	900	150	45	800	-	121

The silver concentrate referred to above is the result of a small flotation circuit which is fed a cut from the secondary ball mill with the purpose of recovering coarse galena (with its contained dyscrasite, pyrargyrite and freibergite) and native silver before they become too finely fragmented in the mill. During periods of high silver feeds, the silver concentrate will assay between 5 and 10% silver, with 25 to 30% lead. This circuit, introduced in 1984, reduced the overall loss of silver to tailings by about seven percent.

Since essentially all of the selenium present in the ore is in solid solution in the copper minerals, the concentrator is ineffective in reducing selenium levels in copper concentrates fed to the copper smelter. The smelter (Macfarlane and co-workers, 1985), however, eliminates about 40% of the selenium by fuming it to the wet-gas cleaning system of the sulfuric acid plant; also, some 2.5% is rejected in the discard slag and about 7.5% is removed from the smelter via the cottrell dust, which is sent to the zinc plant for zinc recovery. This means that almost exactly half of the selenium which enters the smelter appears in the copper anodes prepared for the copper refinery (Newman and Morin, 1987). The "normal" level of selenium in the anodes is about 0.20%; however, during periods when bornite zone ores are being processed in the concentrator, it can rise to 0.50%.

Mainly as a result of the x-ray identification of silver selenide and silver-copper selenide as the major constituents of the anode slimes remaining after the electrolytic refining of copper, there is a long standing mythology in the literature that these are the forms in which selenium occurs in the anodes. Extensive research at Kidd Creek and by CANMET (Chen and Dutrizac, 1987a, 1987b) has shown that this is not the case. Over 99% of the silver is dissolved in the copper as the beta solid solution phase (0.13 to 0.26% Ag). The selenium forms ring-shaped copper selenide inclusions, usually rimming subspherical to tubular copper or copper-lead oxide inclusions (for a very complete discussion of this phenomenon, see Chen and Dutrizac, 1987a). The copper oxide inclusions contain less than 0.05% Ag and only 0.44% Se. The copper selenide contains an average of 0.13% silver and is of very uniform composition in any individual cast. It is a super-stoichiometric form, $Cu_{2.10}Se$ to $Cu_{2.24}Se$, that is presumably a metastable quench species formed during the casting of the anodes. The extent of super-stoichiometry appears to be a function of the absolute amount of selenium present in the blister copper; "normal" levels of 0.2% Se yield a copper selenium atomic ratio of 2.10 in the selenide, whereas "high" levels of 0.4% Se yield ratios which cluster around 2.20.

The importance of selenium, and its relationship to silver in the process mineralogy of the metallurgical complex, emerges in the copper refinery. Since its startup in 1981, the refinery has increased the operating current density in a stepwise fashion from 200 to 220 to 245 A/m² (Sweetin, Newman and Kemp, 1984). Apparently random changes in both the mineralogical composition of the slimes produced on the faces of the anodes and the extent of selenium contamination of the copper cathodes were observed for the various operating current densities and a program of research into the causes of this phenomenon was undertaken.

The result of this has been the discovery that the dominant species formed in the anode slime depends on both the operating current density and the silver to selenium molar ratio in the anode. Predominance diagrams (these are not equilibrium diagrams) for the silver and copper species formed are shown in Figures 1 and 2. They identify the major species, as determined by bulk x-ray diffraction, which are present in each region. The horizontal "phase boundaries" are surprisingly constant and no trace of the adjacent species has been detected at points greater than ± 0.1 ratio units away from a boundary. These are all solid state reactions, since the starting materials are the beta-Cu(Ag) solid solution and ringlike, tubular or subspherical inclusions of $Cu_{2+x}Se$ in the anode.

Metallic silver, as tiny granules, is formed by either a cementation reaction with metallic copper or by a precipitation reaction with cuprous ion. The mean grain size of these silver particles is about 0.5 micron and they are rarely observed to be larger than two microns in diameter; they are thus too small for the microprobe analyses required to define which reaction mechanism is predominant. Despite the fact that fields in which metallic silver is absent are shown on Fig. 1, metallic silver is in fact continuously produced at all current densities (even those as low as 135 A/m²); the diagram merely shows that in these regions the rate of silver redissolution is greater than the rate of accumulation (Sharma, Gupta and Bassi (1984) have shown that the mean lifetime of a silver sphere of two micron radius in copper electrolyte at 60°C is some four hours).

The silver selenide species are produced by progressive replacement in the original $Cu_{2+x}Se$ particles, starting from the outside surface of the grain (Chen and Dutrizac, 1987b), of the copper by silver ions put into solution by the electrolytic dissolution of the beta solid solution in the anodes.

$$Cu_xAg_{1-x} = xCu^{2+} + (1-x)Ag^+ + (1+x)e^- \quad (1)$$

$$Cu_{2+x}Se + xAg^+ = (Cu_{2-x}Ag_x)Se + xCu^+ \quad (2)$$

$$(Cu_{2-x}Ag_x)Se + (1-x)Ag^+ = CuAgSe + (1-x)Cu^+ \quad (3)$$

$$CuAgSe + Ag^+ = Ag_2Se + Cu^+ \quad (4)$$

$$Cu^+ + Ag^+ = Ag^o + Cu^{2+} \quad (5)$$

The extent to which this series of reactions proceeds to completion at any particular current density is a function both of the silver to selenium molar ratio in the anodes and the absolute value of selenium in the blister copper (which appears to define the thickness of the copper selenide shells around the copper oxide inclusions in the anode; very high levels of selenium in anode (0.45%) has been observed to produce thicker selenide shells rather than merely a greater number of thin ones).

At the same time as the silver reaction series is proceeding there is a parallel anodic-oxidative electrolytic reaction series (first inferred by Lange and coworkers, 1981) for the "pure" copper system (Fig. 2). In this, copper is progressively stripped from the original $Cu_{2+x}Se$ particles to produce cupric ions in solution and a series of crystallographically distinct copper selenides. At very low

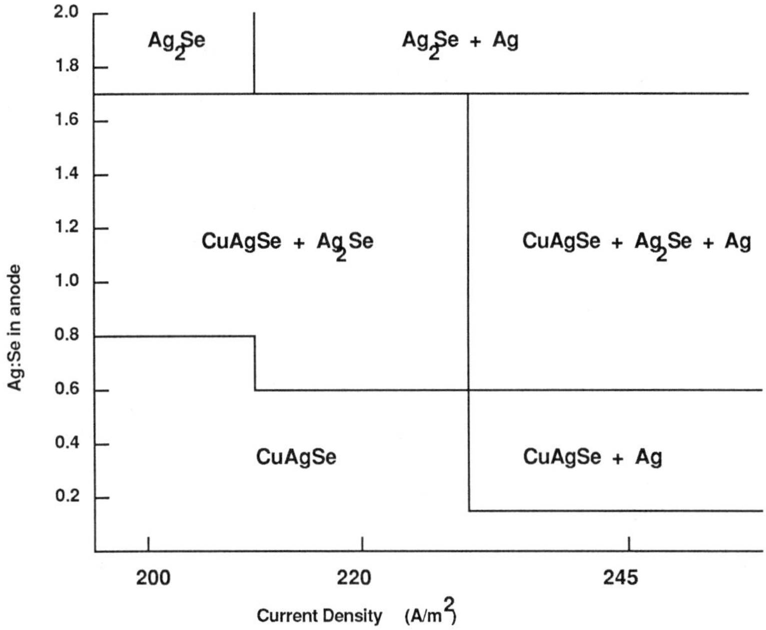

Fig. 1. Predominance diagram for silver

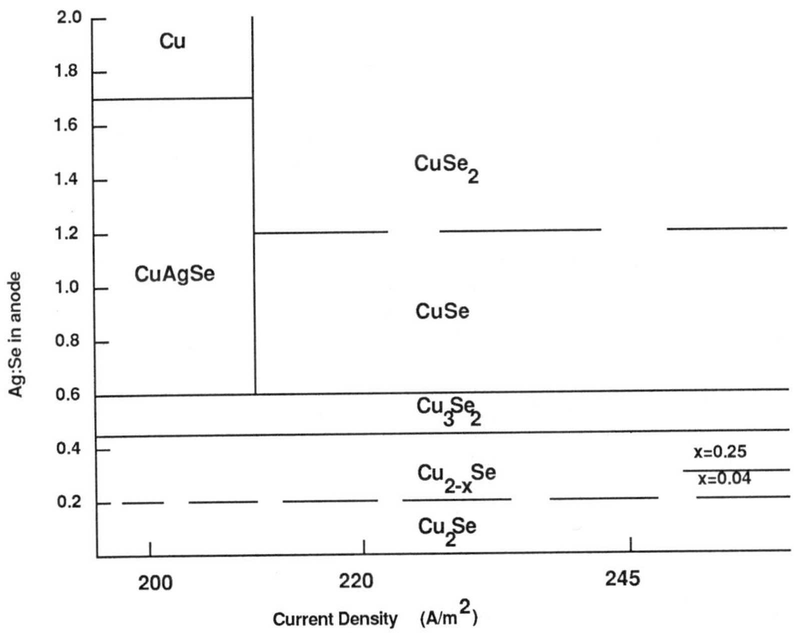

Fig. 2. Predominance diagram for copper

silver levels, the result is apparently stoichiometric Cu_2Se; this has however only been observed in one instance at Ag:Se = 0.180, the lowest value of the ratio which has been produced to date. In the region between Ag:Se molar ratios of 0.20 and 0.45, a family of non-stoichiometric $Cu_{2-x}Se$ species is the limit of the reactions; below Ag:Se = 0.25, x = 0.04 [bellidoite] and above this x = 0.25 [berzelianite]. At progressively higher ratios, the series proceeds through Cu_3Se_2 [umangite] and CuSe [klockmannite] to $CuSe_2$ [krutaite]. As with Cu_2Se, the $CuSe_2$ phase has only been observed in a single instance at 245 A/m^2 and the location of this boundary is somewhat speculative.

The significance of the copper reaction series lies in the fact that the non-stoichiometric $Cu_{2-x}Se$ species are co-platable with copper whereas all of the remaining copper and silver selenides are not. Normally a slime particle must both touch the cathode and be overgrown by copper in order for it to become a contaminant of the final cathode. With the $Cu_{2-x}Se$ species all that is necessary is for the particle to touch the cathode and it is immediately codeposited with the copper; this leads to much higher than normal selenium contamination of the copper cathodes during periods when these species are produced in the anode slime. Their formation is favoured when low silver levels coincide with high selenium levels in the anode copper, and, since this is a description of the result of processing a mixture of bornite zone ore with 'A' ore, this is the basic reason behind the statement that bornite zone ores cause problems at the metallurgical site.

PROCESS MINERALOGY OF SILVER IN THE ZINC PLANT

The leaching circuits and the process mineralogy of iron in the zinc plant have been described in a recent paper (Scott, Donyina and Mouland, 1986); this should be read in conjunction with the following detailed description of the disposition of silver in the circuits. Briefly, the zinc plant was commissioned in 1972 and includes a 770 tonnes per day roasting plant with two Lurgi fluidized bed roasters (offgas is sent to a Monsanto single absorption sulfuric acid plant) and a Jephson (Asarco) design calcine-leach plant with both neutral and high acid leaching sections and a jarosite precipitation circuit. There is also a Jephson cellhouse which feeds twenty four hour plated zinc cathodes to a modern casting plant; a small amount of ammonium chloride flux is used for dross control in the melting furnace. Relatively little has been published on the mineralogy of silver in such roast/leach/electrowin zinc plants and nothing on those with high-silver feeds, such as the 'C' zinc concentrate which can assay as much as 300 ppm silver (Lenchev and Dutrizac (1984), on Matagami Lake Mines concentrate at 34 ppm Ag, is the only detailed discussion).

At Kidd Creek, all of the silver minerals in the feed which are not entirely enclosed by at least 20 microns of quartz are decomposed during roasting and form silver sulfate in the calcine. This silver sulfate is then itself partly decomposed to metallic silver and, if chloride is locally present from dross recycle, to silver chloride. Silver sulfide has never been observed except as tiny relict grains enclosed in quartz. Test leaches (using 3M NaCl or thiourea) show that from 63 to 84% of the silver in the calcine is in the form of silver sulfate. The average results for a series of starting calcines which ranged from 100 to 370 ppm Ag were: 75% silver sulfate, 25% metallic silver; silver chloride was either very faint trace or not detected. The major silver "collector" in the calcine is an aluminosilicate phase, probably the roasted equivalent of chlorite, which always contains variable amounts of silver, usually ranging between 1.4 and 2.0%. There are also occasional grains of a quartz-zinc silicate [willemite] intergrowth with minor (approximately 0.9%) silver. Both of these silicate phases are thought to be impregnated with submicron grains of metallic silver. The majority of the silver sulfate appears to be adsorbed as a thin layer on the outside of zinc oxide or, rarely, zinc ferrite grains; the "inside" of such grains is usually barren of silver. There are however very rare small "worm-like" patches which contain up to 0.2% Ag in the zinc oxide grains; these have minor associated chlorine, and trace sulfur, and are thought to be "fracture penetrations" of the grains by silver chloride.

In the neutral leach residue, all of the silver observed is in the form of a sulfide and occurs as 2 to 10 micron coatings on the outside of residual unroasted sphalerite grains. The majority of the coatings are "pure" silver sulfide, however, there are also occasional thick (10 to 25 micron), somewhat porous, rims of a lead-silver sulfide (approximately $Pb_4Ag(Zn,Fe)_2S_8$, a rather low quality microprobe analysis gave: Pb 57.3%, Ag 7.7%, Zn 6.6%, Fe 3.4%, S 18.9%, sum 93.9%).

In the high acid leach residue, the majority of the silver is in the form of either selenium-rich silver sulfide particles dispersed inside ballike aggregates of elemental sulfur and jarosite or as silver sulfate impregnations in amorphous silica gel masses (20% Ag). Respectively, these are probably the ferric-iron leached equivalents of the silver sulfide coated unroasted sphalerite grains observed in the neutral leach residue and the remains of the silver bearing aluminosilicate grains noted in the calcine. Silver sulfide coated residual sphalerite grains are rarely observed; in almost all cases, these have been attacked to some extent by the ferric sulfate and partly converted to elemental sulfur. There are also minor to trace amounts of silver in solid solution in the jarosite (345 ppm in Na-K jarosite and 1250 ppm in Pb-K-Zn jarosite).

Silver in the daily jarosite residue can vary over a wide range (80 to 500 ppm) and is presently averaging about 200 ppm on a monthly basis. Almost all of this silver has been introduced via the calcine used to neutralize the acid produced during iron hydrolysis and jarosite precipitation. The silver content of the solution entering the jarosite circuit from the high acid leach circuit is only 28 ppb.

Under the present circuit configuration (Scott, Donyina and Mouland, 1986), all of the silver which enters the zinc plant is discarded to tailings with the combined high acid leach and jarosite residues.

PROCESS MINERALOGY OF GOLD

The process mineralogy of gold in the metallurgical site is, in general, very simple. There are two sources of gold input to the copper smelter. The minor source is from a low-gold high silica quartz vein material used as flux in the smelting process and the major source is from a gold-pyrite concentrate. This concentrate is produced from Hoyle Pond ore in a 300 tonne per day crush/grind/float milling operation (Macfarlane, 1987). All of the gold in the Hoyle Pond ore is in the form of elemental argentian gold grains (8% Ag); no gold tellurides have been observed. The gold-pyrite concentrate is continuously fed to the smelting furnace at a rate which is ten percent of the instantaneous feed rate of the copper concentrate. The gold is carried in the matte phase and reports directly to the blister copper as a very dilute solid solution. Losses are minimal, the current average gold content of the discard slag is 45 ppb. No discrete gold-rich phases have been observed in any of the smelter products.

In the refinery, the gold concentrates in the anode slimes and these are currently assaying around 1200 ppm gold. The precise location and form of the gold in the slimes has long been a matter of considerable debate in the industry. Recent work by CANMET on samples of anode slimes from Kidd Creek and other refineries has shown that the gold is present in two distinct forms (Chen, 1987, personal communication). The majority of it occurs as tiny (less than 0.5 micron) particles of metallic gold dispersed in an illdefined copper-silver selenate-arsenate cementing medium which, together with copper sulfate, acts as a loose binder for the copper and silver selenide particles described above. Bulk analyses of separated cementing medium, from ammonium acetate deleaded high-gold anode slime (recovered from the processing of Noranda anodes), yielded variable results which had a strong negative correlation (r=0.98) between the selenium and arsenic assays; the average analysis was: 1.7% Au, 17% Ag, 24% Cu, 11% Se and 7% As. This can be idealized (with gold taken as metallic particles) into a formula resembling $Ag_3Cu_8(SeO_4)_3(AsO_4)_2$. During periods when only Kidd anodes are processed, the gold, silver and arsenic contents of the cementing medium are at a minimum and an idealized formula such as $(Cu,Ag)_2SeO_4$ is postulated. There are also trace amounts of two to four micron particles of a silver copper gold selenide, both as inclusions in the CuAgSe ring-fragments and "loose" in the cementing medium; this material has the approximate formula $Ag_4Cu_4AuSe_5$ and corresponds to no known compound.

REFERENCES

Cabri, L. J., Campbell, J. L., Laflamme, J. H. G., Leigh, R. G., Maxwell, J. A. and Scott, J. D. (1985). Proton-microprobe analysis of trace elements in sulfides from some massive-sulfide deposits. Canadian Mineralogist, 23, 133-143.

Chen, T. T. and Dutrizac, J. E. (1987). Mineralogical characterization of anode slimes from the Kidd Creek copper refinery. In J. E. Hoffman, R. G. Bautista, V. A. Ettel, V. Kudryk and R. J. Wesely (Ed.), The electrorefining and winning of copper, TMS-AIME, Warrendale, Pa. pp. 499-525.

Chen, T. T. and Dutrizac, J. E. (1987). Application of electron microscopy to the electrorefining of copper. In Proceedings of Scanning Electron Microscopy 1987, in press.

Lange, H-J., Geidel, K., Hein, K. and Baum, H. (1981). Untersuchungen zu anodischen Reaktionen des Selens bei der elektrolytischen Kupferraffination. Freiberger Forschungshefte, B228, 123-142.

Lenchev, A. and Dutrizac, J. E. (1984). The behaviour of silver during the roasting of zinc concentrates. Canadian Metallurgical Quarterly, 23, 287-293.

Macfarlane, G. (1987). Conversion of a crushing plant to a gold mill in a copper smelter. 89th annual general meeting of CIM, Toronto.

Macfarlane, G., Storey, A. G., Kelly, V. and Marrs, G. G. (1985). The control of lead and selenium in copper produced at Kidd Creek Mines Ltd., Timmins, Ontario. 24th annual conference of metallurgists of CIM, Vancouver.

Newman, C. J. and Morin, G. (1987). Anode strip casting practice at the Kidd refinery. In International symposium on near-net-shape casting of strip, 89th annual general meeting of CIM, Toronto, pp 1-18.

Pringle, G. J. and Thorpe, R. I. (1980). Bohdanowiczite, junoite and laitakarite from the Kidd Creek Mine, Timmins, Ontario. Canadian Mineralogist, 18, 353-360.

Robinson, D. (1975). Tin and silver mineralogy of a lead, tin and silver rich area of Kidd Creek Mine. BSc. thesis, Queen's University, Kingston, Ontario.

Sharma, I. B., Gupta, R. K. and Bassi, P. S. (1984). Formation of silver selenide in the solid state & its reactivity with $CuSe_2$(s). Indian Journal of Chemistry, 23A, 342-344.

Scott, J. D., Donyina, D. K. A. and Mouland, J. E. (1986). Iron - the good with the bad - Kidd Creek zinc plant experience. In J. E. Dutrizac and A. J. Monhemius (Ed.), Iron control in hydrometallugy, Ellis Horwood Limited, Chichester. pp. 657-675.

Sweetin, R. M., Newman, C. J. and Kemp, D. J. (1984). The Kidd copper smelter and refinery - description and early operation. CIM Bulletin, 77, 84-94.

Wilshaw, N. (1982). The silver mineralogy of Kidd Creek concentrates. MSc. thesis, Queen's University, Kingston, Ontario.

ROLE OF MINERALOGY IN GOLD PROCESSING

Roshan Bhappu and Laszlo Dudas, Mountain States Research and Development, Tucson, U.S.A.

A close scrutiny of various flowsheets utilized for processing different precious metals ores indicate that the mineralogy of the ore dictates the process to be used. Thus, placers and alluvial ores can be processed readily by simple gravity concentration methods while free milling ores containing fine gold require cyanidation and its various modifications such as heap leaching, vat leaching, agitation leaching with carbon-in-pulp and other applicable options. On the other hand, ores with complex mineralogy, such as gold associated with pyrite and base metal sulfides require flotation concentration. Finally, more complex or refractory ores containing gold values in arsenopyrite, tellurides, etc., can only be treated effectively by destruction of the matrix. This paper discusses various flowsheets and plant practices based on the mineralogy of the ores, and operational and economic considerations are presented. Finally, efforts are made to project some newer processing techniques for treating gold ores based on their mineralogical composition.

CHANGES IN THE JERRITT CANYON METALLURGICAL PROCESS AS A RESULT OF GEOLOGICAL CHARACTERISTICS OF THE ORES

D. Birak and K. Deter
Freeport-McMoRan Gold Company, Elko, NV U.S.A.

ABSTRACT

The effectiveness of the current Jerritt Canyon gold recovery process is the result of continual geological and metallurgical investigations. Initially, Jerritt Canyon gold recovery involved separate oxidized and refractory ore process circuits. Oxidized ore was processed in a standard cyanidation-CIL circuit, while refractory ores were processed in a similar CIL circuit following a pre-oxidation and mild chlorination circuit. Currently, all ores are chlorinated. This change, made in 1983, improved recoveries on all ores and was dictated by increased understanding of the physical distribution of carbonaceous and oxidized ores and of the role that sulfide-sulfur and organic carbon components of the ore together play in gold recovery. The effects of each component are highly dependant upon rock type that is being processed.

KEYWORDS

Carbonaceous; Refractory; Pre-Oxidation; Chlorination; Gold Ores.

INTRODUCTION

The Jerritt Canyon operation of Freeport-McMoRan Gold Company is located in the Independence Mountains approximately 80 Km (50 miles) north of Elko, Nevada. It is a 70:30 Joint Venture Partnership between Freeport-McMoRan Gold Company and FMC, Inc. and is one of the three largest single unit sources of newly-mined gold in the United States.

Discovery of the gold mineralization in the area is credited to FMC geologists looking for stibnite. Noticing similarities between the Jerritt Canyon rocks and those of Newmont Gold at Carlin, an exploration program in 1973 identified a gold anomaly now referred to as the Alchem Deposit. In 1976, Freeport Exploration Company was selected as a Joint Venture partner and assumed management of the project.

Freeport geologists studied available data and through reinterpretation, led to the rapid discovery of the Marlboro Canyon deposit, elevating the entire project from a small-tonnage, mineralized area to a potential world-class producing mine.

Two types of gold bearing ores were distinguished: Oxidized ores, amenable to standard cyanidation techniques, and carbonaceous ores, requiring oxidation procedures to render the gold recoverable. Metallurgical tests on the drill samples led to a process of pre-oxidation followed by chlorination, prior to cyanidation.

The pre-oxidation step for the carbonaceous ores consisted of the following procedure (Fig. 1):

The slurry was heated to 70-80°C (160-180°F) in low-air overpressure of 41.3-51.7 x 10^{3} Nm^{-2} (6-7.5 psig). Heat was supplied by steam and total retention time was 16-20 hours in six stages, each tank being 6.71m (22 ft.) diameter by 7.92m (26 ft.) high and containing a working volume of 62,500 gallons. Soda ash was continually added to neutralize the sulfuric acid generated by oxidation of the pyrite, at a rate of up to 20.5 kg/tonne (50 lb/ton). Steam consumption of 3600 kg per hour at 860 x 10^{3} Nm^{-2} (125 psig) and compressed air at 150 m^{3}/min (5000 scfm) was sparged into the slurry. Slurry temperature was raised from ambient by a shell and tube slurry heat exchanger. Slurry from the pre-oxidation circuit tailing at 80°C (180°F) was used to heat the pre-oxidation slurry feed. The pre-oxidation feed would exit the heat exchanger at approximately 54°C (130°F) and be elevated at 77°C (170°F) by contacting the pre-oxidation circuit exhaust gasses in a counter current air contact tower.

Design slurry density was 50 wt.% solids but at normal operating conditions, the circuit could be operated up to 55 wt.% solids. The required residence time to oxidize the sulfide sulfur in the carbonaceous ores would vary from 8 to 20 hours. The purpose of this process was to partially or fully oxidize sulfide to reduce chlorine demand in chlorination.

After cooling to 54°C (130°F) in the shell and tube heat exchanger, the slurry was treated with chlorine gas in four enclosed agitated tanks in series. The tanks, 6.71m (22 ft.) diameter by 7.32m (24 ft.) high, had 62,500 gallon working volumes and were operated under a vacuum of 5cm (2 in.) water to control fugitive chlorine gas and products of carbon oxidation, or off-gas. Off-gas from the tanks was scrubbed with carbonate-bicarbonate solution to complex fugitive chlorine and return it to the slurry as bleach. Two additional tanks were used as decay tanks to allow the hypochlorous acid (HOCl) generated by the hydrolysis of chlorine to decompose to minimal levels (less than 0.05 g/l) prior to adding cyanide. Total retention time was 16 to 20 hours and total chlorine input was 12-23 kg/tonne (30-50 pounds/ton). Slurry exiting the chlorination circuit was cooled to 25-40°C 80-105°F) by water in another shell and tube heat exchanger.

The pre-oxidation/chlorination procedures led to making the Jerritt Canyon area into an orebody and the first production dore' was poured on July 4,1981, some eight years after the initial discovery. Design processing rates were 2500 DMTD (2750 DSTD) with a recovery of 85% on a feed grade of 8.6 grams/tonne (0.250 ounces per ton.) Carbonaceous ore was to be 50% of the feed and require 12-23 kg of chlorine per tonne of ore (30-50 pounds/ton) fed to the pre-oxidation circuit. Ore was defined metallurgically as carbonaceous if its leach ratio (ratio of NaCN-soluable to fire assay gold) was less than 0.60 or greater.

The complex geology of the district has been discussed in prior publications (Birak, 1986; Birak and Hawkins, 1985). During the pre-production period and during early mine operations, Freeport geologists recognized two distinct types of carbonaceous gold ores. The main tonnage of carbonaceous ores initially discovered were hosted in the Roberts Mountains Formation and characterized by extremely fine gold occurring as isolated grains on/or within crystalline pyrites or pyrite aggregates. Additional exploration led to the discovery of carbonaceous ore in the Hanson Creek Formation. The organic carbon material was approximately 0.5 wt.% of the relatively simple limestones, composed primarily of calcite, with lesser amounts of dolomite, sericite, quartz and pyrite. The Roberts Mountains Formation - hosted carbonaceous ores consist of carbonaceous siltstones and limestones with low calcite contents, relatively high dolomite, and high organic carbon levels (.75-1.5 wt %) and high pyrite contents (2-4 wt %).

The carbonaceous levels of the host formations of the Jerritt Canyon District fall into three general visually distinct categories: flat-, shiny-, and sooty-black textures. The flat-black textures are relatively unaltered sedimentary rocks displaying minor tectonic deformation. Shiny-black textures are characterized by graphite coatings on bedding planes and on fault surfaces and represent rocks slightly enriched in organic carbon compared to the flat-black zones and with moderate amounts of shearing and granulation.

The sooty-black textures indicate strongly sheared and carbon enriched rock,with soluble constituents such as carbonates leached out. The flat-black textures rarely contain in excess of 0.5 weight percent organic carbon whereas one to two percent are common in the sooty-black constituents. Sulfide-sulfur levels are often twice as high in sooty carbonaceous rock as in flat-black carbonaceous rock. The most intensely carbonized rock contains a poorly crystalline graphitic residue resulting from vaporization of the volatile constituents of possible petroleum precursors (Birak, 1986). As could be anticipated based on organic carbon contents, the ore grades are relatively enriched in the sooty-black carbonaceous rock with respect to the shiny-and flat-black rock. Gold also is consistently higher in any carbonaceous rock than adjacent oxidized rock, possibly as a function of the ability of carbon to trap hydrothermal gold during the formation of the ore body (Birak and Hawkins, 1985).

The oxidized ore zones occur in both types of host rock and are characterized by extremely hard jasperoids as a result of silicification of the limestones and softer tan-to-brown colored limestone and siltstones. Some hydrothermal oxidation along faults is also evident.

It became rapidly evident that the host formations were extremely faulted, folded and co-mingled, leading to extreme difficulties in segregating the oxidized ore zones from the caronbaceous ore zones. Because of the lack of pre-oxidation treatment on the oxidized ores, no carbonaceous contamination could be tolerated if design recoveries were to be achieved. Various types of mine mapping and grade control techniques were tried in order to insure the oxidized ores, initially described as ore with more than 60% of the gold recoverable by cold cyanidation (leach ratio), could be segregated properly from the carbonaceous ores. As more ore was delineated by geologic interpretation and exploration drilling, it became evident that a precise ore segregation would not be feasible while still providing equal feed tonnages to the two separate mill circuits.

The jasperoid fraction of the oxide ore, roughly 50 weight percent of the oxide ore, caused limitations in the processing rates of the circuit. A very hard siliceous material with a Bond Work Index of 26-28, the jasperoid produced an abundance of 40-70mm (1 1/2-3 in.) critical size particles in the SAG mill circuit that the 100-125mm (4-5") grinding balls were unable to break. Thus the first major metallurgical process alterations were developed.

The milling circuits were altered from a semi-autogenous/ball mill circuit to an autogenous/ball mill/crusher, or ABC circuit to better process the critical size jasperoid pebble produced. This modification came on line in September 1982. The separate processing of carbonaceous ores and oxide ores was discontinued and the pre-oxidation circuit was converted to a chlorine injection circuit. Dual chlorination circuits were thus created to treat all ores. By the end of 1983, this system allowed treatment of increasing amounts of carbonaceous ores at higher throughputs. Sodium hydrosulfide was added to the final decay tank to complex or "kill" any residual hypochlorous acid prior to cyanidation (Fig. 2).

By the end of 1983, with the first major milling and processing improvements in place, tonnage was up to 2700 DMTD (3000 DSTD), chlorine requirements rose to 20 kg/tonne (50 pounds/ton) milled and gold recoveries were 90% on a mill feed of 25-35% Type II ores.

With the advent of chlorination of all ores, ore classifications were redefined as Type I ores and Type II ores, with the Type I being determined by greater than or equal to 20% of the gold recoverable by cold rolling bottle cyanidation tests. Type II ores, less than 20 percent recoverable gold, were further classified by the host rock formations, Hanson Creek or Roberts Mountains. Following blast hole drilling on 14 foot centers, the ore type was determined by averaging assays and leach ratios on samples from the four blast holes defining the corners of a 14' x 14' block. Prior to and following blasting, the mine geologists again examined the ores and identified the proper ore type and host rock formation to better assist the mill in controlling processability.

Statistical analyses of gold, carbon, organic carbon, and sulfur contents were made on the various ore sources and recoveries from them. The research indicated that sulfur was the dominating single geological variable controlling chlorine consumption and therefore affecting overall recoveries.

The equation that best fits this relationship was determined to be:

$$\text{Chlorine (pounds/ton)} = 113.7\ (\text{sulfur }\%) + 35.2$$

As this relationship was valid for roughly only 30% of the ore, it is evident that other geochemical and/or geological parameters will also dictate total chlorine demand. Also evident was the fact that the sulfur/chlorine regression equation behaved differently within the different host rock formations: Roberts Mountain ore requires more chlorine input to achieve similar recoveries than Hanson Creek ores of the same sulfur values. Investigation continues in order to determine the best classification methods for various ores.

Geological evaluations of the remaining ores in 1985 indicated that higher chlorine consumption would be required to make satisfactory metallurgical efficiencies at the then processing rate of 3175 DMTD (3500 DSTD). This led to doubling the liquid chlorine handling facility and an increase of 50% in the chlorine injection circuits, from four to six tanks, the decay tanks being modified with larger impellers and chlorine spargers. Separate agitated tanks were also installed to allow reduction of residual hypochlorous acid with sodium hydrosulfide in the treated slurry prior to cyanidation and use of the prior decay/kill tanks for chlorination (Fig. 3).

Geologic evaluations of the ores to be mined in late 1986 and beyond indicated decreasing amounts of jasperoids in the ore and highlighted the potential for increased throughput well in advance of the ore deliveries to the mill.

The modifications to the chlorination handling, injection and gas scrubbing systems were completed by the start of the Fourth Quarter of 1986, coinciding with the mining of softer ores. Advanced planning allowed for sufficient ore to be delivered to the milling facilities and processing rates reached 3564 DMTD (3929 STPD) for the fourth quarter while recovery increased to 89.7%.

As experience has been gained with the higher chlorine input, ores in excess of 60% Type II are being processed, with chlorine input rates in excess of 40 kg/tonne (100 pounds/ton) and metallurgical efficiencies of 90-93% recovery; all predictable by the geologic evaluations of the host ores.

Further evaluations are continuing on the ores remaining to be mined and on new ore bodies. The projected possibilities of chlorine input in excess of 80 kg/tonne (200 pounds/ton) have led to continuing studies on alternative processing techniques, including autoclaving and roasting. These studies must continue with strong cooperation between the geologic and metallurgical groups to develop and/or alter the techniques required to process any ore with the maximum metallurgical and financial results.

ACKNOWLEDGEMENT

The authors wish to thank Freeport-McMoRan Gold Company for permission to publish this paper. Recognition for C.L. Kinard and D.S. Barr, Vice-Presidents of Freeport-McMoRan Gold Company is given for their support, ideas, and critical review.

REFERENCES

Birak, D. J., (1986). Exploration and Geologic Development of the Jerritt Canyon Gold Deposit, Elko, Nevada, U.S.A.; in proceedings of Gold '86 Symposium, pg. 488-496.

REFERENCES cont'd

Birak, D. J., and Hawkins, R. B. (1985). The Geology of the Enfield Bell Mine and Jerritt Canyon District, Elko, Nevada, in Tooker, E. W., ed., Geologic characteristics of sediment and volcanic-hosted disseminated gold deposits - Search for an occurrence model: U. S. Geological Survey, Bulletin 1646, p. 95-105.

MINERALOGY AND METALLURGICAL PERFORMANCE OF VARIOUS GOLD-COPPER ORES OF THE CHIBOUGAMAU AREA, QUEBEC

A.R. LAPLANTE
DEPARTMENT OF MINING AND METALLURGICAL ENGINEERING
McGILL UNIVERSITY
MONTREAL, QUEBEC, CANADA

ABSTRACT

This paper is intended to show the link between mineralogy and metallurgical performance for various Cu-Au ores of the Chibougamau area. The nature of gold occurence of five different ore types is first briefly described. Its impact on grinding and classification, gravity, flotation and cyanidation response is then presented.

The most salient characteristic of the Chibougamau area Au/Cu ores is their poor response to cyanidation, because of active cyanicides and gold occluded or in solid solution in pyrite. Thus gravity followed by flotation has proven metallurgically more efficient than flotation followed by partial cyanidation of the tail. Cyanidation of the coarse tail (backfill), rather than that of a pyrite concentrate floated from the tails, also appears more attractive, because of much lower cyanide consumptions.

KEYWORDS

Gold/copper ores; applied mineralogy; sampling; assaying; grinding; flotation; gravity concentration; cyanidation.

INTRODUCTION

Mining activity in the Chibougamau area started at the beginning of the century, with the Obalski mine, a very small gold producer. In the early fifties, significant copper deposits were found, and for close to 35 years, the Chibougamau area was a major Quebec copper producer. Gold is generally associated with the chalcopyrite deposits, assaying between 1 and 10 g/t.

The decline of copper prices and reserve grades, and the significant increase in gold prices of the seventies and early eighties turned the highly profitable copper operations into marginally profitable gold producers. This is well illustrated comparing 1974 and 1986 production and price figures. In 1974, Les Mines Camchib Inc. (then Cambell Chibougamau Mines Ltd.) mined ores averaging 1.5% Cu and about 2g/t Au. At the going copper and gold prices, gold was just 27% of the metal values contained in the ore. In 1986, copper in the mill heads assays about 0.8%, and gold 4 to 5 g/t. Gold represents close to 90% of the contained values. Moreover, the real value of copper is actually much lower, as smelting and refining costs add up to nearly 80% of its market value.

The shift in the relative economic importance of copper and gold made it necessary to modify milling practices. Figure 1 shows the yearly (1955-60) and monthly (mid-1961 to 1982) gold recovery as a function of copper grade in the concentrate, in one of the Chibougamau mills. Also shown is the average copper grade of the mill heads, which steadily decreases from 2.34% to 0.95% Cu, from 1956 to 1982. These data are widely scattered; nevertheless, the following regression equation gives a reasonable fit:

$$Rg = 115.05 * F^{0.13} * C^{-0.138} * (27 - C)^{0.054} \qquad \text{Eq. 1}$$

where Rg : gold recovery (%)
F : feed copper grade (%)
C : flotation concentrate copper grade (%)

(Correlation coefficient: 0.87)

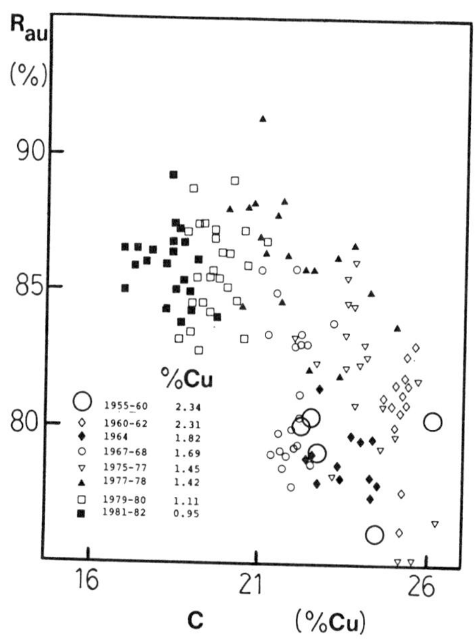

Fig. 1 Metallurgical performance of one of the Chibougamau mills, from 1955 to 1982 (shown is gold recovery, %, vs. concentrate grade, %Cu).

Eq. 1 suggests that decreasing copper grades will severely affect gold recovery

-e.g. a drop of as much as 14% for a drop of feed grade from 2.36% Cu to 0.8% Cu. This can be partially compensated by a drop in concentrate grade, which corresponds to actual plant practice from the beginning of the sixties (24-26% Cu) to the early eighties (17-19% Cu). Eq. 1 predicts that this practice would increase gold recovery by 13%, and effectively compensate for the drop in feed copper grade. However, this practice significantly increases the cost of processing the copper concentrate. During 1986, this cost was slightly above $C200 per tonne of concentrate. For a feed of 0.8% Cu, a decrease of concentrate grade from 25 to 18% corresponds to 0.015 extra tonne of concentrate per tonne of ore milled, or $C3/tonne.

Another drawback of increasing gold recovery by flotation is that smelter payment of gold is both partial and discounted. Smelter contracts normally pay between 95 and 97% of the contained gold, one to three months after delivery. This corresponds to an economic recovery of 92 to 96%.

The above analysis suggests that flotation should not be the sole gold recovery method for the Chibougamau ores. Both cyanidation and gravity concentration have in fact been used at the laboratory and plant scale, over the past few years. Interestingly, gravity recovery has proven quite successful, but cyanidation is no longer practiced.

This paper describes how the behaviour of five Chibougamau ores to the above recovery processes correlates to their mineralogy. The relationship between treatment response and mineralogy is not only predictive, but also inferential. Not only can the mineral processor learn from applied mineralogy, but also can the exploration and mine geologists benefit from the processor's knowledge of ore behaviour. This is because applied mineralogy generates very accurate information, but generally on a limited number of gold grains; surface data are also difficult to translate into actual metallurgical results, because of stereological problems. Processing generates less informative data (e.g. recovery), albeit on very large samples. Thus what processing data loose in intensity of information is partly offset by its statistical reliability.

GOLD MINERALOGY OF SOME TYPICAL ORES OF THE CHIBOUGAMAU AREA

The mineralogy of the gold/copper deposits of the Chibougamau area has been extensively studied over the past thirty years (Miller, 1957; Vollo, 1959; Wilhelmy, 1987). Generally, pyrite is a first-generation sulphide, with chalcopyrite and gold later occurences. It is suggested that copper and gold-bearing hydrothermal solutions infiltrated fractures and precipitated in the pyrite matrix, as shown in Fig. 2. In some cases, pyrite was partially replaced by gold or chalcopyrite; Fig. 3 shows such an occurence, in a loose pyrite matrix, from which they should be readily liberated.

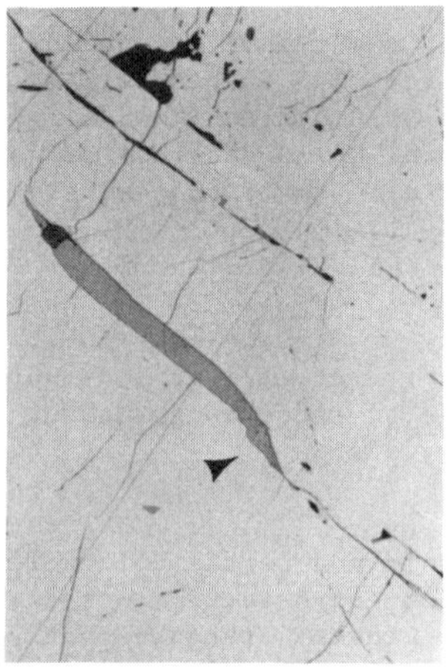

Fig. 2 Occurence of gold and chalcopyrite in a pyrite fracture.

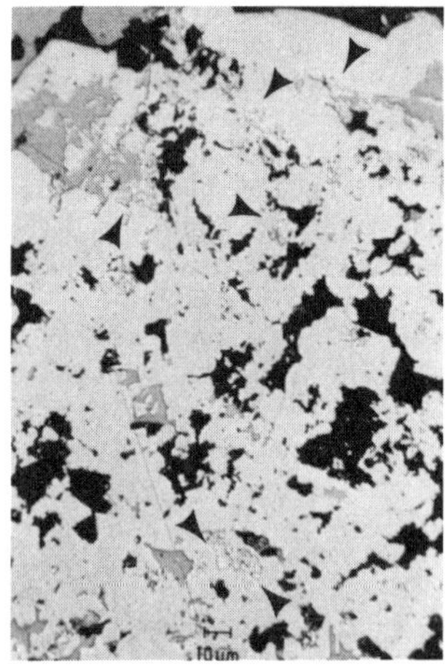

Fig. 3 Gold and chalcopyrite in pyrite.

TABLE 1 Mineralogy of Ores A to E

ORE	Grade of Cu	Au	Gold Grain Size	Gold Associations CuFeS2	FeS2	Fe7S8	Gangue
A	M-H	L-M	F-M	X	X		*
B	M-H	L-H	F	X	X		*
C	L-M	M-H	F-C	X		*	X
D	L-M	M-VH	F-VC	X	X		X
E	VL	M	F-M		X		X

(VL: very low, L: low, M: medium, H: high, VH: very high F: fine, C: coarse, VC: very coarse , X: associated with gold, *: abundant but not associated with gold)

Figure 4 shows another example of gold occurence in a large pyrite grain; the grain appears competent, and some of the smaller (1-3um) gold grains may not be readily liberated or exposed. Pyrrhotite is found in most of the ores, but without intimate association with gold.

The mineralogy of the five ores to be discussed is summarized in Table 1. Of the five ores, pyrite is abundant in four, and pyrrhotite in the fifth (ore C). Chalcopyrite is present in all five ores, from a low grade (less than 0.06% Cu) in ore E to important concentrations in ores A and B (as much as 3%). Gold assays 2-4 g/t for ores A and B, 4 to 7 g/t for ores C and E, and 6-12 g/t for ore D. Generally, gold grain size increases with increasing gold assay; the exception is ore B, with a consistently fine gold mineralization. A detailed mineralogical study of ore D showed two gold grain generations; the first is made of relatively fine grains, equally distributed in pyrite and gangue; the second is made of very coarse grains, mostly associated with pyrite (Wilhelmy, 1987).

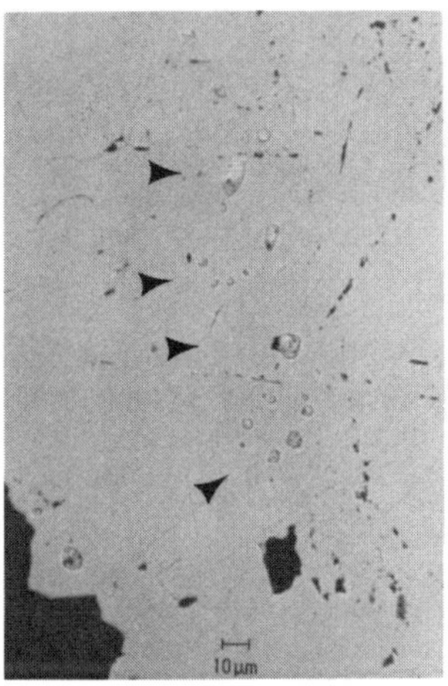

Fig. 4 Fine gold blebs in a pyrite grain.

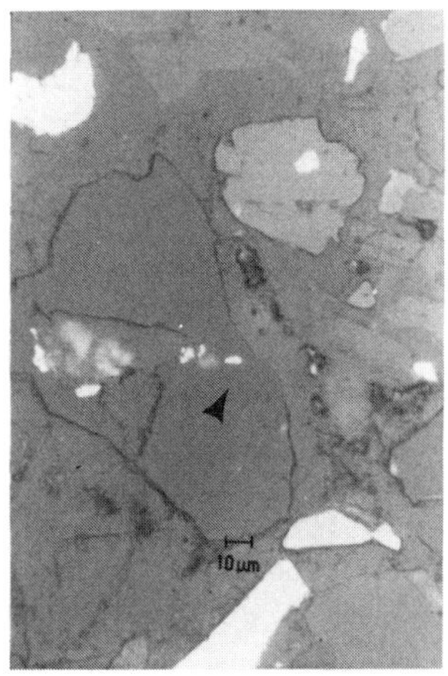

Fig. 5 Fine gold particle in gangue.

Gangue minerals are mostly silicates (quartz, chlorite, clinozoisite) and carbonates (calcite, dolomite, ankerite). Gold is generally not associated with the gangue minerals to any significant extent, except in ore D, and to a lesser extent in ores C and E; Fig. 5 shows a 5um gold particle in a gangue grain, not likely to be liberated. Figures 6 and 7 show two other grains, much more likely to be liberated or exposed, one because of its size (20um), and the other because it is at a grain boundary.

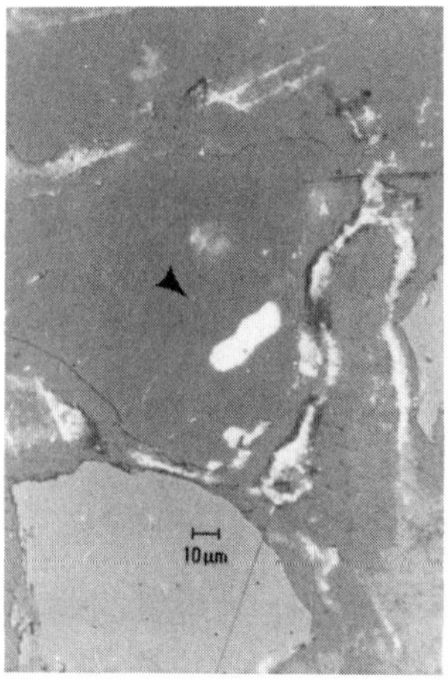

Fig. 6 Gold particle (20um) in gangue.

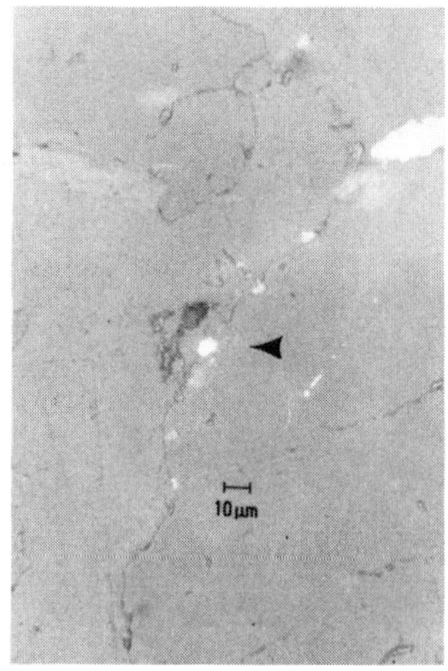

Fig. 7 Gold particle at the boundary of two gangue grains.

RESPONSE TO PROCESSING

Response of the various ores to processing will be examined in four sections: grinding/liberation, gravity, flotation and cyanidation.

Grinding and Liberation

All five ores have similar gangue materials, which largely determine their work index. The work index itself, as determined by dry grinding in a Bond mill (Bond, 1961), shows little variation between the ores. Wet grinding, however, reveals differences due to the different rheological properties of the ground minerals. For example, ore C, which is predominantly chlorite and clinozoisite, displays a significantly higher apparent viscosity than ores A and B, whose main gangue minerals are quartz and chlorite (Wibisono, 1985). As a result, both laboratory and plant grinding require more dilute slurries.

Coarse gold grinds differently in a laboratory mill and a plant grinding circuit with hydrocyclone classification. Figure 8 compares the size distribution of ore C and its gold, ground in a laboratory rod mill and a plant circuit (cyclone overflow). Although the ground ore is much finer in the bench scale test, gold is much coarser. In the plant, gold's malleability is not as critical, because it is recycled to the ball mill until breakage to a suitable size. This size, estimated from Plitt's cyclone model (1976), is equal to 100-150 um (Laplante, 1984); this was confirmed by examination of a gravity concentrate obtained from the plant flotation concentrate (Laplante and coworkers, 1986).

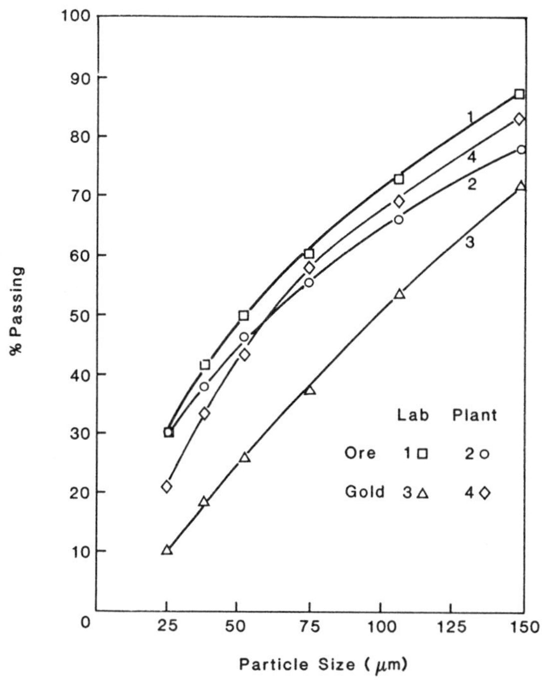

Fig. 8 Gold and overall size distribution for ore C, ground in a laboratory rod mill and in plant (cyclone overflow).

The coarse size of gold, as produced by batch grinding, can create scale-up problems. For example, laboratory cyanidation tests may suggest a longer retention time than needed, because of very coarse gold particles. It may be advisable to skim very coarse gold by gravity prior to cyanidation, to estimate the contribution of very coarse gold particles. If test results suggest a much shorter cyanidation time than for direct cyanidation, installing a gravity circuit with a smaller cyanidation circuit may be warranted. This approach was chosen for the OK Tedi mill (Lammers, 1984)

Figure 9 shows the gold assay of the gravity and flotation tails for ores A, B and C (curve 1) and ore D (curve 2), as a function of size. Curve 1 shows some incomplete liberation, above 420um (35 Mesh). As 95% of the tails is finer than 420 um, little benefit would result from further grinding. Curve 2 shows that ore D is liberated at a much finer size, below 37um; thus tail grade increases with increasing particle size, above 37um. Reproducibility of gold assaying is excellent, both by fire assay and atomic absorption. This suggests that much of the lost gold is very fine, and not associated with massive pyrite, as confirmed by mineralogical studies.

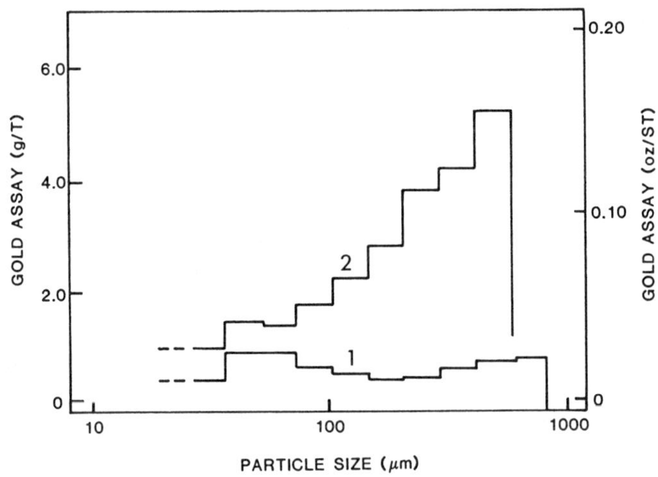

Fig. 9 Gold assay of the gravity/flotation tail for ores A,B and C (curve 1) and ore D (curve 2).

Curve 1 shows an unusual "hump" in the 37-74um range, where sulphides are most abundant. Clearly most of this gold is associated with iron sulphides. This material is largely recovered in the backfill product, whose gold assay is as much as three times that of the fine tails.

Curve 2 (Fig. 9) was obtained on a 2000 tonne lot of ore D, and largely substantiated earlier bench scale work. However, more than 30,000 tonnes of this material has since been processed using the same flowsheet, yielding results much similar to those of ores A, B and C. This exemplifies the difficulty of obtaining representative ore samples, even from underground workings. Similar problems were experienced with early flowsheet development work on ore C. This partly stems from the fact that ores C and D have a low chalcopyrite content, and are not as readily identifiable underground as are ores A and B, more typical of the mining activity of the Chibougamau area over the past thirty years.

Gravity Recovery

This topic has already been broached in the previous section. Although gold size in the flotation feed (cyclone overflow) in relatively fine, gold's circulating load in the grinding circuit is both very significant (as much as tenfold that of solids) and very coarse. Thus the ball mill discharge, or preferably the cyclone underflow, is an ideal feed for gravity separation. The first upgrading step does not have to be very efficient, as most coarse gold particles will be recycled a number of times.

The advantages of gravity concentration, ahead of either flotation or cyanidation, normally outweigh the capital and operating costs of a gravity circuit (e.g. Burt, 1984; Laplante, 1987). Thus gravity recovery is practiced in most South African gold mines (Bath and coworkers, 1973; Loveday and Forbes, 1982). At Les Mines Camchib Inc., the main two advantages are an increased total metallurgical recovery and a higher economic recovery of gold. Economic recovery is higher because gravity concentrates are rapidly smelted and delivered to the Royal Mint in Ottawa, for a pay-off close to 100%. Gold in flotation concentrates, however, is normally paid for at 95 to 97% of its value, from one to three months after its delivery; the pay-off varies between 92 and 96%. As between 20 and 35% of the feed gold is recovered by gravity, the net increase in overall economic recovery is somewhere between 1 and 3%. The overall metallurgical recovery also increases. Eq. 1 predicts, for a feed grade of 0.8% Cu and a concentrate grade of 19% Cu, a gold recovery of 83% by flotation, which would be in good agreement with metallurgical results obtained prior to the start-up of the gravity circuit; total gold recovery was then 86-88%, because a pyrite concentrate was also cyanided. Recoveries are now in the range of 89-91%, for a net improvement of about 7% over flotation alone, and 3% over flotation and cyanidation. The gravity circuit is also much cheaper to operate than the cyanidation circuit, in all respects (manpower, maintenance, reagents and power).

The gravity circuit used at Les Mines Camchib Inc. has been described elsewhere (Laplante and coworkers, 1986; Laplante, 1987). It upgrades the ball mill discharges in four steps, i.e. two pinched sluices, two 76cm and one 21cm Knelson bowl concentrators and one riffleless table. Figure 10 correlates early metallurgical monthly results with the head grade of ores A, B and C, which were jointly processed. As expected, the head grade of ore C best correlates gold recovery by gravity; ore B, despite a higher gold grade, shows little correlation with gravity recovery; for ore A, the correlation is intermediate. The conclusion is that gold grain size, rather than head grade, affects most gravity recovery.

Why would gold recovered by gravity, well liberated, be lost to flotation tails? It is very unlikely that losses originate from particles too coarse to float, because 1) the classifier density effect yield top sizes in a range very amenable to flotation and 2) tail assays would show a "nugget" effect, which they don't. Gold losses most likely come from the inordinately long residence time of coarse particles in the grinding circuit. Upon start-up of one Chibougamau gravity circuit, gold circulating loads dropped from 2000% to 800%, as about 25% of the feed gold was recovered. Thus the recovered gold averaged, prior to gravity recovery, about 50 loops around the grinding circuit -i.e. (2000-800)/0.25. Although it is unlikely that surface aging problems would arise, two other mechanisms would reduce recovery by flotation: firstly, production of very fine gold, difficult to float because of hydrodynamics (as are all fine particles), and secondly, smearing of gold unto other mineral particles, especially gangue ("unliberation").

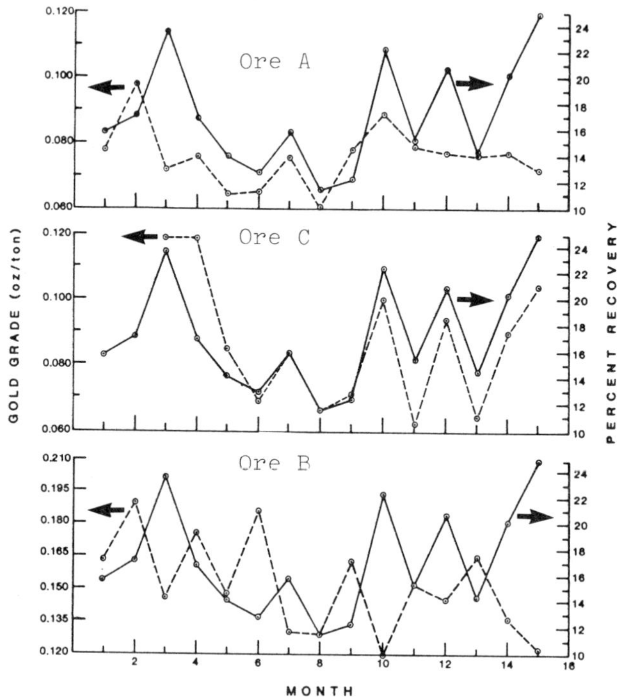

Fig. 10 Correlation between gold recovery by gravity and head grade for ores A, B and C (processed jointly).

Flotation

Approximately 70 to 80% of all gold produced in the Chibougamau area is recovered by flotation, down from nearly 100% a few years ago. Because of intimate association with iron sulphides, concentrate copper grades must be relatively low (18-20% Cu), for good gold recovery. Figure 11 shows that flotation concentrate can be upgraded significantly by flotation, with low copper losses; gold losses, however, are substantial, especially above 25% Cu.

Gold-pyrite associations are not found in all Chibougamau ores. Ore C is rich in pyrrotite, but has little pyrite. Gold shows little association with pyrrhotite, but is intimately associated with chalcopyrite. When an iron sulphide concentrate is produced, either at bench or plant scale, gold assays the same as the gangue (0.7g/t). Ores A and B, on the contrary, will yield significant upgrading in iron sulphide concentrates produced by flotation of the tails (up to 3-4g/t). Because of these mineralogical differences, ores A and B respond differently than ore C to different collector combinations. Table 2 (Laplante and coworkers, 1986) shows laboratory flotation results, for a collector combination which favors iron sulphides recovery, and a second one, of a shorter chain xanthate and a longer chain dithiophosphate, which favours the selective flotation of gold and chalcopyrite. As expected, the first combination outperforms the second for ores A and B (as corroborated by plant results). Ore C, however, responds better to a more selective flotation, which stresses the recovery of chalcopyrite and gold middlings, rather than that of iron sulphides.

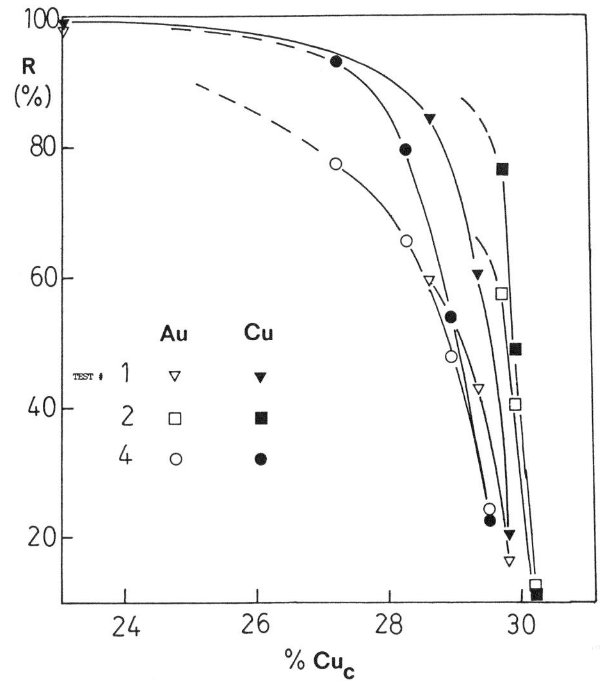

Fig. 11 Grade-recovery curves for three plant flotation concentrates (feed: 22-24% Cu).

TABLE 2 Mass and Gold Recovery (%) in a Laboratory Rougher Concentrate

	Collectors A		Collectors B	
	%Rec. Mass	%Rec. Au	%Rec. Mass	%Rec. Au
Ores A,B	15.6	87.5	11.3	81.9
Ore C	3.7	83.1	5.9	90.4

(Coll. A: Na amylxanthate, 35g/t, and Cyanamid Aerofloat 208, 15g/t; coll. B: Na isopropyl xanthate, 35 g/t, and Cyanamid Aerofloat 3477, 15g/t)

When the same mill must treat many ores which may respond differently to flotation, separate processing may be considered. However, it is advisable to process ores jointly, with careful blending (e.g. Madh Adh Dhahab, Lewis, 1984). This minimizes capital costs and simplifies day-to-day operations. It is especially important to avoid, whenever possible, separate processing of different ores from the same mine, as it is nearly impossible to achieve effective segregation.

Cyanidation

Of the two cyanidation circuits of the Chibougamau area, the first, which treated for a few years ore E, and a pyrite concentrate from ores A and B, is now shut down. Some ores from the area, low in copper content, were also custom milled successfully in this circuit. The second circuit, of larger capacity, was recently installed to process an auriferous pyrite concentrate, scavenged by flotation from a copper flotation circuit. After operating for a few months, the circuit was shut down, as gold recoveries were much lower than anticipated.

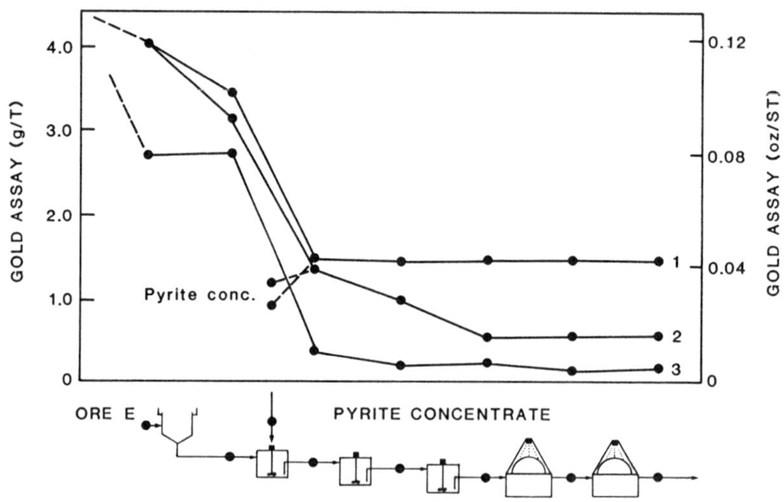

Fig. 12 Gold assay of the solid phase in a Chibougamau cyanidation plant.

Cyanidation of Au/Cu ores can be much more difficult than that of free-milling ores. Often, high cyanide consumptions preclude direct cyanidation; most of the copper, and a significant fraction of the gold, must then be removed by flotation. Cyanidation is then performed on either all, or part of the tail (usually a pyrite concentrate produced by flotation). This is not without problems, as often with iron sulphides are floated fine or slime chalcopyrite, which was not recovered earlier because of its slow flotation kinetics. Its high specific surface area makes it a strong cyanicide, even at low concentrations; at Les Mines Camchib Inc., significant increases in cyanide consumption could be detected if the pyrite concentrate assayed more than 0.08% Cu.

A second problem, also related to copper-bearing cyanacide, is that of the accurate detection of free cyanide. Copper ions in solution should not affect gold cyanidation, provided the concentration of free cyanide is above 0.05% (0.5g/L) (Stamboliadis and coworkers, 1976). The usual method to measure free cyanide, silver nitrate detection, also measures part of the cyanide tied with copper, due to partial dissociation of the cuprocyanide complex. Thus a free cyanide titration with silver nitrate of 0.05% can correspond to much lower free cyanide concentrations. Figure 12 shows what can then happen: curve 1 was obtained as a pyrite concentrate (from ores A and B) assaying 0.10% Cu was fed to the first reactor of the cyanidation circuit processing ore E. No gold dissolution takes place after the first reactor, presumably because of a severe deficiency in free cyanide. Curve 3 shows how the circuit normally performs, when no pyrite concentrate is processed; very rapid gold dissolution takes place, up to 95% complete.

Curve 2 (Fig. 12) illustrates another problem associated with the cyanidation of pyrite flotation concentrates. Even with adequate concentrations of free cyanide, only part of the gold associated with the pyrite concentrate is recovered. Laboratory tests show that only a 65% gold recovery is achieved; very fine grinding (99% -37um) does not increase recovery significantly. Very fine blebs of gold in massive pyrite (Fig. 4), or gold in solid solution (whose presence in the Chibougamau ores is suspected) would explain this refractoriness.

Another problem associated with cyanidation of flotation concentrates is that of the chemical environment. Collectors adsorbed onto gold can retard dissolution (Mrkusic and coworkers, 1968). Collectors and thiosalts are also known to adsorb onto activated carbon, and can seriously affect the efficiency of CIP circuits. This problem may have played a key role in the demise of the most recent cyanidation circuit of the Chibougamau area, inasmuch as there was no acid cleaning of the carbon. An 'elegant' solution is to destroy, with SO2, collectors and free cyanide prior to dissolved gold recovery in a CIP circuit, as practiced at Equity Silver (Semple, 1987).

An alternative to the cyanidation of flotation concentrates is that of the coarse tail, often recovered as backfill. Table 3 summarizes laboratory tests on the coarse and fine fractions of ores A, B and C vs. that of ore D. Cyanide consumption is high for the fines of ores A, B and C, and moderate for the three other products. Although the fine fractions yield better gold recoveries, more gold is recovered from the coarse fractions, because their gold grades are about three times those of the fine fractions. In all cases, cyanidation times are relatively short (7-10 hours).

Backfill cyanidation offers the added advantage of an easy solid-liquid separation. Most of the solids can be rejected in a counter-current cyclone circuit, such as the one used at Golden Sunlight (Dayton, 1984).

TABLE 3 Summary of Laboratory Cyanidation Tests with Ores A,B,C,D

	Gold Assay Metric (g/t)	Gold Assay British (oz/st)	Gold Recovery (%)	Cyanide Consumption (kg/t)
Ores A,B,C				
(+37 um)	0.58	0.017	50	1.4
(-37 um)	0.34	0.010	65	3.9
Ore D				
(+53 um)	2.7	0.08	55	1.4
(-53 um)	1.0	0.03	75	1.9

CONCLUSION

The mineralogy of most of the Chibougamau Au/Cu ores is such that the classical approach of direct cyanidation is both metallurgically and economically non-feasible. Flotation, which has been used for over thirty years as the sole mean of copper recovery, remains the workhorse of the three mills in the area. The emergence of gold as the most important valuable species to recover has necessitated some adjustments to the processing route. Both gravity and cyanidation have vied as competing technologies for secondary gold recovery. Surprisingly, gravity has emerged as the clear winner; the mineralogy of most of the local ores should have at least in part predicted such an outcome. Most applied mineralogy work, however, was performed after flowsheets were developed and tested at full scale. Its findings largely corroborate and explain plant results, and could in part be inferred from metallurgical behaviour.

The demise of cyanidation in the Chibougamau area is another reminder that gold recovery needs not follow Elsner's equation each time. Cyanidation will remain for quite a few years the dominant route to gold recovery. Other routes must, however, be given their due place, which can often be predicted from applied mineralogy. This highlights the need for using it as part of the flowsheet development program, rather than only as part of a 'post-mortem' exercise.

ACKNOWLEDGEMENTS

The author wishes to thank Les Mines Camchib Inc., Les Mines Northgate Inc. and Les Ressources du lac Meston Inc., for invaluable help in preparing this paper, and permission to release information. The author is also indebted to Dr. Jean-Francois Wilhelmy, mineralogist at Le Centre de Recherches minerales (Quebec City), for providing Figs. 2-7, and many helpful discussions regarding the mineralogy of the Chibougamau Au/Cu ores.

BIBLIOGRAPHY

Bath, M.D., A.J. Duncan and E.R. Rudolph (1973). Some factors influencing gold recovery by gravity concentration. South Afr. Inst. Min. Metall. J., June, pp. 363-384.

Bond, F.C. (1961). Crushing and grinding calculations, parts 1 and 2. Brit. Chem. Engng., Aug., Vol. 8, No. 8, pp. 543-548.

Burt, R. (1984). Gravity Concentration Technology. Ed. Elsevier, Vol. 5, Developments in Min. Proc., Amsterdam, 605 p.

Dayton, S.H. (1984). Golden Sunlight sheds warming rays on Placer U.S.. E/MJ, MAY, PP. 34-39.

Lammers, J.M. (1984). The design of the OK Tedi Project. First Intern. Symp. on Precious Metals Recovery, Reno, June, Paper IV, 15 p.

Laplante, A.R. (1984). Sampling and mass balancing for gold ores. First Intern. Symp. of Precious Metals Recovery, Reno, June, Paper V, 20 p.

Laplante, A.R., A. Wibisono and A. Cauchon (1986). Development of a flowsheet for the S-3 ore. Can. Min. Proc. Regional Meet., Chibougamau, Sept., pp. 155-193.

Laplante, A.R. (1987). The use of gravity concentration for gold recovery. McGill Univ., Prof. Dev. Sem. on the Proc. Gold Ores, April 30-May 2, 19 p.

Lewis, P.J. (1984). Metallurgical evaluation and process development for the Mahd Adh Dhahab gold deposit. First Intern. Symp. on Precious Metals Recovery, Reno, June, Paper XV, 28 p.

Miller, R.J. (1957). Geology and ore deposits of the Cedar Bay mine area, Chibougamau district, Quebec. Laval University thesis.

Loveday, B.K. and J.E. Forbes (1982). Some considerations in the use of gravity concentration for the recovery of gold. South Afr. Inst. Min. Metall. J., May, pp. 121-124.

Mrkusic, P.G., H.A. Simonsen and M. Sciarone (1968). The effect of aeration, before cyanidation, of pulps of gold ore containing xanthate. South Afr. Nat. Inst. Metall. Report No. 380, Aug., 38 p.

Plitt, L.R. (1976). A mathematical model of the hydrocyclone classifier. CIM Bull., Dec., pp. 114-123.

Semple, P.G. (1987). Equity Silver Mines scavenger circuit. 18th Ann. Meet. Can. Min. Proc., Ottawa, Jan. 20-22., Sess. B, New Operations, in print.

Stamboliadis, E., J. McHardy and T. Salman (1976). The effect of copper ions on the cyanidation of a gold ore. CIM Bull., March, pp. 128-130.

Vollo, N.B. (1959). The geology of the Henderson copper deposit, Chibougamau area, Quebec. McGill University thesis.

Wibisono, A. (1985). Development and testing of a flowsheet for the S-3 ore from Les Ressources Camchib Inc. McGill University Thesis, 162 p.

Wilhelmy, J.-F. (1987). Applied mineralogy and the processing of gold ores. Presentation at the McGill Univ. Prof. Dev. Sem. on the Processing of Gold Ores, Montreal, April 30-May 2.

DIAGNOSTIC LEACHING IN THE DEVELOPMENT OF FLOWSHEETS FOR NEW ORE DEPOSITS

J.A. Tumilty, A.G. Sweeney, L. Lorenzen

Anglo American Research Laboratories, Johannesburg, South Africa

ABSTRACT

Diagnostic leaching, is a simple series of sequential leaches developed to elucidate the deportment of gold in various matrices. The mineralogical examination quantitatively determines the minerals present in a matrix, and the diagnostic leach sequence is designed accordingly. Each leach destroys a specific mineral, and subsequent cyanidation determines the mineral's gold content. The differences in the kinetic and thermodynamic stability of the various minerals allows leach selectivity.

Various examples were chosen to explain the diagnostic leach steps and the interpretation of the results. Prospect A has a low gold grade (1,4 g/t) and low ore reserve and is best treated by heap leaching. Prospect B has a high gold grade (14,5 g/t) but approximately half of the gold is locked in sulphides and may be best treated by a flotation step followed by bacterial leaching. Prospect C has a gold grade of approximately 9 g/t but very high cyanide consumption due to copper mineralisation and the most attractive route is to bulk float and smelt the copper concentrate and obtain credit for the gold and arsenic content. Gold in a flotation tailings sample from Plant A (4,3 g/t gold) may be reduced by modifying the milling circuit to prevent overgrinding of the heavy minerals, for example by use of fine screening. Plant B flotation tailings could be cyanided, followed by flotation, roasting of the concentrate and cyanidation of the calcine. Diagnostic leaching of these samples gives a clear picture, which when combined with experience, indicates the logical process route.

The optimum particle size reduction is also specified by this technique. Sulphides are liberated in the ore processing stage earlier than the silicates; thus the residual silicate gold content is a good indicator of the optimum matrix grind.

KEYWORDS

Cyanidation; diagnostic-leaching; flotation; mineralogy; ore-deposits.

INTRODUCTION

The first step in the investigation of a gold deposit should be an examination of the matrix mineralogy and the determination of mineral abundance. The diagnostic leach technique can then be used to determine the gold deportment. Diagnostic leaching combines sequential selective leaches of an ore or concentrate with interstage cyanidation. The process is varied according to the matrix mineralogy and aims to destroy selectively the minerals associated with gold.

Estimation of gold deportment has been limited to gravity separation and assay of the density fractions, as the gold is not normally visible. The information is of more use to the geologist and the mining engineer. The data required, to select process routes available, determine the gold liberation at particular grinds and specify possible recoveries expected in e.g. flotation are greatly simplified when a diagnostic leach is performed.

Diagnostic leaching was designed to answer these questions and in doing so it opened up the field of mineralogical analysis to the gold metallurgist. Tumilty (1986) investigated Witwatersrand deposits utilising this method successfully. This paper describes the application of the method to samples from plants and prospects and illustrates process implications.

A number of ore deposits from various parts of the world were treated.

EXPERIMENTAL AND DISCUSSION

The concept of diagnostic leaching is very simple. The least stable mineral present in the matrix of the sample is first solubilised in aqueous acidic media. The residue is filtered, washed and cyanided to extract the gold liberated. The gold leached by the cyanide gives an accurate record of the amount of gold associated with that mineral. The cyanidation residue can be subjected to a more oxidative acid leach and then re-cyanided. The 'recipe' for each leach is determined by the mineral suite, as determined by the preliminary optical examination.

The diagnostic leaching technique is still in its infancy. It will be developed further to give the metallurgist more information by adjustment of type and severity of the leach conditions. The diagnostic leach eliminates a whole lot of testwork but very careful interpretation must be made of the results.

At the end of the diagnostic leach sequence the metallurgist is left with an almost complete record of the deportment of gold. An overlap of extent of leaching does occur but is generally limited to 85% of the required mineral and 15% of the next most stable mineral. The metallurgist can now use the information to formulate a metallurgical flowsheet to treat the ore or to solve problems at an existing gold plant. The diagnostic leach can deal with any type of intermediary gold plant product. The information gained can give a useful indication of process efficiency.

For example, most of the precipitated gold in slimes dam material fed to a flotation plant will report to the float tails - determined by diagnostic leaching. Another example is the determination of gold locked by ferrites produced from base metal sulphides in fluo-solids roasting of flotation concentrates.

In this paper various examples were chosen to explain the diagnostic leach steps and the interpretation of the results. Prospect A has a low gold head grade (1,4 g/t) and low ore reserve while Prospect B has high gold head grade (14,5 g/t) but approximately half of the gold is locked in sulphides. Prospect C has a gold grade of approximately 9 g/t but very high cyanide consumption due to copper mineralisation. Flotation tailings samples from Plant A containing 4,3 g/t gold and Plant B containing 0,57 g/t gold were also investigated to determine a suitable flowsheet for gold recovery.

PROSPECT A

The ore sample (crushed to less than 25 mm) was received at Anglo American Research Laboratiries (AARL) for mineralogical and metallurgical examinations. The ore assayed about 1,4 g/t Au, 0,62 g/t Ag and 0,1 per cent sulphide sulphur.

The mineralogical composition of the ore in decreasing order of abundance was quartz (major), feldspar (predominant), mica, goethite, chlorite, pyrite, arsenopyrite.

The diagnostic leach routine was then designed to selectively leach the minerals indicated by mineralogical examination. This routine consisted of-

Cyanidation - to determine leachable gold.

HCl treatment followed by cyanidation - to determine gold in goethite.

HNO_3 treatment followed by cyanidation - to determine gold in sulphide minerals, especially pyrite and arseno-pyrite.

TABLE 1. Diagnsotic Leaching Results for Prospect A

	Calc. head	Free gold leachable (NaCN only)		Gold in iron oxide labile sulphides & coatings (HCl & NaCN)		Au in remaining sulphides incl. pyrite (HNO_3 & NaCN)		Residual gold in quartz	
Test No.	Au g/t	Ext.		Ext.		Ext.		Ext.	
		g/t	%	g/t	%	g/t	%	g/t	%
A	1,49	1,23	82,55	0,13	8,72	0,09	6,04	0,04	2,68
B	1,36	1,10	80,88	0,13	9,56	0,09	6,62	0,04	2,94
Average	1,43	1,17	81,72	0,13	9,14	0,09	6,33	0,04	2,81

The ore was milled to the usual Witwatersrand grind of 70 per cent minus 74 micrometres and diagnostically leached in duplicate, Tests A and B in Table 1. At this grind, the cyanidable gold represented 82 per cent (1,17 g/t) of the contained gold. 15 per cent (0,22 g/t) of the gold was associated with goethite and sulphide minerals with 3 per cent (0,04 g/t) of the gold being locked in quartz (see Table 1).

Gold dissolution proceeded very rapidly with 66 per cent gold extraction being achieved in one hour.

Flotation of the sulphides to recover the gold content would be marginal for this ore thus the lower recovery of 82 per cent must be accepted. Silicate gold content was low at 0,04 g/t, thus grinding to this fineness may not be necessary.

Due to the low head grade (approx. 1,4 g/t) and low ore reserve (approximately 0,5 million tons) the treatment capital and operating costs must be low to maximise return on investment. These factors and the high leachable gold content indicate this ore-body should be treated by heap leaching.

Further testwork should be conducted to determine the gold extraction for different crushed sizes (up to 25 mm).

PROSPECT B

A sample of milled (70 per cent less than 74 micrometre) ore was subjected to a sequence of diagnostic leaches in order to establish the deportment of gold and silver.

The lixiviants were chosen to destroy the gold and silver bearing minerals sequentially. Inter-stage cyanidation indicates the deportment of gold and silver, see Table 2.

TABLE 2. Diagnostic Leaching of Prospect B

Lixiviant sequence	Mineral leached	Au Extr. (%)	Ag Extr. (%)
NaCN	Free gold and silver and chalcocite	42,70	52,3
Na_2CO_3/HCl/ NaCN	calcite, dolomite gypsum and galena	13,50	17,00
H_2SO_4/ NaCN	Sphalerite, tetrahedrite and labile copper sulphides	32,00	21,00
HNO_3/ NaCN	Pyrite, arsenopyrite, marcasite	8,20	8,10
Residue	Quartz or gangue	3,60	1,60
	Head	14,5 g/t	6,22 g/t

The testwork indicated that 43 per cent of the gold was cyanidable while 46 per cent was associated with acid solubles and labile sulphides. A ferric leach followed by cyanidation resulted in 12 per cent of gold and 10 per cent of the silver, reporting to the residue.

The variation of the ore-body mineralogy required further diagnostic leaching of a representative sample.

A second sample of this ore-body indicated the diagnostic leach results in Table 3. This sample had a similar quantity of free gold but much more gold in pyrite.

TABLE 3. Further Diagnostic Leaching of Prospect B

Lixiviant Sequence	Mineral leached	Gold Extracted		Silver extracted	
		g/t	Cum. %	g/t	Cum. %
NaCN	Free gold, silver and chalcocite	4,90	35,8	88	49,7
HCl	Galena, pyrrhotite and some sphalerite	0,25	35,8	39	71,7
NaCN		0,54	39,7	21	83,6
$FeCl_3$ (90°C)	Sphalerite, galena labile sulphides tetrahedrite	0,25	39,7	12	90,4
NaCN		1,92	53,8	2	91,5
HNO_3	Pyrite and all remaining sulphides	0,94	60,7	0,5	91,5
NaCN		4,85	96,1	7	95,5
Residue	Mainly quartz	0,54	100	8	100

Head 13,7 g/t Au 177 g/t Ag

*Free silver is defined as that silver soluble in cyanide and includes some present as sulphide and chloride.

The base metals in the diagnostic leach solutions were determined and utilising simultaneous equations and the mineralogy data, the gold was appointed to the minerals as described below.

A large proportion of the copper (approximately 60%) leached during the initial cyanidation, indicating a labile copper sulphide such as chalcocite.

The HCl leach solubilised 90% of the lead while the ferric chloride solubilised 90% of the zinc. The iron (85%) was solubilised in the nitric acid leach.

Thus by determining the minerals solubilised and with simple mathematics the gold distribution in the mineral matrix is specified.

The gold in quartz is 0,54 g/t, considering the grind, additional milling costs to recover this gold would be prohibitive.

The routes available for consideration could be sulphide flotation followed by roasting or pressure leaching. Roasting of sulphides containing tennantite-$(Cu,Fe)_{12}Sb_4S_{13}$-could generate bed sintering problems. Roasting and pressure leaching are high capital cost unit processes and the size of the deposit would not support the high capital expenditure.

Ore bodies of this type require a flotation step with treatment of the concentrate by new processes such as bacterial leaching. Other processes which create ferric for leaching may also be suitable.

The copper leached during the cyanidation dictates a high cyanide consumption and the solubilised copper will be adsorbed if eg a CIP unit process is utilised for gold recovery. The CIP would be problematic, thus some other method would have to be utilised.

The viability of the project will be determined by the economics associated with all these problem areas.

PROSPECT C

A bulk sample was split and a portion crushed to minus 1650 micrometres, milled, and submitted for mineralogical examination, head assay and diagnostic leaching.

The mineralogical investigation revealed that the sample consists essentially of pyroxene, quartz, amphibole, feldspar, calcite, ore minerals and minor to trace amounts of chlorite, mica and clay minerals, see Table 4A. The heavy minerals were: pyrite, arsenopyrite, bornite, chalcopyrite, magnetite, hematite, metallic bismuth and bismuthinite, see Table 4B.

TABLE 4A. Estimated Major Mineral Composition

Mineral	Mass per cent in sample
Pyroxene	35
Amphibole	12
Quartz (major))	
Feldspar (minor))	25
Mica (trace))	
Calcite	12
Chlorite	2
Ore minerals	14
Clay minerals	1
	100

TABLE 4B. Estimated Heavy Mineral Composition

Mineral	Mass per cent of	
	Total sample	Ore minerals
Pyrite	1,1	9
Arsenopyrite	7,6	54
Bornite	1,0	7
Chalcopyrite	1,7	12
Magnetite	2,0	14
Hematite	0,4	3
Metallic bismuth, bismuthinite other sulphide mineral and gold	0,2	1
	14,0	100

The head assays were 9,11 g/t gold, 16,9 g/t silver and 1,33 per cent sulphide sulphur (available at 1400°C after a carbonate leach). Calculated head values of 10,0 to 10,8 g/t Au obtained from diagnostic leaching testwork indicated the difficulties in obtaining a representative head sample.

TABLE 5. Diagnostic Leaching of Prospect C

Leaching sequence	Au	Ag	Fe	Cu	As	Ca
Assay Head	10,1 g/t	20,1 g/t	12,65%	9209 g/t	4,14%	10,65%
Calc Head	9,71 g/t	22,15 g/t	11,1%	9305,9 g/t	4,16%	6,2%
NaCN (5 kg/t)	53,6%	-	0,01%	21,12%	0,05%	0,28%
NaCN (10 kg/t)	33,5%	54,3%	0,30%	23,91%	0,06%	0,08%
HCl/NaCN	5,6%	35,8%	5,75%	0,59%	6,02%	17,21%
H_2SO_4/NaCN	5,1%	5,3%	10,72%	18,05%	0,03%	3,13%
HNO_3/NaCN	1,7%	3,7%	20,50%	36,26%	91,59%	28,48%
Residue	0,5%	1,1%	62,72%	0,08%	1,95%	50,82%

An initial leach indicated high cyanide consumption. 58 per cent of the gold was directly leachable. A second test at a grind of 83 per cent passing 74 micrometres and using 20 kg/t NaCN for free gold determination showed that 88 per cent of the gold was directly leached. This cyanidation used aggressive conditions to show a point i.e. how much of the gold was free, thus care must be taken on the interpretation of the results. The cyanide consumption in this testwork does not indicate the expected plant cyanide consumption. 8 per cent of the gold was associated with sulphides and 4 per cent with quartz. A third diagnostic leach, see Table 5, was performed using two-stage cyanidation, which clearly demonstrated that the copper mineralisation was consuming large amounts of cyanide and hindering gold dissolution. NaCN addition of 15 kg/t in the two stages leached 87 per cent of the gold. A dilute hydrochloric acid leach followed by cyanidation liberated a further 6 per cent of the gold, either from calcite or the bismuth mineralization. Around 7 per cent of the gold was associated with other sulphides and only 0,5 per cent was associated with quartz.

The diagnostic leaching results show that the majority of the locked gold is associated with the copper mineralisation. Thus with direct cyanidation sodium cyanide consumptions will be high. Further investigations were conducted to determine the feasibility of producing separate copper and arsenic concentrates for treatment by various methods.

Using the differential flotation technique, a rougher concentrate assaying 200 g/t Au, 450 g/t Ag, 23 per cent Cu and only 1 per cent arsenic can be produced at recoveries of 70 per cent gold, 88 per cent silver and 97 per cent copper.

Adjustment of the pulp pH to 5,5 is necessary for efficient arsenic flotation and this would require a high acid addition (10-12 kg/t). However, there are indications that between 18 and 19 per cent of the gold in the original feed could be taken into solution by cyaniding the copper flotation tailings.

The high cyanide consumption and high copper tenors render direct cyanidation unattractive. Further, the copper adsorption in the CIP process prohibits the use of this unit process.

The economics of producing an arsenic concentrate need to be carefully assessed since it would appear that a copper float plus tailings leach may well produce 89 per cent Au recovery.

Differential flotation and roasting of the copper concentrate and cyanidation of the tailings are not attractive either.

The most attractive route is to bulk float and smelt the copper concentrate and to obtain credit for the gold and arsenic content.

PLANT A

A sample of reclaimed flotation tailings was received to evaluate the gold deportment, gold leaching characteristics using cyanide and gold recovery from cyanide, possibly using activated carbon.

Gold distribution in various size fractions is presented in Table 6.

A diagnostic leaching routine was designed to determine the gold deportment.

64,6% of the gold is cyanidable gold, see Table 7. The conditions used for the diagnostic leach are "aggressive" thus this is probably the maximum gold extraction obtainable in cyanidation. Leaching testwork using lower cyanide additions showed a gold extraction of 52 per cent.

The cyanidation residual gold (1,52 g/t) is distributed in the calcite/dolomite fraction, in sulphide minerals and in gangue. The diagnostic leach results indicated that the flotation circuit at this plant is inefficient in terms of gold recovery, metallic gold should be recoverable in a float circuit. Poor recovery of these gold fractions is probably due to under grinding of the quartz and over grinding of the sulphides. Modifications to the milling circuit at the plant could be effected by utilising a linear screen, instead of cyclones, after the mill and returning the oversize to the mill.

Thus recovery of gold from existing tailings, with regrinding after cycloning, by CIP offers the most attractive route.

TABLE 6. Size Distribution in Flotation Tailings

Size fraction (micrometre)	Cum. Mass (%)	Au content (g/t)	Cum. Dist. Au %	S/S content %
+147	6,95	5,39	8,95	
-147 +104	24,96	5,21	31,39	0,41
-104 + 74	44,63	4,42	52,17	
- 74 + 53	58,29	3,85	64,74	
- 53 + 44	63,43	3,65	69,34	2,20
- 44	100,00	3,52	100,00	

TABLE 7. Diagnostic Leach of Plant A Flotation Tailings

Lixiviant sequence	Aim	Au Extraction	
		%	g/t
NaCN	Leachable gold	64,6	2,77
HCl/NaCN	Gold in calcite/dolomite	11,7	0,50
HNO_3/NaCN	Gold in sulphides	14,6	0,63
-	Gold in gangue	9,1	0,39
	Total	100,0	4,29

PLANT B

Tailings from Plant B were received to determine a possible treatment to recover the gold in their tailings. The sample as received was screened into size fractions and each fraction was diagnostically leached to determine the deportment of gold.

TABLE 8. Distribution of Gold in Plant B Tails

Tailings	Au head g/t	% of total gold in minerals					
		Leach-able Au	Base metal Sulphides	FeS_2	Adsorbed on carbon	Silicates	Mass %
-208+74 um	0,84	27,6	24,1	37,9	6,8	3,6	18,3
- 74+43 um	0,96	39,4	13,0	27,6	15,9	4,1	13,2
-43 um	0,42	53,0	9,7	23,9	11,5	1,6	68,6
Overall	0,57	46,6	12,8	26,8	11,5	2,3	100,0

The largest proportion of the gold in the less than 74 micrometre fractions is free gold at 51% of the total gold while the largest proportion in the greater than plus 74 micrometre size fraction is pyrite gold at 38% of the total gold.

The overall gold head grade is low at 0,57 g/t with 0,26 g/t free gold (47% of total). Pyrite gold is the next largest portion (27% of total). Base metal gold and gold adsorbed on carbonaceous material are about 12% of the total gold each. Only 2,3% of the gold was locked in the silicates.

The diagnostic leach indicates that direct cyanidation with CIP for gold recovery offers the most attractive route.

If the tailings reserves are large and there is a ready market for sulphuric acid - then cyanidation followed by flotation, roasting of the concentrate and cyanidation of the calcine - would be attractive.

CONCLUSIONS

The diagnostic leach consists of a set of simple experiments which can determine the deportment of gold accurately in various mineral matrices. The data obtained from a modest amount of testwork enables the derivation of possible process routes for treatment of the ore-body for gold recovery.

The routes chosen depend on such factors as the economics of the project, location of the ore-body and ore-body size.

Diagnostic leaching and experience also determine whether there is a need for further testwork and indicate what testwork is necessary.

ACKNOWLEDGEMENT

This paper is published by permission of the Anglo American Corporation of S.A. Ltd.

REFERENCES

Tumilty, J.A., C.G. Schmidt, (1986) In C.E. Fivaz (Ed.), Extractive Metallurgy of Gold, Vol. 2, The South African Institute of Mining and Metallurgy, Johannesburg. Part 6, pp. 541-553.

EFFECT OF MINERALOGY ON THE MCLAUGHLIN AND OTHER HOMESTAKE METALLURGICAL PROCESS DEVELOPMENTS

R.S. Kunter, Russell M. Honea and Robert D. Lear
Homestake Mining Company
Golden, U.S.A.

Gold in Homestake's McLaughlin Mine is occluded in several fine grained sulfide species including pyrite, marcasite, pyrargyrite, and miargyrite. Active carbon in this orebody, asphaltic in character and originating from the seabottom sediments, causes "preg robbing" from gold bearing cyanide solutions when present.

Metacinnabar and cinnabar are present in substantial quantities in the upper zone of the orebody but decrease significantly with depth. Stibnite is present at times in massive quantities but is generally not auriferous. Arsenopyrite is present in lesser amounts than the stibnite. The presence of both arsenic and antimony in the system has given rise to a complex but rarer mineral suite also present in the orebody. Chalcopyrite, sphalerite, galena, and millerite are all present in locally anomalous amounts.

Roasting of this ore or its concentrate would have been a traditional and technically feasible approach to metallurgical amenability to cyanide leaching. The fine grained nature of the sulfides caused some ferritic sintering which somewhat reduced the overall gold extraction. Additionally, roasting would have required a state of the art gas cleaning and SO_2 adsorption plant in an environmentally sensitive area of California comprised of three air quality control districts.

The hydrometallurgical pressure oxidation process eventually developed and installed by Homestake had a number of process and environmental advantages. The gold extraction by cyanide was higher than by roasting. The arsenic and antimony are disposed of in the solid residues as ferric-arsenates and ferric-antimonates which are stable compounds. Mercury is primarily contained as a mercuro-jarosite which does not readily leach. Copper, zinc, nickel, cobalt, and iron are solubilized during the pressure oxidation and are removed by washing and are therefore not present as cyanicides during the subsequent cyanidation with a substantial decrease in cyanide consumption. All of the sulfides in the ore act as fuel during the oxidation and provide the heat energy necessary to keep the process at temperature. The process tails are in a very stable oxidized condition and are highly buffered from changes in the basic pH range.

An understanding of the mineralogy of the McLaughlin deposit and the effects of various metallurgical processes on this mineralogy was concurrently pursued during the metallurgical development which led to the pressure oxidation process installed.

Homestake is currently developing a hydrometallurgical concentrate leach for the North Amethyst project at the Bulldog Mountain operation in Creede, Colorado. Bulk sulfide concentrates produced from certain areas in the operation had very high cyanide consumptions. Chemical and mineralogical investigation indicated that the chemical species was ferro-cyanide and that during the leach cyanide was reacting with metastable pyrite showing very evident etched grain boundaries. Subsequently, the problem was alleviated by cleaning the concentrate with cyanide and sodium silicate and thereby depressing the pyrite into the cleaner tails. The concentrate was then leached with the high levels of cyanide required because of the silver content. The cleaner tails containing the pyrite was leached with low levels of cyanide which gave good precious metal recovery but eliminated the attack on the pyrite with an overall much reduced cyanide consumption.

Technical Sessions

Session No.4

Plant Practice 2

REDUCING POWER AS THE CIRCUIT CHEMISTRY MONITOR AT ECHO BAY MINES LIMITED, LUPIN GOLD OPERATION

J. Fulcher and K. Kim

Lupin, N.W.T., Canada, XOE 1MO

ABSTRACT

In the spring of 1986, a change was made in the preaeration circuit to solve a mechanical problem. By doing so, it created an unexpected upset in the chemistry, which took approximately four weeks to correct. The reducing power measurement was found to be invaluable in monitoring the circuit chemistry.

In the preaeration circuit, the cyclone overflow was initially sent to the preaeration tanks prior to the thickener and then on to the leach tanks. Since the change, the cyclone overflow is sent to the thickener prior to the preaeration tanks.

The benefits from this change include tremendous reduction in downtime, in maintenance costs, and in reagents. As well, the operators find the new circuit arrangement much easier to control.

KEYWORDS

Reducing power; preaeration; thickener; lead nitrate; air lifters; suspended solids; cyclone overflow; leach; filtrate; cyanide; pH; density.

INTRODUCTION

Echo Bay Mines' Lupin operation processes 1700 tons of ore per day with an average head grade of 0.320 ounces of gold per ton of ore and a 95.3% recovery. The Lupin mill consists of seven circuits: crushing, grinding, preaeration, cyanidation leaching, filtration, precipitation and refining. (Fig. 1)

In the spring of 1986, a change was made in the preaeration circuit to alleviate mechanical mixing problems. This change created an unexpected upset in the chemistry, which took approximately four weeks to correct.

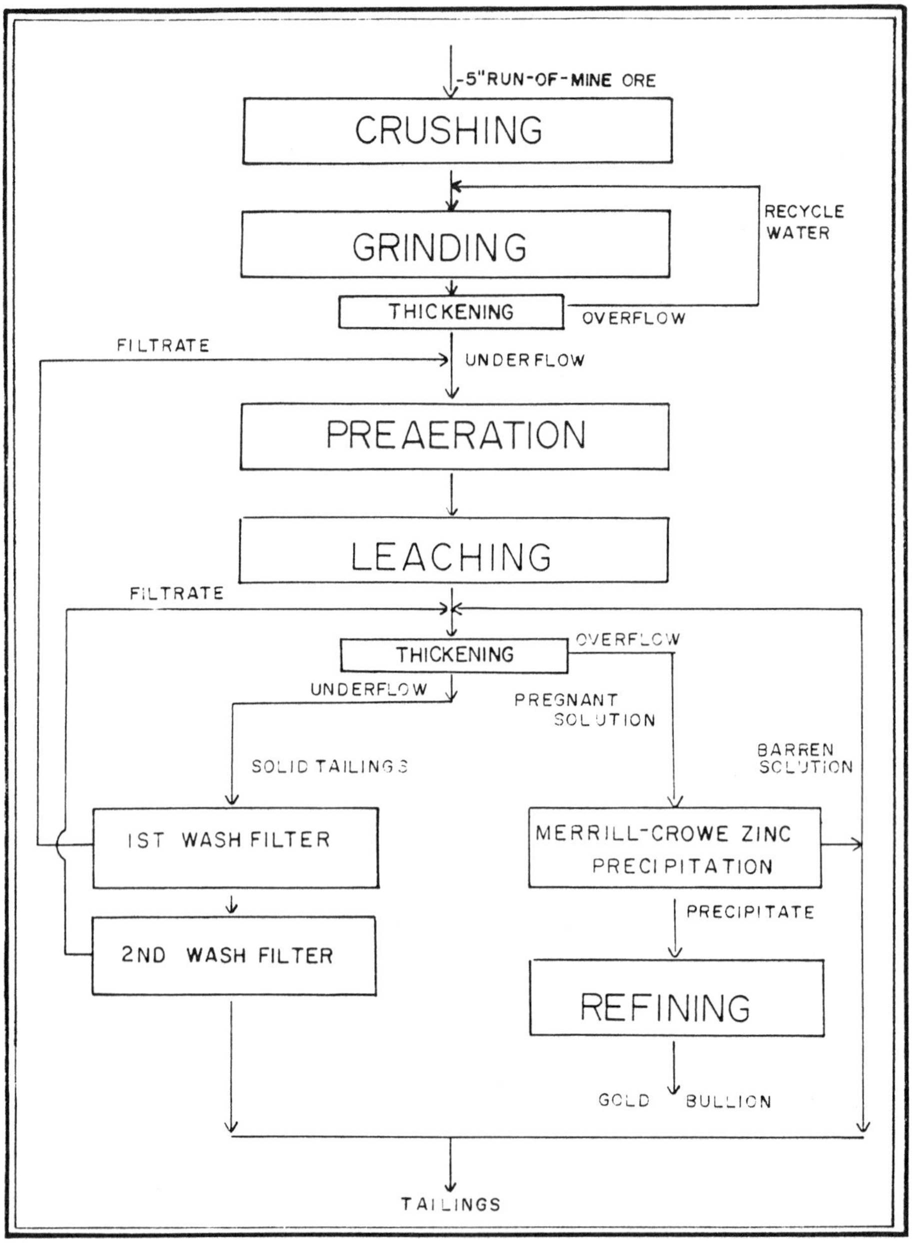

Fig. 1. Process flow sheet for Lupin mill, Echo Bay Mines Limited.

BACKGROUND

Initial feasibility study results (1982) indicated that the preaeration (PA) would commence with two 40 foot tray thickeners (Th) to thicken the cyclone overflow (COF) from 35% solids to 50% solids.

The thickened pulp would then be aerated in three consecutive preaeration tanks, 18 feet in diameter and 30 feet high. The aerated pulp would then be filtered in a single stage filter. The filter cake would be repulped with barren solution and then fed to the leach circuit.

The actual installation included three preaeration tanks that were 24 feet in diameter by 26 feet high, which received the cyclone overflow from grinding directly. In practice, aeration took place at 32% solids. The aerated pulp was then thickened in a single thickener, 50 feet diameter, to about 60% solids, pumped to leach, and was then diluted to 45% solids, using the barren solution.

The aeration tank mechanisms consist of a horizontal rake that moves the pulp toward the center where a single center airlift raises the pulp and distributes it at the top via a launder. The airlifts are provided with a supply of compressed air at 40 psi. As built, these aeration mechanisms could not be successfully operated. Due to the light density in the tank, differential settlement which resulted in a middling type of product collecting in the bottom portion of the tanks caused the mechanism to over torque and stop. A power bump also resulted in the instant settling of the pulp and buried the rakes.

In a year and a half, all three mechanisms ran simultaneously for a total of only three weeks. The third tank was finally abandoned, and converted into a recycled water tank, as part of the expansion program. In order to solve the settling problem, Echo Bay designed, built, and installed additional airlifts for the #1 and #2 tanks. (Fig. 2) The success of these modifications resulted in the installation of a new recycle water tank, and the conversion of the third preaer tank from recycle water to preaeration.

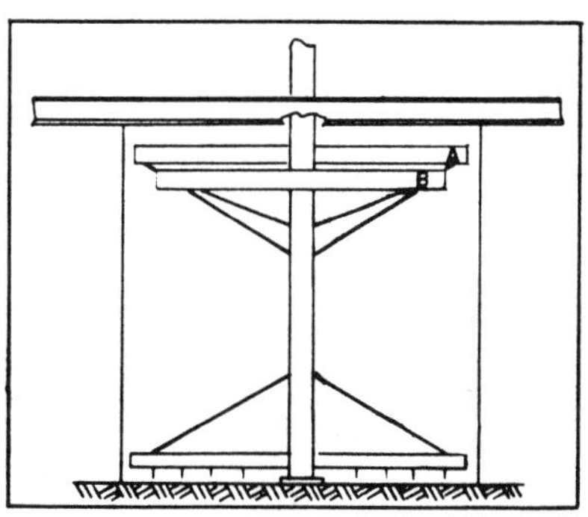

Fig. 2. Preaeration tank mechanisms: (A) original and (B) additional airlifters

With this mechanism design change, the interfacing of the grinding circuit and the preaeration tanks was more forgiving but still very sensitive. The main problem stemmed from settling of solids in the preaeration tanks due to a light COF density. The density in the preaeration tanks must be kept above 33% solids. With lighter densities, the solids settled out creating a bed of solids 3 to 5 feet in depth. This would cause the mechanisms to over torque and stop, approximately once per month, causing havoc in the mill, as well as increase in reagent consumption, clean up and maintenance costs. This situation placed undue pressure on the grinding operators to be alert to the very tight operating parameters on the COF density. (Fulcher, 1985)

To fully solve this mechanical problem of the preaeration tanks dependance on the COF density and with previous knowledge that these tanks work most efficiently around 45% solids, a circuit change was made. On March 31, 1986, the COF was rerouted to feed into the thickener in the density range of 28 - 33% solids. The thickener underflow with a density of 60% solids is pumped to the preaeration tanks with dilution to 45% solids. From these tanks, the slurry is pumped to the leach tanks. The additional airlifts were not required with the greater percent solids and were removed during the circuit changeover.

THEORY

Suspended Solids

The stability of suspended solids is dependent on three factors: density and type of suspension, and shape of the medium particles. A low density gives rise to a low stability. Hence diluent suspensions have high settling rates. (Fig. 3) Addition of fine particles into diluent suspensions increase stability, and the spherical particles have higher settling rates than the angular particles. (Burt, 1984)

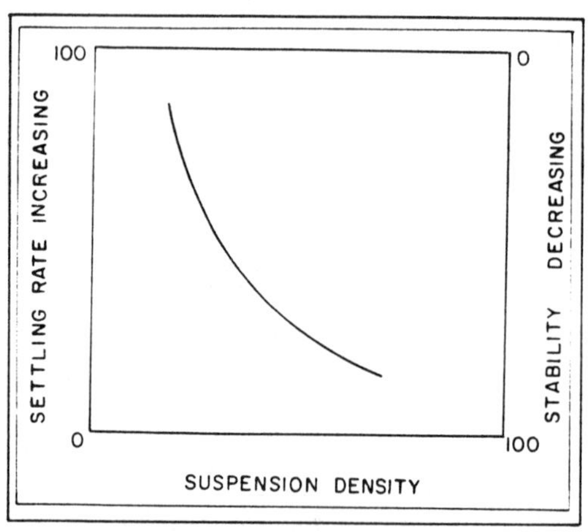

Fig. 3 Approximate relationship between settling rate and suspension sensitivity.

Chemistry of Preaeration

The Sulphide minerals in the ore are the contributing factors in forming ferrous sulphide (FeS), which causes problems in the cyanidation process. FeS competes with gold for the consumption of oxygen and cyanide.

$$2FeS + 2NaCN + \frac{3}{2} O_2 + 3H_2O = 2NaCNS + 2Fe(OH)_3 \quad (1)$$

In order to prevent this competition, FeS is oxidized in the preaeration circuit before cyanidation. FeS reacts with oxygen to form ferrous sulphate (FeSO) and then is converted to the stable, insoluble ferric hydroxide (Fe(OH)). (Hedley, 1963)

$$FeS + 2O_2 = FeSO_4 \quad (2)$$

In an alkaline solution

$$FeSO_4 + Ca(OH)_2 = Fe(OH)_2 + CaSO_4 \quad (3)$$

$$2\,Fe(OH)_2 + \frac{1}{2}O_2 + H_2O = 2Fe(OH)_3 \quad (4)$$

The addition of lead nitrate ($Pb(NO_3)_2$) aids in this oxidation. The exact mechanism of lead nitrate is unclear at this time.

One method of measuring the compilation of aeration is the reducing power (RP) measurement. The RP is usually monitored on the barren solution and the normal value of RP in this solution is about 500. However, during the circuit change, the RP measurements were made through the plant.

The literature search on the RP produced only a few paragraphs on this topic. The following was published in "The Analysis of Cyanide Solutions," (CIL, 1936).

Principle. The reducing power is measured by a titration method. The cyanide solution in a high concentration of sulfuric acid is titrated with potassium permanganate solution. Reducing agents include thiosulphate, sulphide, thiocyanate, ferrocyanide and any metals forming complex cyanides such as copper, zinc and lead, which react with permanganate. This method only measures the substances which react with permanganate and not all the substances which consume oxygen. Thus, this method only gives an approximate measure of constituents which consume oxygen.

Interferences. The presence of thiosulphate interferes and causes a recurring end point.

Method. 10 ml of cyanide solution is diluted to 200 ml with distilled water. 25 ml of 25% sulphuric acid is added. A titrate is carried out with 0.1 N potassium permanganate solution to the first pink end point, which should stay pink for 60 seconds. 0.1 N potassium permanganate solution is made by dissolving 3.16 g in 1 liter of distilled water. The R.P. number is determined multiplying the titration by ten.

While the exact chemistry of the RP measurement is not known, there is some helpful information. The RP can be decreased with the addition of lead nitrate. As the RP increases, the consumption of cyanide will increase, and as the pH increases over 11, the RP increases dramatically.

CHEMISTRY PROBLEMS AFTER THE CIRCUIT CHANGE

The circuit change was made to solve a mechanical problem, but in doing so, created a chemical problem. Upon the physical completion of the circuit change of thickening (Th) prior to preaeration (PA), there was a metallurgical upset in the mill for approximately a month. The reducing power (RP) increased drastically, and consequently the cyanide consumption increased. The RP increased from 500 to 1500 during the following ten days. The consumption of sodium cyanide (NaCN) increased by about 75% within a week. No other changes had been made other than physical piping changes. The severity of the problem was not initially recognized because two of the three preaeration tanks were out of the circuit for rebuild. The preaeration circuit (chemistry and retention time) was only at 30% capacity. By the end of the second week the cyanide consumption increased by about 100% to 3.00 lb/ton. This was unacceptable even though the recovery had not suffered but had remained constant.

During the following weeks seven metallurgical scenarios were carried out to bring the NaCN consumption back to normal levels.

IN PLANT TESTS

Original Operating Parameters Before the Change Over

COF ⟶ PA ⟶ Th ⟶ Leach + Recycled Filtrate+$Pb(NO_3)_2$+NaCN

	COF	PA	Th	Recycled Filtrate+$Pb(NO_3)_2$+NaCN
% Solids	28-33%	28-33%	58-60%	42-45%
pH	9.0	9.0	10.5	10.5
RP				500 at barren

Cyanide Consumption 1.6 lb/ton

Problem: The operational difficulties with the preaeration tank mechanisms were unacceptable.

Observation: The preaeration rake mechanisms sanded out at least once a month.

Solution: The density in the preaeration tanks was increased by thickening prior to preaeration.

Immediately After the Change Over

COF ⟶ Th ⟶ PA (one tank) ⟶ Leach + Recycled Filtrate+$Pb(NO_3)_2$+NaCN

	COF	Th	PA	Recycled Filtrate+$Pb(NO_3)_2$+NaCN
% Solids	28-33%	58-60%	58-60%	42-45%
pH	9.0	10.5	10.5	10.5
RP				1500 at barren

Cyanide Consumption 3.0 lb/ton

Problem: The cyanide consumption was unacceptably high.

Observation: The RP and pH were very high. During this time, the ph in the thickener rose up as high as 12 causing the thickener water to turn green in colour, but the pH was lowered immediately, and the colour disappeared.

Reasoning: Only one preaeration tank was on line so it was thought that this initial upset was due to the reduced capacity in the preaeration circuit.

Solution: The second and third tank rebuilds were completed as quickly as possible, and the tanks were put back into service immediately.

Scenario No. 1

(three tanks) Recycled

COF ⟶ Th ⟶ PA ⟶ Leach + Filtrate+$Pb(NO_3)_2$+NaCN

	COF	Th	PA	Leach
% Solids	28-33%	58-60%	58-60%	42-45%
pH	9.0	10.5	10.5	10.5
RP		300	400	1500 at barren

Cyanide Consumption 3.0 lb/ton

Problem: The cyanide consumption was unacceptably high.

Observation: The RP measurements were monitored on the thickener and preaeration tanks.
The problem still existed. There was no affect on the chemistry with an increase in the retention time.

Reasoning: The preaeration step may be the cause of the problem.

Solution: The preaeration circuit was bypassed and the thickener underflow was fed directly to the leach tanks.

Scenario No. 2

Recycled

COF ⟶ Th ⟶ Leach + Filtrate+$Pb(NO_3)_2$+NaCN

	COF	Th	Leach
% Solids	28-33%	58-60%	42-45%
pH	9.0	10.5	10.5
RP		250	800 at barren

Cyanide Consumption 2.5 lb/ton

Problem: The cyanide consumption was still too high, and the preaeration circuit was not being utilized.

Observation: The RP decreased somewhat, resulting in a slight decrease in the cyanide consumption.

Reasoning: The high density of thickener underflow and preaeration circuit may be one of the factors in the upset.
The filtrate would dilute the thickener underflow and existing cyanide may help reduce the RP.

Solution: Two preaeration tanks were brought back on stream, and the filtrate line was extended to the first preaeration tank.

Scenario No. 3

	COF →	Th →	Recycled PA+Filtrate →	Leach+$Pb(NO_3)_2$+NaCN
% Solids	28-33%	58-60%	42-45%	42-45%
pH	9.0	10.5	10.5	10.5
RP		250	500	1000 at barren

Cyanide Consumption 2.5 lb/ton

Problem: The cyanide consumption was too high.
Observation: The RP was increasing, especially in the preaeration tanks.
Reasoning: Lead nitrate worked towards reducing the RP.
Solution: Lead nitrate addition line was diverted from the leach tank #1 to the preaeration tank #1.

Scenario No. 4

	COF →	Th →	Recycled PA+Filtrate+$Pb(NO_3)_2$ →	Leach+NaCN
% Solids	28-33%	58-60%	42-45%	42-45%
pH	9.0	10.5	10.5	10.5
RP		250	400	700 at barren

Cyanide Consumption 2.0 lb/ton

Problem: The RP and cyanide consumption were decreasing but not significantly enough.
Observation: With addition of lead nitrate into the preaeration circuit, the RP decreased immediately.
Reasoning: The filtrate may not have significant effect on the RP in the preaeration tanks.
Solution: The filtrate was redirected back to the leach tank #1.

Scenario No. 5

	COF →	Th →	PA+$Pb(NO_3)_2$ →	Recycled Leach+Filtrate+NaCN
% Solids	28-33%	58-60%	58-60%	42-45%
pH	9.0	10.5	10.5	10.5
RP			600	750 1000 at barren

Cyanide Consumption 2.75 lb/ton

Problem: The cyanide consumption was still too high.
Observation: The RP started to increase again.
Reasoning: The filtrate is needed in the preaeration circuit to help reduce the RP.
Solution: The filtrate was once again pumped back into the preaeration tank #1.

Scenario No. 6 (Return to Scenario No. 4)

COF ⟶ Th ⟶ PA+$Pb(NO_3)_2$+Recycled Filtrate ⟶ Leach+NaCN

	COF	Th	PA+$Pb(NO_3)_2$+Recycled Filtrate	Leach+NaCN
% Solids	28-35%	58-60%	42-45%	42-45%
pH	9.0	10.5	10.5	10.5
RP		250	400	700 at barren

Cyanide Consumption 2.0 lb/ton

Problem: The cyanide consumption was too high.
Observation: The operation conditions have been returned to the levels achieved in Scenario 4.

Scenario No. 7

Observation: Due to an operator error, the pH in the thickener jumped to 12.0 and the green colour in the thickener water appeared for the second time.
Solution: At this time, the lime addition was shut off to the thickeners and let the pH drop to 9 and brought back up to 9.5. The pH in the preaeration circuit was 10.2 and the pH in the leach was back up to 10.5.

COF ⟶ Th ⟶ PA+$Pb(NO_3)_2$+Recycled Filtrate ⟶ Leach+NaCN

	COF	Th	PA+$Pb(NO_3)_2$+Recycled Filtrate	Leach+NaCN
% Solids	28-35%	58-60%	42-45%	42-45%
pH	9.0	9.5	10.2	10.5
RP				500 at barren

Cyanide Consumption 1.6 lb/ton

Observation: With these changes, the RP decreased to 500 at the barren solution and the cyanide consumption dropped to 1.6 lb/ton of ore, which was the rates prior to the circuit change.

Present Operating Parameters after the Change Over

COF ⟶ Th ⟶ PA+$Pb(NO_3)_2$+Recycled Filtrate ⟶ Leach+NaCN

	COF	Th	PA+$Pb(NO_3)_2$+Recycled Filtrate	Leach+NaCN
% Solids	28-35%	58-60%	42-45%	42-45%
pH	9.0	9.5	10.2	10.5
RP		150	200	400 500 at barren

Cyanide Consumption 1.6 lb/ton

Reflecting back on the ordeal, the importance of pH in the thickener was not immediately realized, because the lime addition into the thickener was only for aiding the flocculation process and not for control of any chemical reactions.

SUMMARY

The reducing power measurement was found to be a valuable guide in monitoring the circuit. The key operating variables that emerged from this flowsheet change were pH, diluent and lead nitrate.

The advantages resulting from the circuit change are the stability of the preaeration tanks (no rapid settling) and the decrease in the lime consumption.

The two advantages that have resulted from the circuit change are: 1) The preaeration tanks have not settled out since the circuit change; and 2) There is a trend towards a decrease in the consumption of lime. These two factors have managed to save thousands of dollars for the operation.

OTHER HIDDEN BENEFITS

The density (% solids) of the COF is no longer a critical factor and can vary between 28% and 35%.

FUTURE STUDIES

A number of questions have been generated from this experience:

1) What does RP exactly measure?
2) What is the chemical reaction of $Pb(NO_3)_2$ in the preaeration and leach circuits?
3) What causes the green water in the thickener at high pH? Chemical reaction?

ACKNOWLEDGEMENT

Authors would like to acknowledge Messrs. P. Parashyniak, Mill Superintendent, F. Hamonic, General Mill Foreman, and G. Shaw, Shift Foreman, for working on this project and would like to thank the management of Echo Bay Mines Ltd. for allowing us to publish this paper. We would also like to thank Ms. C. Hart, Mill Clerk, for typing this manuscript.

REFERENCES

Fulcher, J. (1985). The Milling of Gold Ore at the Lupin Operation, Echo Bay Mines Ltd., Paper presentation 87th CIM Annual General Meeting, Vancouver.

Burt, R. O. (1984). Gravity Concentration Technology, Elsevier, New York.

Hedley, N. and Tabachnick, H. (1968). Chemistry of Cyanidation, American Cyanamid.

(1936). The Analysis of Cyanide Solution, CIL.

DESIGN AND PLANT OPERATING PRACTICE IN CANADIAN HEAP LEACHING OPERATIONS

W.J. Witte
Witteck Development Inc.,
Mississauga, Canada

In the past three years heap leaching has become an accepted, low cost process for the recovery of precious metals in Canada. Many existing major Canadian producers as well as Junior mining companies and entrepreneurs are investigating and implementing the technology into present and future operations. Laboratory test programs and pilot heap leaching has resulted in the design and installation o full scale heap leaching operations. A description of the design and operating parameters required to implement heap leaching of Canadian ores under severe climatic conditions and environmental restraints is presented. The design of a cost effective heap leach operation must be adapted from the experience of operations in relatively arid climates to handle net precipitation gain, wind and freezing conditions. A general description of operating criteria for heap leaching in Canada will be given.

MODERNIZATION OF THE CAMPBELL RED LAKE MILL IN 1987

Scott Roberts - Mill Superintendent, Campbell Red Lake Mines Limited, Balmertown, Ontario
John Starkey - Senior Metallurgist, Kilborn Limited, Toronto, Ontario

ABSTRACT

This paper describes the modernization project now underway at the Campbell Red Lake Mines Limited Balmertown mill. It gives an overview of the process changes and presents details of the new carbon-in-pulp circuit, and the new automated grinding and backfill plants.

The project was undertaken in 1986 to decrease unit operating costs, to increase the recovery of backfill and gold and to upgrade mill technology by adding a central distributed control system. Start up for the new facilities is scheduled for August, 1987.

KEYWORDS

Carbon-in-pulp; automation; distributed control; rod and ball mills; gold gravity concentration; cemented backfill; flyash; ARF screens.

INTRODUCTION

In 1986, Campbell Red Lake Mines Limited (CRLM), started planning a project to improve and modernize the milling operation at their Balmertown Ontario plant. Kilborn Ltd. was then hired to study the concepts and to engineer the new facilities.

This mill originally started production in 1949 at 273 t/d (300 TPD) and has grown over the years, in stages, to the present capacity of 1000 t/d (1100 TPD) of ore grading 21 g/t (0.6 oz per ton) gold.

The process used consists of crushing to 6mm ($\frac{1}{4}$ inch), grinding to 78% -200 mesh, gravity concentration of the free gold, flotation recovery of an arsenical gold concentrate, roasting and calcine leaching, flotation tail leaching, backfill recovery and Merrill-Crowe recovery of the leached gold as a precipitate for refining to doré bullion. Understandably, some of these areas will not accommodate additional production. Grinding, Merrill-Crowe, and backfill production systems are at their limit. The plant is manually controlled and is clean and well operated.

Operating costs relate mainly to three areas: 1. Operating labour and supplies; 2. Parts and labour to maintain 2 - 1.3m (4 ¼ foot) crushers, 5 ball mills, many Dorr air lift agitators, disc and string filters, and related equipment; and 3. Alluvial sand to augment backfill production.

The major objectives of the modernization project are listed below.

1. Modernize plant control by adding a distributed control system and a central control room.
2. Provide milling capacity for up to 1100 t/d (1200 TPD) at design grind using a new rod-ball grinding circuit.
3. Increase backfill production by grinding mine waste and constructing a new backfill plant.
4. Decrease solution losses in the tailings leach by replacing string filters with a carbon adsorption system.
5. Reduce operating costs in major areas listed above.

This paper will discuss this project by comparing the existing flowsheet with the new. It will also show the basis for equipment and flowsheet selection. Finally, it will conclude with a slide presentation showing construction progress during the period April, 1986 to July, 1987.

SI units of measure used in this paper are calculated from Imperial data with the result rounded to two significant figures.

PROCESS DESCRIPTION

A simplified flowsheet is presented on the following page (Fig. 1), to indicate the existing process in block diagram form. Revisions are indicated as additions while deleted items are shown with an asterisk.

Crushing Plant

The CRLM surface crushing plant consists of a jaw crusher and 2 - 1.3m Symons cone crushers, one standard and one shorthead. The shorthead crusher is fed by a surge bin belt feeder and operates in closed circuit with a 2.1m x 3.0m (7' x 10') double deck Tyler screen with a 6mm (¼") slotted opening lower deck. The crushed ore is conveyed to fine ore bins on an 460mm (18") conveyor belt which runs over the "B" circuit fine ore bins. Plows are used to feed the "B" bins while 2 additional short conveyors are used to place ore in the "A" circuit bins.

Changes to the crushing plant are minimal. Capacity will be increased from 98 to 127 t/h (108 to 140 tons per hour) by opening the crusher settings and screen to produce minus 16mm (5/8") feed to the fine ore bins. The screen undersize 460mm conveyor will be replaced with a 510mm (20") belt running on the same idlers but at a faster speed. In order to retain the opportunity to run the "B" grinding circuit on waste for backfill, the plows are to be retained. A new vari-speed drive for the conveyor will allow the belt to run at up to 1m/s (200 feet/min.) against the plows for waste handling and 1.5m/s to deliver ore. This conveyor discharges to a new 610mm (24") conveyor, which feeds a 2900 t (3200 ton) new steel walled fine ore silo.

The most difficult part of this conversion will be the installation of the tail pulley for the new conveyor because existing conveyors must operate until the new grinding circuit is ready to run. Planning has been based on conversion to the new conveyor with a minimum shutdown, not to exceed 2 shifts.

FIG. 1

CAMPBELL RED LAKE MINES LTD.
SIMPLIFIED FLOWSHEET – SHOWING REVISIONS

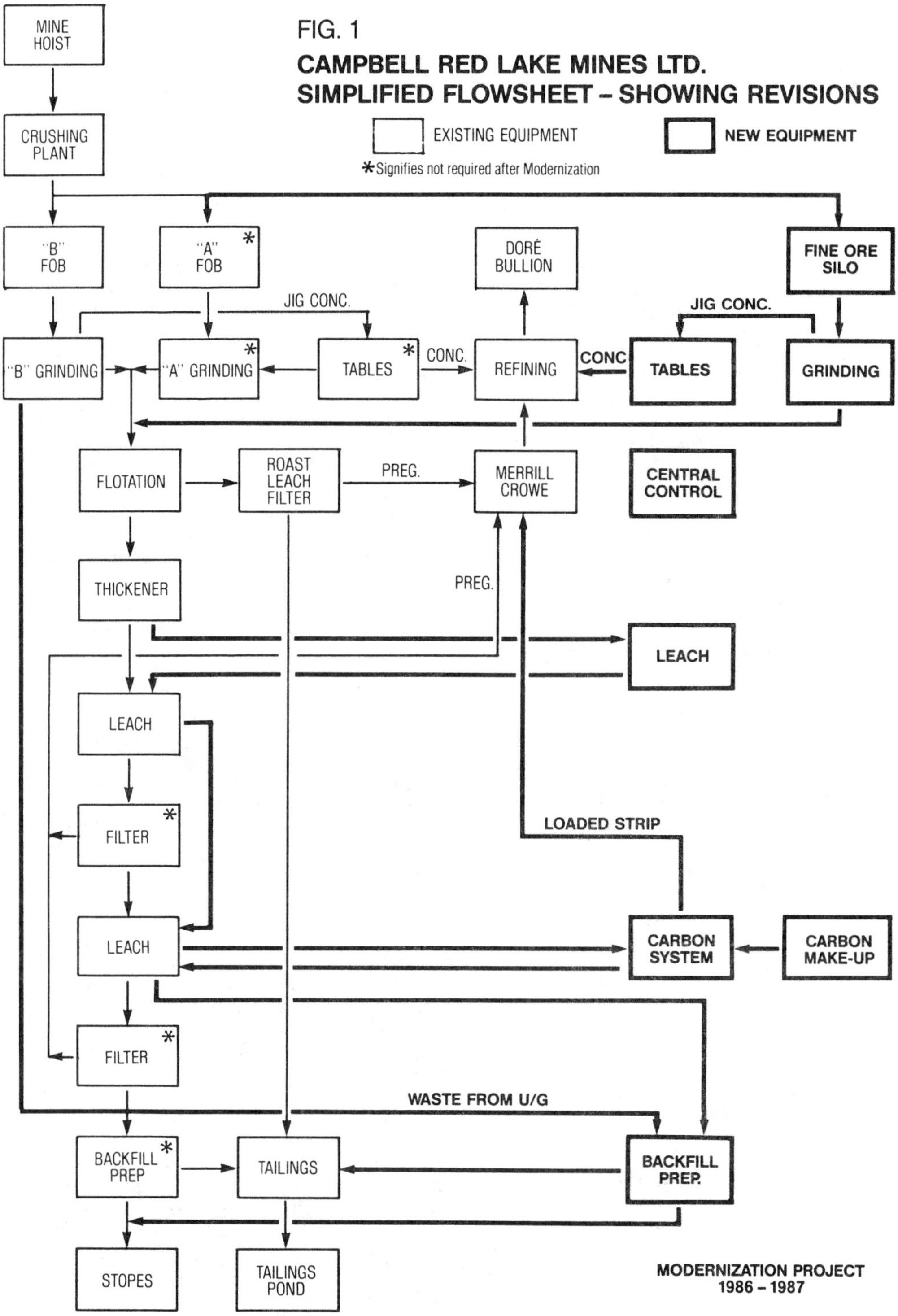

MODERNIZATION PROJECT
1986 – 1987

Grinding and Gravity Concentration

The present grinding plant consists of 5 ball mills arranged in two parallel circuits with primary and secondary mills and cyclone closed circuit classification on the feed to the secondary mills. These mills are listed below.

Circuit	Mill Size	KW	HP	Duty
"A"	2.1mΦ x 3.7m (7'Φ x 12')	223	300	Primary
	2.1mΦ x 1.8m (7'Φ x 6')	112	150	Secondary
	2.1mΦ x 1.8m (7'Φ x 6')	112	150	Secondary
"B"	2.1mΦ x 3.4m (7'Φ x 11')	223	300	Primary
	2.4mΦ x 3.7m (8'Φ x 12')	260	350	Secondary

Although the installed power is 930 kw (1250 HP), a review of actual power used suggests that less than 820 kw (1100 HP) is available for useful grinding work. As a result, the grind achieved is occasionally coarser than design.

Two duplex 16x24 jigs are used in each grinding circuit on secondary mill discharges to recover free gold. About 40% of the gold in CRML ore is recovered from these jigs through tabling and direct refining to doré bullion.

The new grinding - gravity circuit (see Fig. 2), has been designed to operate independently of the old systems, except for the tie-in of the feed conveyor which will cut out the possibility of using the old "A" circuit after start-up.

Ore will be withdrawn from the 2900 t fine ore silo by 2 - 910mm (36") wide slot feeders, and discharged to a 610mm mill feed conveyor. Grinding will be done in a 2.7m diameter x 3.7m (9'Φ x 12'), 300 kw (400 HP) rod mill and a 3.8m diameter x 4.6m (12 1/2'Φ x 15') long, 930 kw (1250 HP) ball mill, both supplied by MPSI. Selection of a rod mill was important because it will allow the crushing plant to be opened up as noted above.

Four 250mm (10") diameter Krebs primary cyclones (one spare) will classify mill discharges which flow into a common pumpbox. Cyclone overflow is prepared for flotation by removing wood chips on a Delkor linear screen located above the flotation feed pumpbox. This application of the Delkor screen (developed by Anglo American Corp.), is, to our knowledge, the first in North America. Cyclone underflow, flows by gravity to 3 -24 36 Minpro duplex jigs (one is spare), via a 3-way static distributor designed specially for this purpose.

Jigs tailings will then be dewatered in two 510mm (20") diameter Krebs cyclones (1 spare), prior to grinding in the large ball mill. Flotation middlings will also be reground in this mill after classification in one of two 150mm (6") diameter cyclones.

Jig concentrates will flow intermittently by gravity to a 5.7m^3 (200 ft^3) storage launder located above the first of two Holman James tables. A second storage launder with 2m^3 (70 ft^3) capacity will hold the rougher table concentrate and recycled middlings from the second table. The tabling operations will be done on day shift only.

Removal of wood chips from Campbell ore has been an instructive exercise. First, wood chip removal is mandatory prior to any carbon circuit because wood chips interefere with the carbon handling and can cause problems in the reactivation kiln. Second, it is now evident that removing wood chips prior to flotation improves both the flotation and roasting/calcine operation.

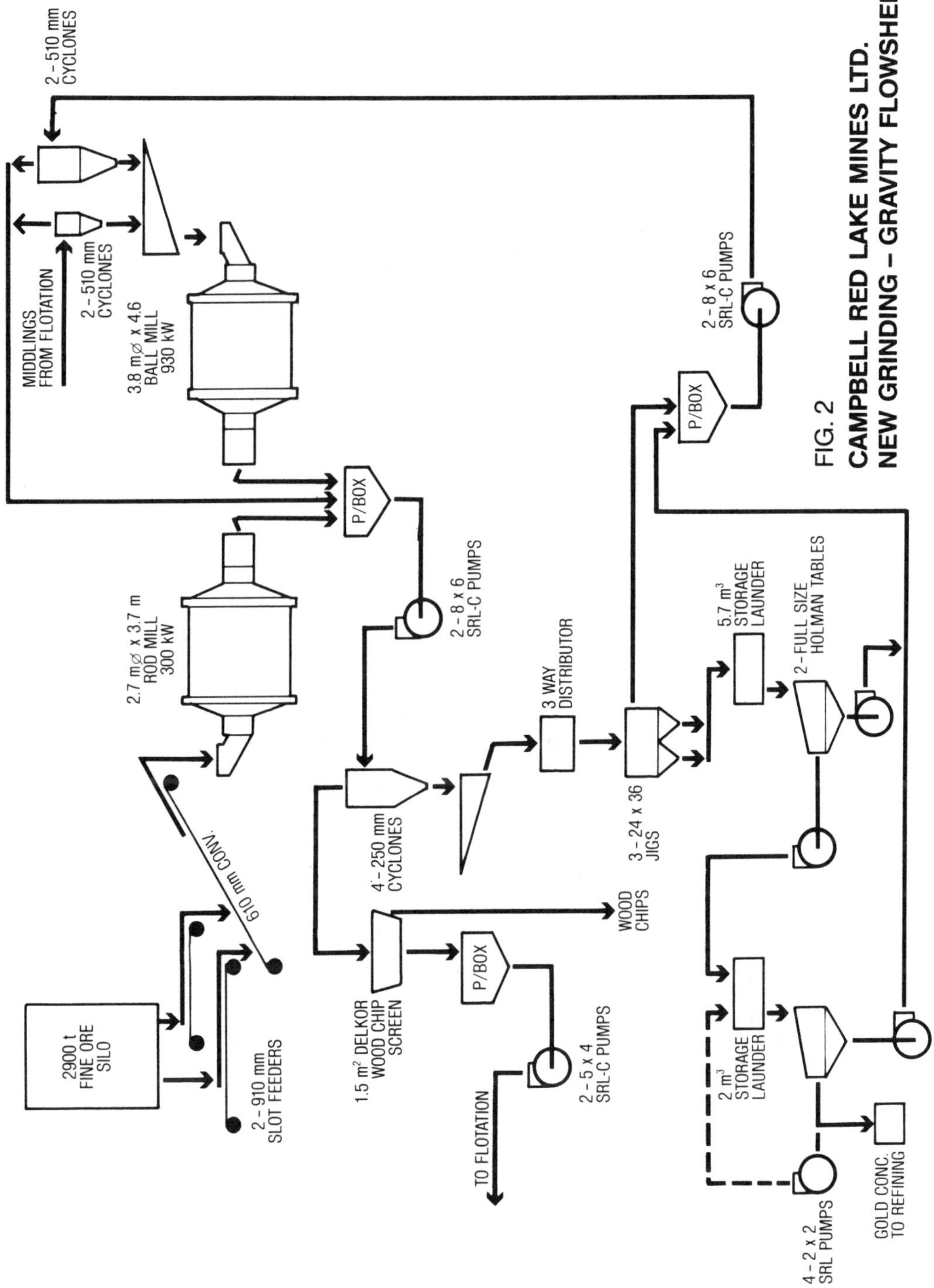

FIG. 2
CAMPBELL RED LAKE MINES LTD.
NEW GRINDING – GRAVITY FLOWSHEET

Sulphide Treatment and Gold Refining

Portions of the milling circuit will remain unchanged when this project is complete. The following areas will not be modified: 1. sulphide flotation in 2 banks of 10-2.8m^3 (100 ft^3) Denver cells, 2. flotation tail thickening (27.4m diameter), 3. roasting, 4. calcine leach, 5. arsenic recovery, 6. pregnant solution storage, 7. Merrill-Crowe system and, 8. gold refinery. Most of these facilities have either been renovated or installed new during the last ten years.

The sulphide flotation circuit was installed during the 1979 expansion and the tailings thickener in 1982. Sulphide concentrate comprises only about 5 weight percent of Campbell ore, so the roasting and calcine treatment systems do not treat large tonnages. Pregnant solution handling, prior to 1986, was a serious problem because adequate filter washing overloaded the Merrill-Crowe system. However, eliminating the tailing leach filters and replacing the pregnant solution with much higher grade strip solution will eliminate this problem entirely. Finally, a new refinery was built in 1983 and will be more than adequate to handle any changes in gold production.

Tailing Leach Circuit

The present flotation tailings leach circuit consists of 6 - 283m^3 (10,000 ft^3) and 7 - 113m^3 (4,000 ft^3)wood stave leach tanks with Dorr air lift agitators. The first stage of this circuit is 566m^3 (20,000 ft^3) of aeration in 2 - 7.3m (24') diameter x 7.3m tanks, followed by primary leaching in 680m^3 (24,000 ft^3) of tankage and filtration on two Eimco disc filters. This filtration stage is now optional. Secondary leaching is then done in 1246m^3 (44,000 ft^3) of tankage with final recovery of pregnant solution being done on 3 - 3.5m (11 ½') diameter x 4.9m (16') Eimco drum filters with string discharge.

The modified leach circuit now under construction, will also use 13 tanks, 8 -7.3m (24') x 7.3m and 5 - 5.5m (18') diameter x 5.5m. Total volume will now be 2832m^3 (100,000 ft^3), including 566m^3 for aeration, 1700m^3 for primary leaching and 566m^3 for carbon-in-pulp recovery of the gold. In order to make room for the carbon facilities, 3 existing tanks are being removed but they are being replaced by 4 new, outdoor, 7.3m (24') diameter x 7.3m steel tanks with 22 kw (30 HP) Prochem agitators.

Figure 3 shows the flowsheet for the entire tailing leach operation. Of special interest is the new carbon-in-pulp circuit, consisting of 5 modified 5.5m diameter x 5.5m woodstave agitators. The screens in these tanks will be referred to as Airless Reverse Flow (ARF) carbon retention screens. A diagram illustrating the principal of operation is given in Fig. 4. These screens were developed by CRLM in order to utilize the existing Dorr mechanisms and to improve on the operating performance of carbon screens used in other plants.

The ARF screens were tested first in a 4.9m (16') diameter x 6.1m (20') tank and next in an 5.5m diameter x 5.5m tank. The tests were highly successful for two reasons. First, two paddles rotate inside a circular launder, creating a wave which causes a backwash effect. This prevents carbon particles from being pinned to the screen surface. Second, discharge from the rotating launder arms physically wash near the screen face, again preventing carbon accumulation.

The trial tank mechanisms were run at 3 RPM using normal bottom rakes, at 8 RPM using paddle agitators, and with no rakes or paddles, but 4 air sparge pipes instead for agitation. CRLM prefers the latter method because it is more flexible to operate; that is, agitation air or rotation speed can be increased, without excessive power draw.

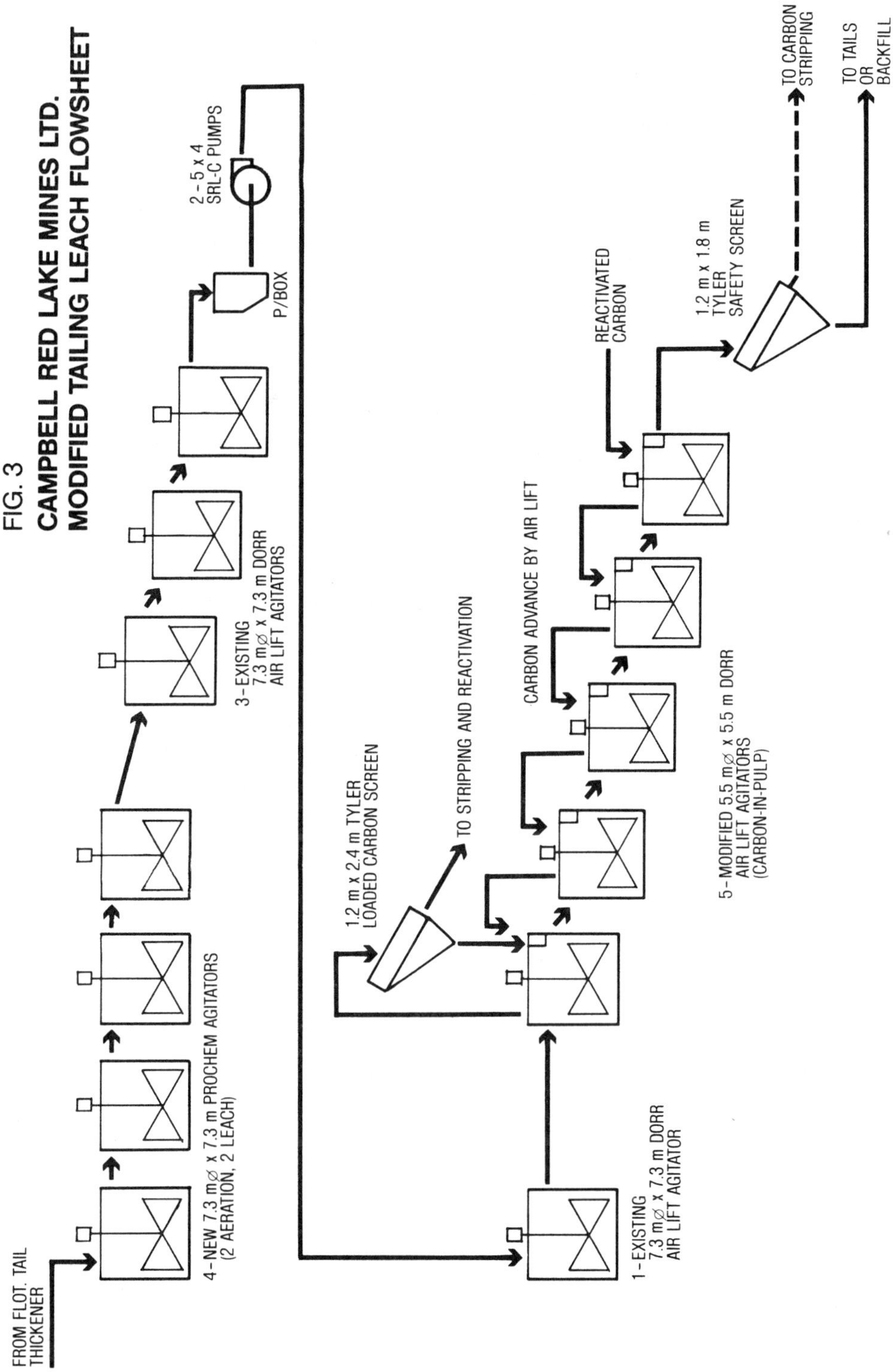

FIG. 3
CAMPBELL RED LAKE MINES LTD.
MODIFIED TAILING LEACH FLOWSHEET

FIG. 4

CAMPBELL RED LAKE MINES LTD.
AIRLESS REVERSE FLOW CARBON RETENTION SCREENS

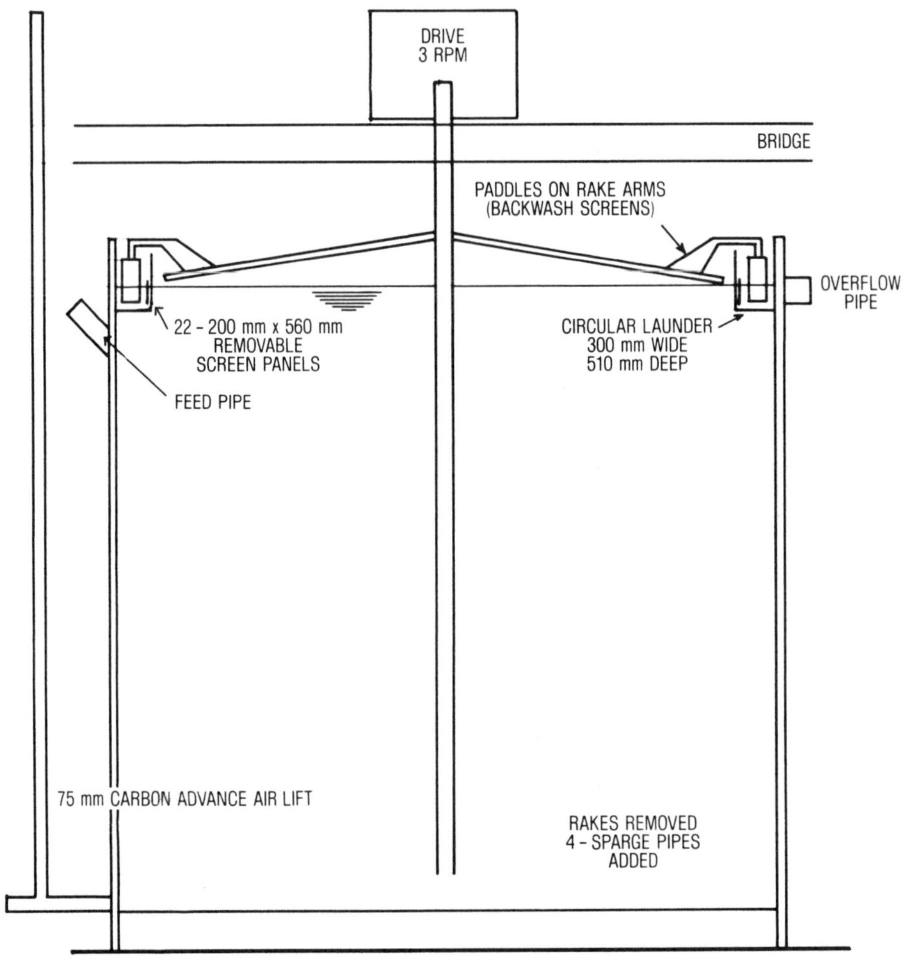

CROSS SECTION VIEW OF EXISTING 5.5 m∅ x 5.5 m DORR AIR LIFT AGITATOR IN WOOD STAVE TANK SHOWING MODIFICATIONS

Carbon Handling Circuit

The new carbon handling facilities can be described very simply as a 0.9 t (1 ton) per day system, using conventional pressure stripping with acid wash and carbon reactivation facilities. (See Fig. 5.) This capacity is at least 3 times greater than required to recover the tailing leach gold, but as long-range plans call for conversion of the calcine circuit to C.I.P. as well, so ultimately, it will be used. In the meantime, stripping and reactivation can be done every three days on day shift because there is storage ahead of each stage for the full one ton of carbon.

The carbon will be screened before and after reactivation on 16 mesh 610mm x 1830mm (2' x 6') Tyler screens. It will be screened before the kiln in order to dewater the carbon and not to waste energy on material too fine to use, and after, so as not to add abraded fines to the process which ultimately would result in gold loss to tailings. For the same reason, the barren reactivated carbon will be transferred to the fifth C.I.P. tank, using fresh water instead of eductor water because the eductor water recirculated from other transfer operations will contain some fine carbon particles.

The kiln selected for this plant is a 0.56m (22") diameter x 6.4m (21') long Lochead-Haggerty indirect heated electric unit.

New carbon will be received in drums and added as make up after attrition scrubbing in a 0.9mm (3') diameter x 1.2mm (4') conditioner, and screening on the 16 mesh sizing screen referred to above. This operation ensures that only the most stable carbon particles will be used in the process.

Design of the strip solution system is based on the following parameters:

Size of Tank	13 Bed Volumes - 4m (13') diameter x 4m
Flow Rate	2 BV per hour (7m^3/h)
Strip Temperature	135 deg. C
Strip Pressure	4700 k Pa (100 psi) Max.
Assay of Loaded Carbon	5100 g/t (150 oz per ton) of carbon
Assay of Loaded Carbon	68 g/t (2 oz per ton) of carbon

Moving of carbon from stage to stage will be done using 50mm (2") diameter eductors powered by a pump drawing recirculated eductor water from a 2.7m (9') diameter x 3m (10') eductor water tank. This tank will be constructed with a baffled bottom and submerged feed wells. The purpose here is to allow any coarse carbon to settle out before it is drawn into the pump suction. Ultimately these settled carbon particles will be removed through a manual drain and, depending on the actual assay, may be sent to a copper smelter for recovery of gold values.

When the carbon circuit is operational, the tailing string filters will be shut down while the stripping circuit is run in. Finally, the intermediate disc filters will be by-passed and recovery of all the gold from the tailing leach circuit will be done by the C.I.P. circuit.

FIG. 5
CAMPBELL RED LAKE MINES LTD.
NEW CARBON HANDLING FLOWSHEET

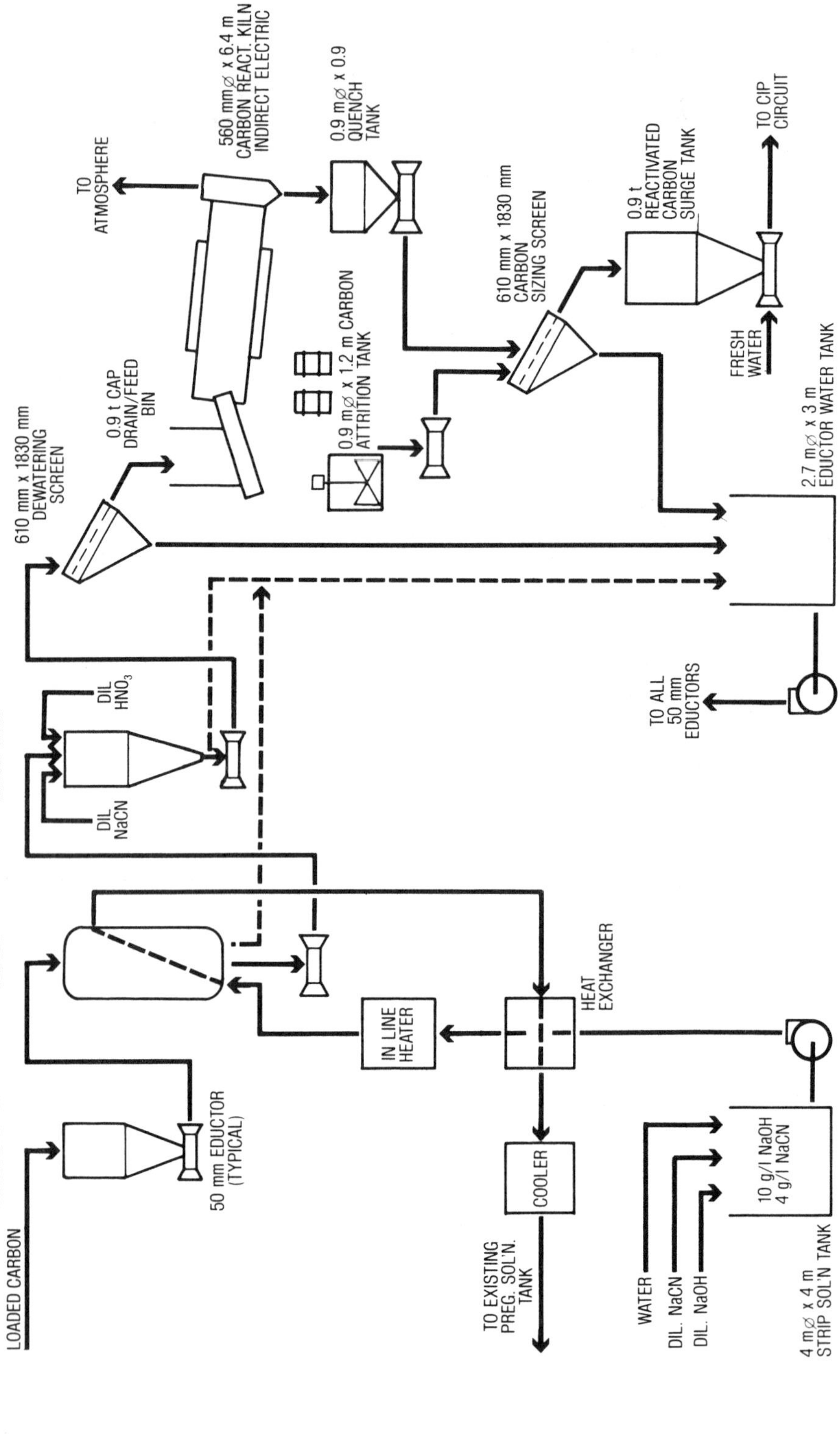

Backfill Plant

The existing backfill plant consists of three main parts. First, 2 stage cyclone recovery of tailing sand in 300mm (12") diameter Dorrclones; second, storage of the second stage underflow in two backfill storage tanks (settled solids); and third, cement addition to recovered solids for distribution to underground. This operation is augmented by the addition of expensive screened alluvial sand as slurry to the backfill storage tanks. The existing backfill system is not a cost effective operation for two reasons. First, because some cement is wasted due to manual control and second, because current fill requirements greatly exceed the amount which can be produced from mill tailings.

The new plant flowsheet is pictured in Fig. 6 and was designed to improve the performance of the old system. In order to maximize recovery of solids, new cyclones were designed to fully exploit the recovery of solids from the new and slightly finer initial grind. The first stage is now 6 - 150mm (6") diameter Krebs cyclones, while second stage cyclones are 3 - 250mm (10") diameter Krebs units. Careful discussion with the underground engineers indicated that a significant portion of the fill could be low percolation rate material, 75mm (3") per hour, for deferred fill applications. By by-passing the second stage cyclones, the performance is expected to change from 40% weight recovery of 200mm (8") per hour percolation rate material to approximately 50% weight recovery of 75mm per hour backfill.

A second method for augmenting backfill production will be to grind all the mine waste in the "B" circuit primary mill. It is felt that the fill produced in this circuit will cost about one fifth of what alluvial sand will cost over the long-term.

Backfill storage will be increased from 910 t (1000 tons) to 1230 t (1350 tons) by installing a twin outdoor storage tank with 640 t (700 ton) capacity to replace a smaller 320 t indoor tank which will become obsolete.

The final part of the new backfill circuit is the cement/fly ash addition equipment. Two identically sized storage tanks will hold 90 t of cement and 60 t of fly ash respectively. Withdrawal from each tank will be fully automated via a bin activator, screw pre-feeder, and weigh belt feeder system, each feeding a 1.8m (6') diameter x 2.1m (7') agitated backfill mix tank. The fly ash addition will be used in ratios to cement of up to 1:1 in order to cut the cost of cement addition by replacing cement with flyash. The system has also been designed to accurately control the ratio of fill to cement at ratios varying from 5:1 to 50:1.

Slurry withdrawal from the storage tanks will be done by water additions to the tank cones and a transfer pump box. A transfer pump will then deliver a density/flow controlled slurry to the backfill mix tank.

Process Control

An important part of the present project is the addition of a new environmentally controlled central control room and a Foxboro distributed control system. Initially, this system will control the new grinding and backfill systems and the existing roasting/arsenic recovery plant which has now been automated. Future plans are to automate the entire plant operation and to this end, all electrical facilities are now being upgraded with the addition of I/0 racks in the MCC rooms. Controls and instrumentation will be added later.

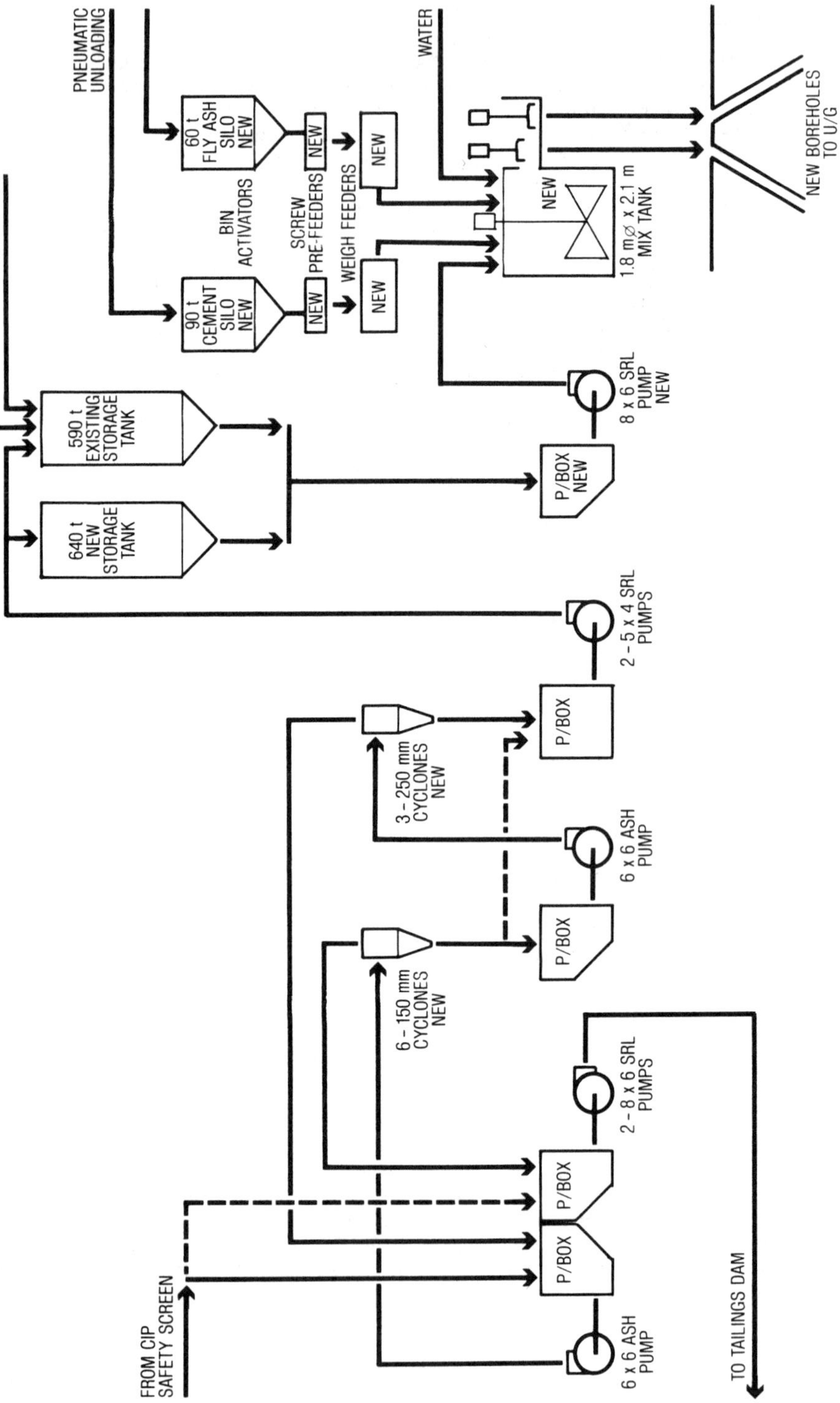

FIG. 6
CAMPBELL RED LAKE MINES
NEW BACKFILL FLOWSHEET

The heart of the automation package consists of a Foxboro 'Spectrum' distributed control system for analogue control, fully integrated with an Allen-Bradley PLC-2/30 programmable controller for motor control, sequencing and interlocks. The system includes a multistation processor which provides the memory and central logic for process display, monitoring, alarms and control functions. The multistation processor is complete with main ram memory of 1526K Bytes to handle the process control and management. The multistation processor is both configurable and programmable, and serves dual functions as an operator interface, as well as a process computer.

Three multistation interactive graphic consoles provide high resolution data display and convenient data entry. Process parameters, system data, operator-oriented displays such as plant overviews, group displays, real time and historical trending, shift and daily reports and equipment start/stop and status are easily accessible from the console.

The flexibility of the distributed control system will permit the addition of a single control loop or whole process areas as required, with minimum disruption to the working plant.

The long-term potential for this system is very attractive because it will allow CRLM to improve labour productivity and possibly improve metallurgical performance through faster discovery of malfunctioning equipment and quicker response to other process changes.

SUMMARY

It will be several months before the full assessment of this project can be done. However, CRLM is confident that the objectives are conservative and that most if not all of the anticipated benefits will be achieved or exceeded.

For years Campbell Red Lake Mines Limited has been the premier low cost gold producers in Canada. The present modernization project will ensure that this trend will continue and that CRLM will maintain its position as a leader in the field of gold production.

Because of the recent surge of gold mining activity in Canada, the industry is rapidly changing because the newer plants generally use the new technologies for processing and control. Obviously, not every plant can afford to modernize, but when the payback warrants it, good management demands that these projects be implemented.

ACKNOWLEDGEMENTS

The authors wish to thank the management of Campbell Red Lake Mines Limited and Kilborn Limited for the permission and resources to prepare and present this paper. Also, we wish to thank Mr. Ken Wright, Chief Metallurgist, Dome Mines Limited for his work in developing the Airless Reverse Flow screen concept and his helpful comments prior to preparation of this paper.

CARBON STRIPPING; THE BATEMAN AARL SYSTEM COMPARED

Mark K. Suddaby
Bateman Metallurgical Laboratories
Sparks, Nevada, U.S.A. 84111

EXPANDED SYNOPSIS; The Bateman - Anglo America Research Laboratories carbon elution system is described. The relative merits of this system are discussed with respect to other systems and incorporating various site specific factors.

A sample graph showing the number of bed volumes of eluant (B.V.) for varying concentrations of sodium, calcium, and magnesium in the strip solution required to achieve the same stripping efficiency is attached. The AARL process is compared with the most common pressure stripping process used in North America, known as the pressure Zadra process.

Reference is made to the AARL strip plant recently installed at Pegasus Gold Corporation's Zortman - Landusky Operation in Montana - U.S.A.

The differential responses of cadmium, silver, gold and mercury to the AARL strip process are described.

Water balance impact for AARL is compared with that of the pressure Zadra system. In positive water balance environments, such as high rainfall low evaporation heap leach situations, and also where good water quality if unavailable, the possibilities of reverse osmosis water purification are briefly described.

A continuous 1 TPD AARL pilot strip plant has been operated by Bateman in Australia and the potential future application to large scale AARL plants is described.

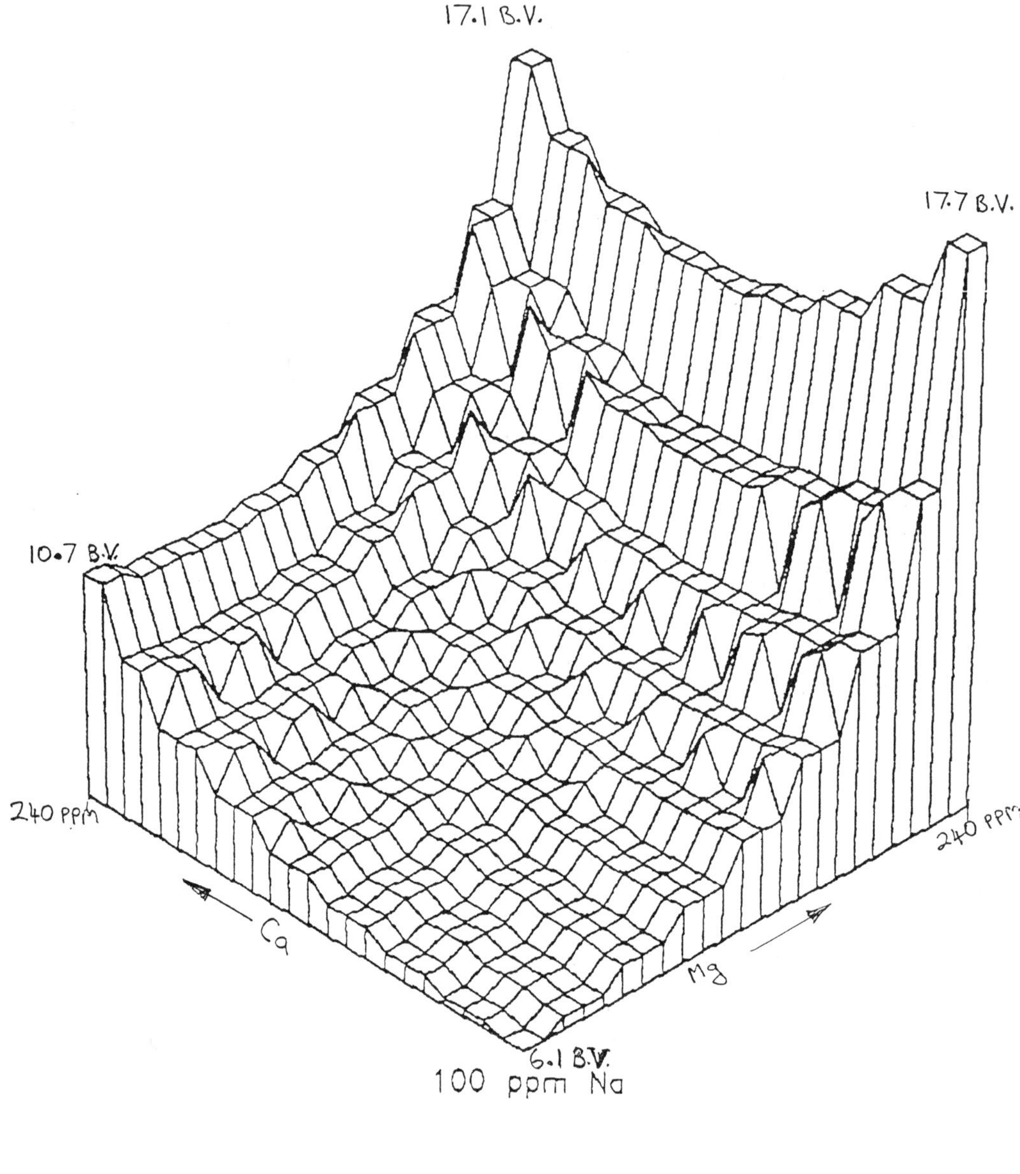

17.1 B.V.
17.7 B.V.
10.7 B.V.
240 ppm
Ca
Mg
240 ppm
6.1 B.V.
100 ppm Na

ENVIRONMENTAL CONSIDERATIONS IN GOLD FLOWSHEET DEVELOPMENT

R.W. Lawrence and P.B. Marchant

Coastech Research Inc.
869 West Third Street, North Vancouver
B.C. V7P 1E2

ABSTRACT

Environmental impact and permitting requirements in gold flowsheet development are often overlooked until very late in the project. However, environmental considerations should be addressed following initial flowsheet definition and be dealt with concurrently with metallurgical unit operation specification. Environmental factors, which are highly site specific, play a significant role in the project feasibility study and process selection.

This paper describes the important environmental laboratory test procedures and process considerations in gold flowsheet development. Discussions include waste rock and effluent classification and treatment, water management, and acid mine drainage potential and prediction. Selection criteria for cyanide destruction alternatives, including alkaline chlorination, SO_2-air, hydrogen peroxide, and acidic regeneration are also discussed.

KEY WORDS

Gold processing; environmental testing; flowsheet design; cyanide destruction; acid mine drainage; environmental impact; tailings disposal; water quality

INTRODUCTION

It is not the purpose of this paper to provide the latest update on environmental protection legislation and guidelines concerning the disposal of mine waste materials. Rather, the purpose is to provide an overview of the factors that the metallurgist should take into consideration when designing a gold flowsheet.

There is now a large body of opinion which stresses the importance of environmental impact assessment in pre-planning as a tool to aid decision making in any mine and mill development. In many instances, environmental issues are considered secondary to the design process. At best, environmental planning and flowsheet design steps are enacted on parallel but separate paths. Often this leads to a situation where technical and economic feasibility can be demonstrated but overall non-feasibility is the result due to opposition from government and public bodies and interest groups, or long term reclamation difficulties. We are not advocating that environmental issues set the criteria for the engineer and metallurgist. However, we strongly suggest that environmental studies are integrated with engineering feasibility studies at the earliest possible stage.

The metallurgist should be aware of (a) the nature of the environment with which his mill will be interacting and what specifically needs to be protected, and (b) the ways in which the mill operation and its waste products can affect the environment. Such knowledge will allow realistic cost factors to be included in the design process and expedite the environmental permitting effort through to mill start-up. A mill design produced with full consideration of the potential for environmental impact might minimize or prevent costly remedial measures both during operation and long after abandonment.

GOLD FLOWSHEETS AND ENVIRONMENTAL IMPACT

Gold can exist in a large number of ore types exhibiting a wide variation in mineralogy. Consequently, many processing options exist for the treatment of ore and the recovery of gold. For the purpose of this paper, processes are limited to mill applications in which gold is the principal value recovered and ore-types are limited to free-milling and the refractory sulphidic ores. Gold recovery by heap leaching techniques will not be specifically referred to, although many of the environmental considerations discussed for the mill situation will be applicable.

A general flowsheet showing processing options for the recovery of gold from free-milling and refractory ores is provided in Fig. 1.

In the flowsheet, ore is reduced in size by crushing and grinding and is then processed according to its mineralogical characteristics. The ground ore might proceed directly to the gold extraction step or it might desirable to concentrate the gold and/or its host minerals in a concentration process prior to extraction. Typically this might by flotation or by gravity techniques The recovered concentrate then passes to the gold extraction step and the tailing is disposed of.

For refractory ores, a pretreatment step is required to liberate the gold from the sulphidic host minerals such as pyrite and arsenopyrite. Pretreatment is usually an oxidation process such as roasting, pressure oxidation or biooxidation. Refractory ores might be concentrated by flotation prior to oxidation. The pretreated ore or concentrate then reports to the gold extraction step. The oxidation process also produces by-products which have to be further treated before disposal.

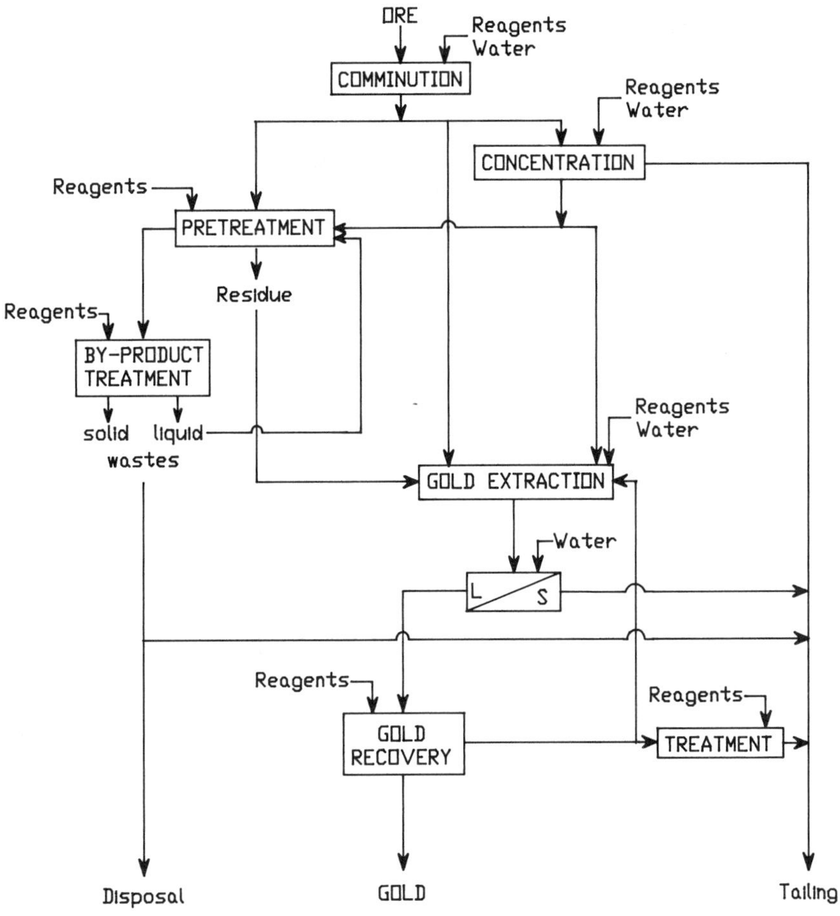

Fig. 1. General gold flowsheet.

Gold extraction is typically carried out by dissolution in a cyanide solution. Gold pregnant solution refining is typically by metallic displacement from solution or by activated charcoal contact with slurry and carbon refining. Following gold recovery, the barren solution is recycled and/or treated to remove residual cyanide before discharge to the receiving environment.

All of the unit operations described have a potential environmental impact for which due consideration must be given in the design stage. Each operation utilizes water drawn from the environment or from recycle streams from other parts of the process. Most operations require the input of various chemical reagents and all operations produce a potential impact, either on the workplace environment or the natural environment, or both, in the form of noise, dust, fumes, and liquid and solid waste products. The potential impact from each operation is summarized below.

Comminution

Crushing and grinding of ore produces noise and dust. Given the remoteness of most mines from human habitation, the concerns of the impact of noise and dust is mainly applicable to the workplace environment. Sometimes, reagents such as cyanide are added to the grinding circuit to initiate gold extraction. Since there is no direct discharge of comminuted product to the environment, the addition of reagents is, again, principally a workplace environmental issue.

Concentration

Since flotation is the principal concentration step in the flowsheet being described, the major sources of potential environmental impact in concentration are the chemical reagents, which might impact both the workplace and natural environments, and the flotation tailings discharged to the environment. Reagent handling and storage will be discussed below. The impact of tailing is related to:

(a) the potential source of acid mine drainage,
(b) the potential source of soluble metal values in seepage and surface run-off,
(c) the potential for increased turbidity in discharge waters.

Pretreatment

Pretreatment of gold ores and concentrates is usually carried out by oxidation to break down the gold-bearing sulphide minerals such as pyrite and arsenopyrite. Whichever oxidation method is used, the products of oxidation require further treatment to allow safe environmental disposal. For roasting, the major consideration is the production of sulphur dioxide and arsenic gases which cannot be directly discharged into the atmosphere. For aqueous oxidation systems such as pressure oxidation and biooxidation, the products of sulphur oxidation are found in solution as sulphuric acid and as metal sulphates.

In all cases, the products of oxidation have to be further treated to produce environmentally stable disposable forms. These products may be disposed of with the flotation and gold extraction tailings or they may require separate disposal or shipment to a remote disposal site.

Gold Extraction

The gold extraction method usually employed is dissolution in an alkaline cyanide solution although alternative extraction reagents such as thiourea (in acidic solution) can be considered. In any case, reagent storage and handling are considerations for protection of the workplace environment. The waste product from gold extraction is the solid ore particles which are disposed of with the flotation tailing. These solids will likely contain some residual extraction reagent. The solid tailing fraction presents similar environmental consideration to that discussed in the concentration section.

Gold Recovery

Gold recovery from cyanide solution might be carried out by adsorption on carbon either during (CIL) or following (CIP) the extraction step, or by precipitation with a metal such as zinc as in a Merrill-Crowe system. Whichever system is used, a final solution is produced which will contain residual reagent, some of which can be recycled and the remainder is discharged with the tailings stream. Before discharge, the residual reagent needs to be removed to reduce concentrations to acceptable limits.

Tailings Disposal and Water Balance

There are a number of tailings disposal options available such as sealed ponds, decant ponds, drainage beaches, lakes, spiggoting (sub-aerial deposition), marine disposal, backfill, dry land-fill, and offsite storage. The selection of the best disposal method is site specific and depends on population proximity, topography, hydrology, local regulations, tailing characteristics, and property operating policy.

The tailing disposal method selected might have a significant effect on mill process design, especially where tailing decant solution is recycled to the process or the tailing solution is discharged directly to the regulated environment.

Following definition of a disposal method, a property water balance, based on property hydrology and process materials balance, can be determined. A detailed property water balance is essential for process design and reagent control as well as for permit specifications.

Reagent Shipping and Storage

Reagent shipping and storage is an important consideration in any property development and should be addressed during the metallurgical process design stage. The definition of reagent shipping and storage practice as well as contingency plans will be an integral part of property operating permits and must be described in detail early in permit procedures. Often the reagent supplier(s) will provide much of the detail regarding reagent handling and should be included in the permitting procedures as a technical resource.

ENVIRONMENTAL TESTING

The purpose of laboratory testwork on mill products is to determine what components of the waste streams might have a deleterious impact on the environment. Questions that need to be addressed are:

(a) what could be produced as a result of chemical and biochemical reactions taking place in the wastes,
(b) what is the effect on water quality as a result of waste discharge, and
(c) what is the toxicity of the wastes and their decomposition products to fauna and flora in the receiving environment.

A number of accepted test procedures can be carried out to categorize wastes according to the above criteria. These include the prediction of acid mine drainage tests, metal dissolution tests and bioassay tests.

Baseline Database

The first step in assessing the potential environmental impact of a gold mill is to establish a predevelopment database of the existing environment. This is normally carried out as part of the Environmental Impact Assessment or Study and includes an extensive study, including seasonal variations, of the many aspects of the environment such as:

Air: meterology and climate, air quality, noise,
Hydrology: river and lake systems, aquifers ground-water, flow rates,
Water Quality: temperature profiles, turbidity, metal concentrations, dissolved oxygen, conductivity, pH, hardness,
Fisheries: species, species abundance, bathymetry, habitat evaluation,
Wildlife: Mammals, birds, habitat evaluation,
Soils and Vegetation.

Establishment of the database, which is normally carried out by specialist environmental companies or consultants, is an essential part of the permitting process and in documenting the environmental performance of the operating mill.

Acid Mine Drainage Prediction

The formation of acid mine drainage (AMD) resulting from the natural chemical and biochemical oxidation of sulphide minerals such as pyrite contained in solid wastes is perhaps the most serious impact that mining can have on the environment. Many gold ores contain pyrite so that in many cases, flotation tailing and the residual solids from incomplete or partial oxidation pretreatment and from the gold extraction step, contain significant quantities of oxidizable sulphur. If oxidation of this sulphur produces sulphuric acid in excess of that which the material can consume, then acid mine drainage will occur.

Several methods are available to predict the formation of acid mine drainage. Methods include static tests, which examine the balance between acid producing components (primarily pyrite) and acid consuming components (carbonates and other rock sources capable of neutralizing strong acids), and kinetic tests which attempt to model acid producing and consuming reactions over time. All methods have recognized limitations and the development of improved predictive techniques is currently of great interest. An excellent review of AMD predictive techniques has been provided by Ferguson and Erickson (1987).

Typical results obtained in our own laboratory of acid-base accounting tests conducted on two waste rock samples using procedures modified from those recommended by the EPA (Sobek and others, 1978) are shown below.

TABLE 1 Acid Base Accounting for Two Waste Rock Samples

Sample	Sulphide (%)	Acid Potential (kg $CaCO_3$/t)	Paste pH	Neutralization Potential (kg $CaCO_3$/t)	Net NP (kg $CaCO_3$/t)
A	0.65	20.3	8.7	70.5	120.2
B	4.35	135.9	7.1	110.6	-25.3

The results given in Table 1 show that the net neutralization potential of a waste material (Net NP) is dependent on both the sulphide content and the neutralization capacity of the sample. In this example, Sample A has a positive Net NP and can be classified as a material unlikely to be a source of AMD. In contrast, Sample B, despite having a neutralization capacity greater than that of Sample A, has a sufficient sulphide content to give a negative Net NP value and therefore has the potential to be a source of AMD. The EPA procedures (Sobek and others, 1978) define a potential acid generating material as any rock or earth material having a negative Net NP of 5 tonnes of calcium carbonate equivalent or more per 1000 tonnes of material.

Confirmation of the acid generating potential of a waste material can be carried out by means of laboratory oxidation tests in which the chemical and biological reactions, largely responsible for the acid-producing reactions, are established and the quantity of acid produced assessed. Special consideration of the disposal of mill wastes which are shown to be potential sources of AMD must be given. Failure to take precaution in the disposal and/or use of waste materials in the construction of, for example, tailings dams, dykes and haulage roads, can result in very expensive remedial measures at a later date if acidic effluents and runoffs occur. Generation of AMD within wastes can take several years to develop but, once initiated, is extremely difficult to control.

Extractable Metal Tests

Extractable metal tests are carried out on mill products to assess the possible extent of contaminant release to the environment. The primary objective of extractable metal tests is to provide a relative classification of waste products to indicate the degree of disposal precautions necessary for long term storage. Typically the elements of interest include antimony, arsenic, cadmium, cobalt, copper, chromium, lead, mercury, nickel, vanadium, zinc, and sulphur (as sulphide). Specifically, tests are designed to:

(a) characterize leachate produced from a waste to evaluate if it will meet dissolved cation compliance levels,
(b) determine whether seepage and drainage from waste and tailings would be harmful to the aquatic environment, and
(c) evaluate the efficiency of a waste disposition scenario or stabilization technique.

Tests to determine the extractable metal contents of mill waste materials take many forms. Regulatory authorities usually have their own required procedures. These often involve a single leach test followed by analysis of the resulting leachate to determine the extractable concentrations. For example, in British Columbia, classification of waste products from industrial processes is determined by an acetic acid extraction procedure specified under the Special Waste regulations (B.C. Reg., 1984).

Sequential extraction tests can also be used which, although do not currently provide criteria which can be tied to the regulatory guidelines, can be used very effectively to predict the effect of waste disposal on water quality. Such tests use a sequence of leaching agents to extract components in order of decreasing solubility. This provides knowledge of the chemical forms of soluble metal species or of metals associated with particulates or colloids which is useful in estimating bioavailability, mobility and chemical reactivity of the waste components (Hickey and Kittrick, 1984). Sequential extraction procedures can simulate a variety of environmental conditions to which waste materials may be exposed. Such tests have the potential to predict not only the short-term elemental releases but also what might become mobile in a longer time frame.

Bioassay

The objective of bioassay testing is to indicate the relative and absolute toxicity level of an effluent stream (before and/or after treatment). Typically bioassays are conducted using rainbow-trout underyearlings, brown trout, bass, insect larvae, fish eggs, indigenous flora (Daphnia), and other site specific organisms.

Generally a bioassay test is conducted to determine the concentration of the effluent stream at which acceptable biological survival and support is demonstrated. Note that it is extremely important to conduct representative control tests in parallel with effluent bioassays for meaningful interpretation of the results. For example, if city tap water or deionized water is employed in metallurgical testing, a water control sample, a process water control sample, and an effluent stream sample should be tested in parallel.

Samples for bioassay are typically derived from locked cycle testing, to indicate the effect of solution recycle, if any, on effluent toxicity. However, where severe leaching conditions are employed (eg. oxidative sulphide pretreatments) bioassays should be conducted early in the process selection testwork to indicate deleterious byproducts that might affect overall process economics.

Cyanide Destruction

It is often necessary to incorporate a cyanide treatment step in a gold flowsheet for one or more of the following reasons:

(a) to comply with regulations to discharge liquid effluent,
(b) for removal of cyanide solution "fouling" prior to recycle,
(c) to regenerate unused dissolved cyanide for further precious metals dissolution.

The methods of cyanide treatment available are listed in Table 2. Generally the treatment methods can be categorized:

(a) volatilization (natural or chemical),
(b) oxidative destruction,
(c) precipitation of inert metallo-cyanide byproducts.

The most common cyanide treatment method in gold milling is natural degradation by ponding. Where net positive water balance conditions exist, pond control becomes critical and is often difficult to permit. A double pond system, for proper seasonal operation, provides excellent cyanide removal, including thiocyanate, but requires a large surface area and dyke construction to provide sufficient residence time. Often an impervious pond liner is required for regulatory approval.

Where cyanide laden tailing decant water is recycled for service in a sulphide flotation mill a double pond system is acceptable. However, this practice might pose a higher metallurgical risk than most operations will accept. Tailing water recycle without cyanide treatment might also lead to a circulating load of `fouled' species which might be deleterious to gold extraction. Locked cycle testing during flowsheet development is essential to determine the effects of recycle. In some ponding cases a back-up cyanide destruction process (usually hydrogen peroxide) is incorporated to polish the tailing decant water prior to recycle or discharge to the environment.

TABLE 2 Methods of Cyanide Removal

NATURAL DEGRADATION

- Volatilization
- Biodegradation
- Oxidation, Photodecomposition

OXIDATION PROCESSES

- Alkaline Chlorination
  - chlorine gas
  - hypochlorites
  - electrolytic (in-situ) generation
- SO_2-air (copper sulphate)
- Ozonation
- Hydrogen peroxide (copper sulphate)

ACIDIFICATION/VOLATILIZATION/RENEUTRALIZATION-AVR

ADSORPTION PROCESSES

- Ferrous sulphide
- Ion Exchange and Acidic Regeneration
- Ion Flotation

ELECTROLYTIC PROCESSES

- Regeneration or Destruction

CONVERSION/PRECIPITATION

- To thiocyanate
- To ferro-/ferricyanide

The most common oxidative cyanide destruction processes used in operations are: alkaline chlorination, SO_2-air, and hydrogen peroxide. Alkaline chlorination is not recommended for applications where significant thiocyanate and ferrocyanide occur in solution, due to high reagent consumption and poor product quality. Alkaline chlorination is generally not competitive with the alternatives suggested above unless chlorine is readily available as a byproduct or waste stream.

Hydrogen peroxide treatment has become a widely accepted cyanide destruction alternative. Peroxide systems offer fast reaction kinetics, ease of operation, simple control instrumentation, adaptation to wide swings in feed solution chemistry, and a dependable product quality. However, peroxide applications are generally limited to treatment of solutions (barren bleed, tailing decant) due to high reagent consumption with solids present.

SO_2-air treatment is an excellent alternative to hydrogen peroxide destruction where a slurry stream is the cyanide effluent (eg. CIP, CIL tailing). However, an SO_2-air system is more capital intensive than peroxide and is more difficult to control than peroxide, often resulting in `spikes' of non-compliant effluent and therefore should be employed where sufficient dilution of the product is available to smooth the resultant effluent quality. The SO_2-air process is also applicable to solution treatment situations, especially where high dissolved copper is encountered, and is often very competitive with peroxide economics. Again, control complexity and product predictability are important considerations for selecting SO_2-air or peroxide. In general, SO_2-air systems have demonstrated lower operating costs when compared with peroxide based systems.

The selection of SO_2-air versus hydrogen peroxide often enters feasibility analyses where a choice between zinc precipitation and carbon technology is available. In general, it is much cheaper to treat a bleed solution stream rather than a slurry effluent stream, using either process. Similarly, when analysing the sensitivity of incremental cyanide concentration during leaching on incremental gold/silver extraction the incremental cyanide destruction costs must also be incorporated in the analysis.

Note that during metallurgical testing a batch hydrogen peroxide test is usually sufficient for design purposes, however, continuous reactor testing is critical for SO_2-air process design and costing purposes.

Cyanide regeneration using ion exchange and acidification-reneutralization methods has received considerable attention in the last ten years. Todate this technology has suffered from poor economic return and highly complex operating strategies, resulting in few commercial installations. Generally, if there is sufficient regenerable cyanide species in an effluent stream to warrant an ion exchange plant then optimization of the cyanide addition rate is the first step in reducing cyanide effluent treatment costs. Another commercial cyanide removal process available is the Prussian Blue process where the effluent cyanide species are complexed to form inert metallo-cyanide precipitates. This process is particularly attractive where ferric/ferrous sulphate solutions are readily available as a byproduct (eg. sulphide pressure leachate or bioleachate, acid mine drainage).

A detailed understanding of process solution chemistry and receiving environment restrictions, as well as locked cycle test data are important when selecting a cyanide removal alternative during gold flowsheet development. Additional processes listed in Table 2 are for interest only and have not been satisfactorily proven technically or economically in operations.

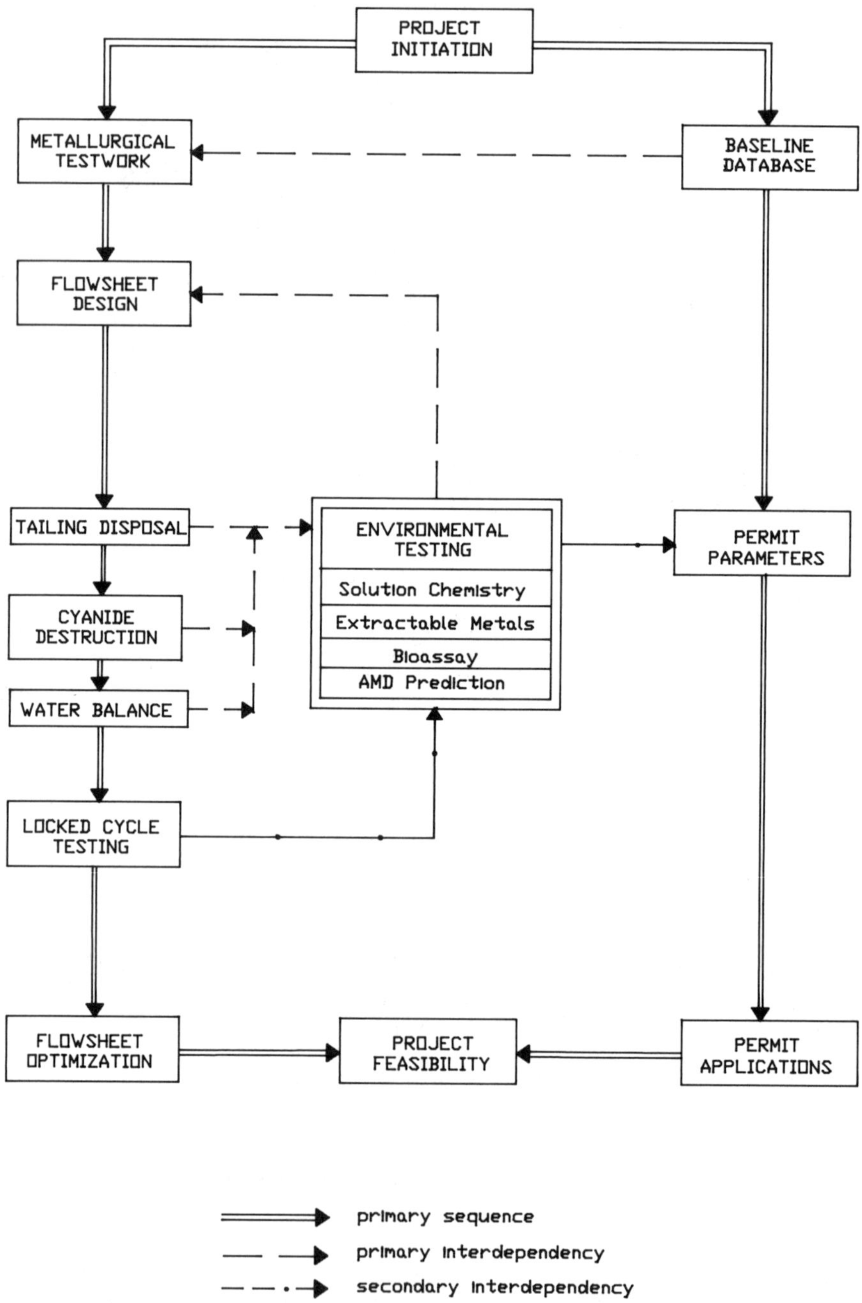

Fig. 2. Project sequencing and interdependencies.

CONCLUSIONS

The discussion above has illustrated the essential considerations of the effect of gold flowsheet alternatives on property environmental impact. To properly determine the feasibility of production the environmental impact of the process, both short and long term, must be clearly defined and incorporated in the feasibility study.

Figure 2 shows a simplified project sequence indicating the interdependency of environmental testing on flowsheet design and consequently flowsheet optimization and permit parameters. The sequence and integrated test procedures suggested, as described previously, would provide an environmentally responsible metallurgical flowsheet as well as expedite the permitting procedure by providing the necessary permit data and process design details within the feasibility study.

Again, it was not the purpose of this paper to define environmental protection legislation and guidelines concerning the disposal of mine waste materials but to highlight the environmental considerations for the process metallurgist during flowsheet feasibility analysis and to suggest a method to integrate environmental design within the metallurgical process design sequence.

ACKNOWLEDGEMENT

The authors thank R.A. Hawes, Norecol Environmental Consultants and K.D. Ferguson, Environment Canada, for helpful discussions.

REFERENCES

B.C. Reg. 42/84, B.C. Gazette, Part II, March 6, 1984

Ferguson, K.D., and P.M. Erickson (1987). An overview of methods to predict acid mine drainage. Proceedings Acid Mine Drainage Seminar, Halifax, Nova Scotia.

Hickey, M.G., and J.A. Kittrick (1984). J.Environ. Qual.,13(3), 372-376.

SAMPLING AND SAMPLE PREPARATION OF GOLD AND SILVER ORES

G. Zárate
Project Head
CIMM, Casilla 170, Santiago 10, Chile

ABSTRACT

Practical and statistical considerations concerned with sampling and sample preparation of bulk materials are reviewed. Then the application to gold and silver ores is discussed including for that sake typical examples and common mistakes in sampling and sample preparation. As a result some recommendations are given.

Finally the problems posed by the presence of coarse gold in the gold ores is analyzed. And some recommendations are given to solve these problems.

KEYWORDS

Gold and silver; sampling: sample preparation; chemical analysis; coarse gold

INTRODUCTION

Generally the sampling and sample preparation procedure of bulk materials, namely ore, concentrates and tailings must accomplish several basic requirements in order that samples be representative of sampled stream. These conditions are related with the procedure itself and the equipment. In this work the application to gold and silver ores is discussed including the problems posed by the presence of coarse gold.

PRACTICAL CONSIDERATIONS

Practical and statistical consideration in sampling and sample preparation were reviewed in detail by the author in previous papers. A brief summary follows.

Sampling

In order to define an appropiate procedure it is necessary to define or estimate the following parameters : required level of precision, ore quality variation, size and number of increments. These parameters can be experimentally determined or can be assumed based on previous knowledge.

Concerning the sampling equipment, this must accomplish several requirements in terms of geometry, velocity, opening, layout, capacity and location.

Sample Preparation

Sample preparation stage is very important in a sampling system, specially for gold and silver ores. Within this, the sample division is the main step and it is closely related to the particle size through the division rule. The best division methods are riffle division, fractional shovelling and increment division and the equipment must not contaminate either have losses of material.

Other steps in sample preparation are crushing, grinding, pulverizing, screening, mixing, filtration and drying. These steps don't have to contaminate either produce fine losses, which can reach up to 8 % depending on the equipment.

STATISTICAL CONSIDERATIONS

The determination of the content of gold, silver or any other quality characteristics in ores, concentrates, etc. has generally an associated error. Error in this context is defined as the difference between the measured assay and the actual value of the lot which the sample is assumed to represent. It has two components:

Precision : dispersion of error distribution (σ).
Bias : difference between the mean value of error distribution and the actual value.

A good sampling system is such that the value of precision is minimun and the bias is not statistically significant.

APPLICATION TO GOLD AND SILVER ORES

General case. Ores without coarse gold. For evaluating a sampling system the coefficient of variation, CV, related to total precision through the relationship $CV = (\sigma t/x) \times 100$ bias, d, are used. The required level of CV is established beforehand based on technical or commercial data. For example the japanese standard for ores proposes the levels of CV shown in Table 1, according to the monetary value of the lot.

TABLE 1 : Level of Precision (CV %)

Monetary value of Lot (US$)	Level of Precision (CV %)
1.750	5,0
5.000	2,5
50.000	1,0
500.000	0,5
> 500.000	0,25

Note : Original table in yen. 1 US$ = 200 yen

Following some typical examples of sampling systems applied to gold and silver ores are presented noting the common mistakes introduced when the sampling and sample preparation requirements are not accomplished. These are discussed in terms of precision and bias for sampling and sample preparation.

Sampling examples: precision. When the main contribution to the total precision is sampling, the problem is generally solved increasing the number of increments whenever is possible or preparing daily or weekly composites. For determining the appropriate number of increments the following relationship is used:

$$\sigma^2 M = \frac{\sigma^2 w}{n} \qquad (1)$$

Where

$\sigma^2 M$: sampling variance
$\sigma^2 w$: quality variance
n : number of increments

A typical example of the effect of the number of increments in the coefficient of variation is shown in Table 2 for a gold ore. The values of CV have been calculated using the equation (1) from experimental results.

TABLE 2 : CV for Different Number of Increments. Lots of 20 ton. Average Gold Content : 4,2 g/t

Number of increments	CV (%)	Dispersion ($\pm 2\sigma t$)*
9**	8	0,66
18	5,7	0,46
27	4,6	0,40
36	4,0	0,33

* σt : total precision of sampling system
** Experimental result

Bias is generally caused when using automatic cutters whose edges are not parallel radial and horizontal or whose opening are lower that those required according to the maximum particle size. Depending on the sampled material and the equipment the systematic errors can be significant. In Table 3 results for copper tailings are included. It is believed that for gold and silver ores similar results can be obtained.

TABLE 3 : Systematic Errors in Automatic Sampling of Tailings

Material	Cutter Charact.	Actual Content (% Cu)	Measured Content (% Cu)	Bias (% Cu)
Copper Tailings	Vertical edges and not parallel opening	0,07	0,09	+ 0,02
Copper Tailings	Id.	0,2	0,18	- 0,02

Systematic errors are also introduced when sampling is done only in a part of the stream. Merks compared partial stream cuts from a rod mill discharge against full stream cuts concluding that partial stream cuts are deficient in heavy particles. These results are shown in Table 4.

TABLE 4 : Partial Stream Versus Full Stream Cuts (4)

Parameter		Copper	Gold	Silver
Mean :	Full	3,18	2,00	41,1
	Partial	3,16	1,46	39,7
Mean	Difference	- 0,02	- 0,54	- 1,4
Student's	t-value	0,66	4,15	3,21

Sample preparation examples: precision. When sample preparation variance is the main component of total variance, this is reduced by applying the rule of division. A typical scheme of gold ore sample preparation, according to Mc Lean is shown in Figure 1. This kind of procedure rises significant errors, in terms of both precision and bias, because it doesn't follow the rule of division.

For instance studies carried out at CIMM show for a system that goes from 30 kg to 1 kg of - 3/8" gold ore a poor precision as it is indicated in Table 5. The level of precision is improved by including a crushing step of all the sample to 28 Tyler mesh. Both methods are shown in Figure 2.

Generally speaking a more appropiate procedure for gold and silver ores is shown in Figure 3. However the final procedure must be defined for each particular case.

TABLE 5 : Precision Level for Correct and Incorrect Sample Preparation Schemes for Gold Ores

Average Grade	Dispersion (± 2σt)	
	Correct Scheme	Incorrect Scheme
4,1 g/t Au	± 0,21 g/t Au	± 1,5 g/t Au
3,2 g/t Au	± 0,15 g/t Au	± 1,0 g/t Au

Bias

Concerning bias, using an incorrect sample preparation system, such as that shown in Figure 1, may introduce systematic errors. A summary of results obtained at CIMM for two different gold ores is shown in Table 6.

TABLE 6 : Bias on Incorrect Sample Preparation System. Average Results for 20 Experiences

Ore	Actual Grade (g/t)	Measured Grade (g/t)	Bias (g/t)	Obs.
Gold	2,1	2,5	+ 0,4	Division from 50 kg (-2") to 1 kg (-2")
Silver	620	600	- 20	Division from 20 kg (-1/4")to 1 kg(-1/4")

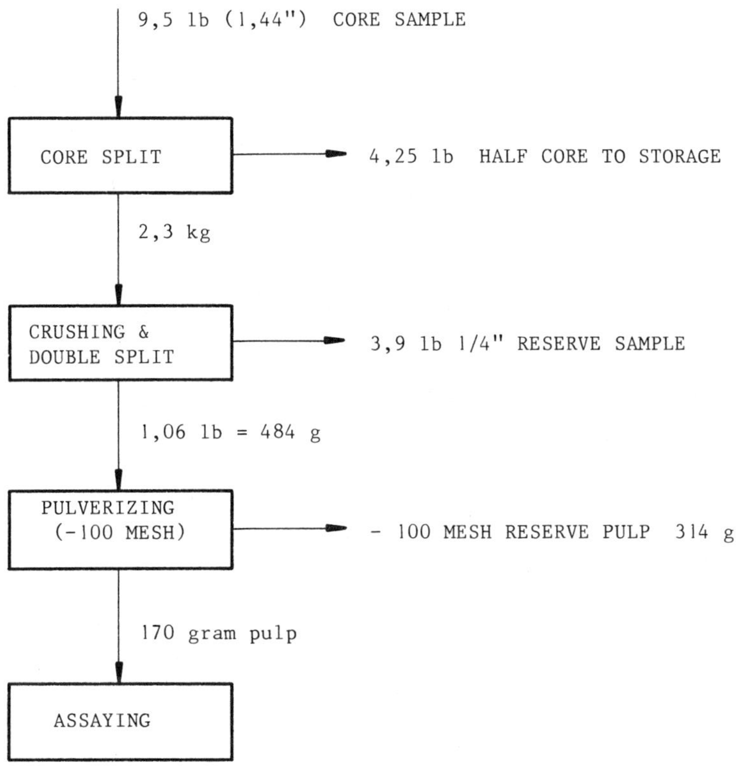

FIGURE 1 : CONVENTIONAL GOLD SAMPLING PROCEDURE

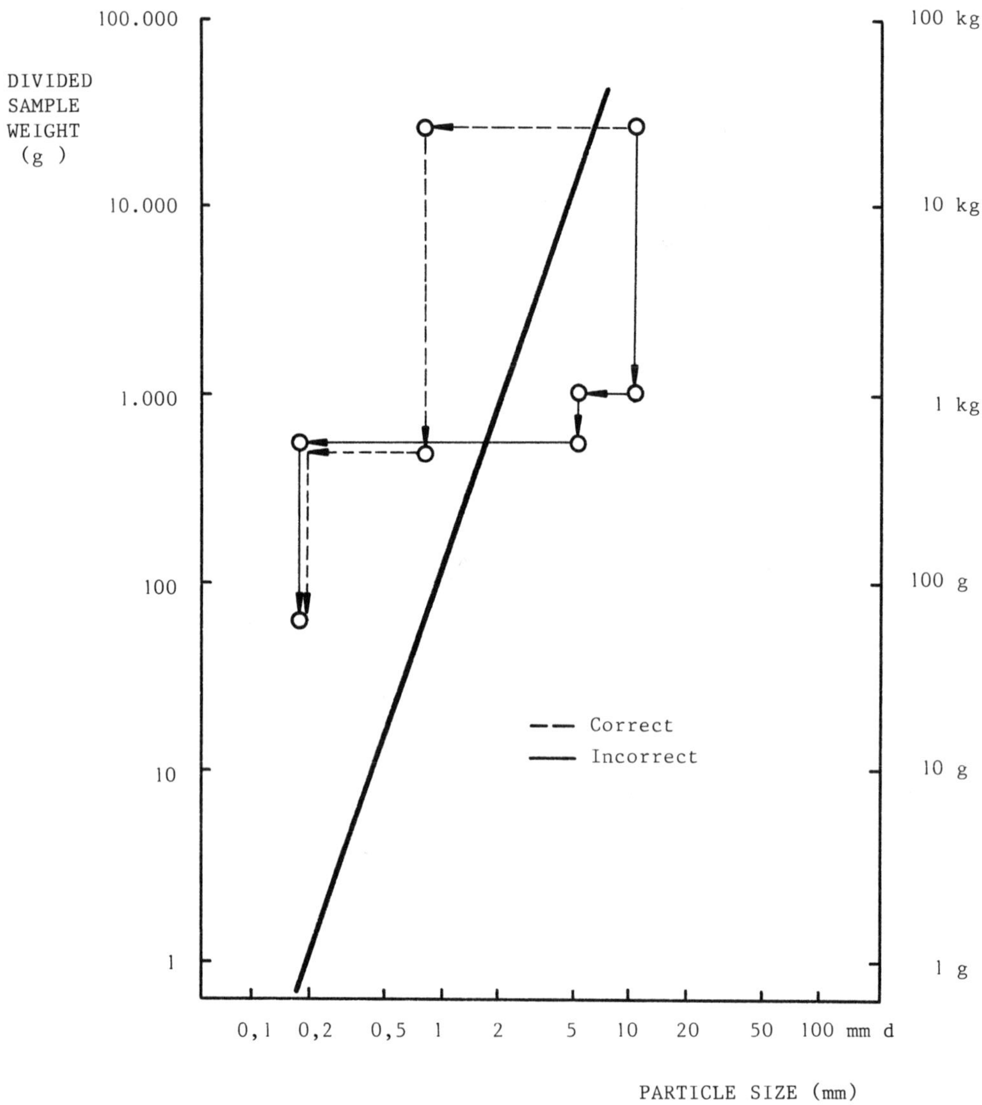

FIGURE 2 : EXAMPLE OF CORRECT AND INCORRECT SAMPLE PREPARATION METHODS

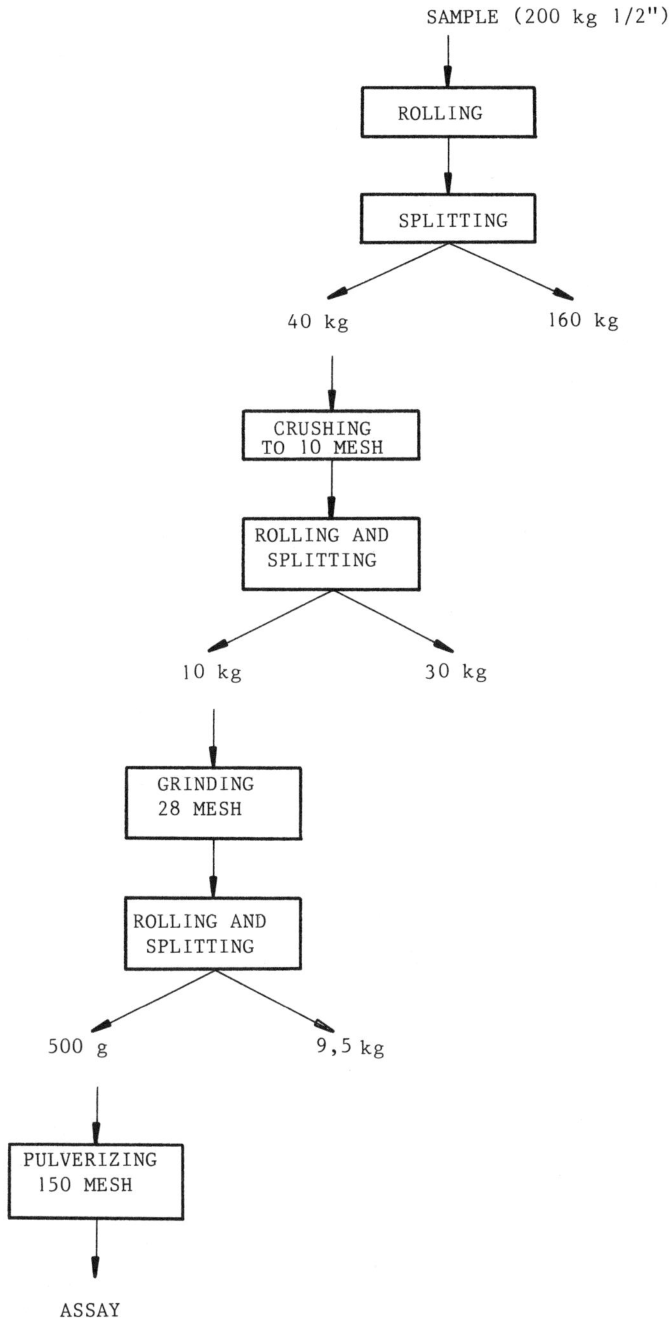

FIGURE 3 : EXAMPLE OF CORRECT SAMPLE PREPARATION SCHEME. GOLD AND SILVER ORES

Another steps that may introduce systematic errors are : pulverizing with loss of fines and drying at high temperatures. In the first case, depending on the pulverizing equipment, the fine losses may reach up to 8 % (2). In the second case studies carried out at CIMM showed that for drying temperatures much higher than 100°C the differences in the final grade may reach up to 20 %, relatively, depending on the dried material and the analyzed element.

Final sample preparation. In the case of gold and silver Mc Lean says that the final sample for chemical analysis must be pulverized under 200 Tyler mesh if gold and silver is fine. But generally this is ground under 150 Tyler mesh..

Particular Case: Ores With Coarse Gold

There are three important aspects to be considered when sampling ores with coarse gold. One is related to the sampling theories which are based on normal distribution while ores with coarse gold follow skewed or bimodal distribution. However this problem may be overcome by working with average values.

The second aspect is related with the sample preparation stage. This is clearly shown in results published by Tippin et all who carried out a sampling study at the Manhattan Mine in Nevada. Gold in this mine appears mostly as particles larger than 100 Tyler mesh. Tippin et al compared gold analysis for different sample preparation methods including also acalculated assay from the processing of 10 ton ore in a pilot plant facility. The methods are schematically shown in Figures 4 and 5 and the results summarized in Tables 7 and 8.

TABLE 7 : Sampling Results at Manhattan Mine.Different Sample Preparation Methods (6)

Method	Gold (oz/t)	CV (%)	Standard Deviation (oz/t)
Grinding-10 Tyler Mesh	0,230	110	0,252
Grinding-200 Tyler Mesh	0,142	72	0,102
Cyanidation	0,094	44	0,042

TABLE 8 : Sampling Results at Manhattan Mine. Pilot Plant Evaluation (6)

Sample	Gold (oz/t)	CV (%)	Standard Deviation (oz/t)
Tailings	0,0142	13	0,0019
Concentrate	4,1009	3*	0,1065*
Calculated Head	0,1511	14*	0,021*

* Estimated by author

From these tables it can be seen a great variation in results, even after a grinding under 200 mesh which corresponds to particle sizes below 74 μm.

In another study at the same mine, Dixon et all based on the previous results defined a sample preparation program (see Figure 6) which has three techniques

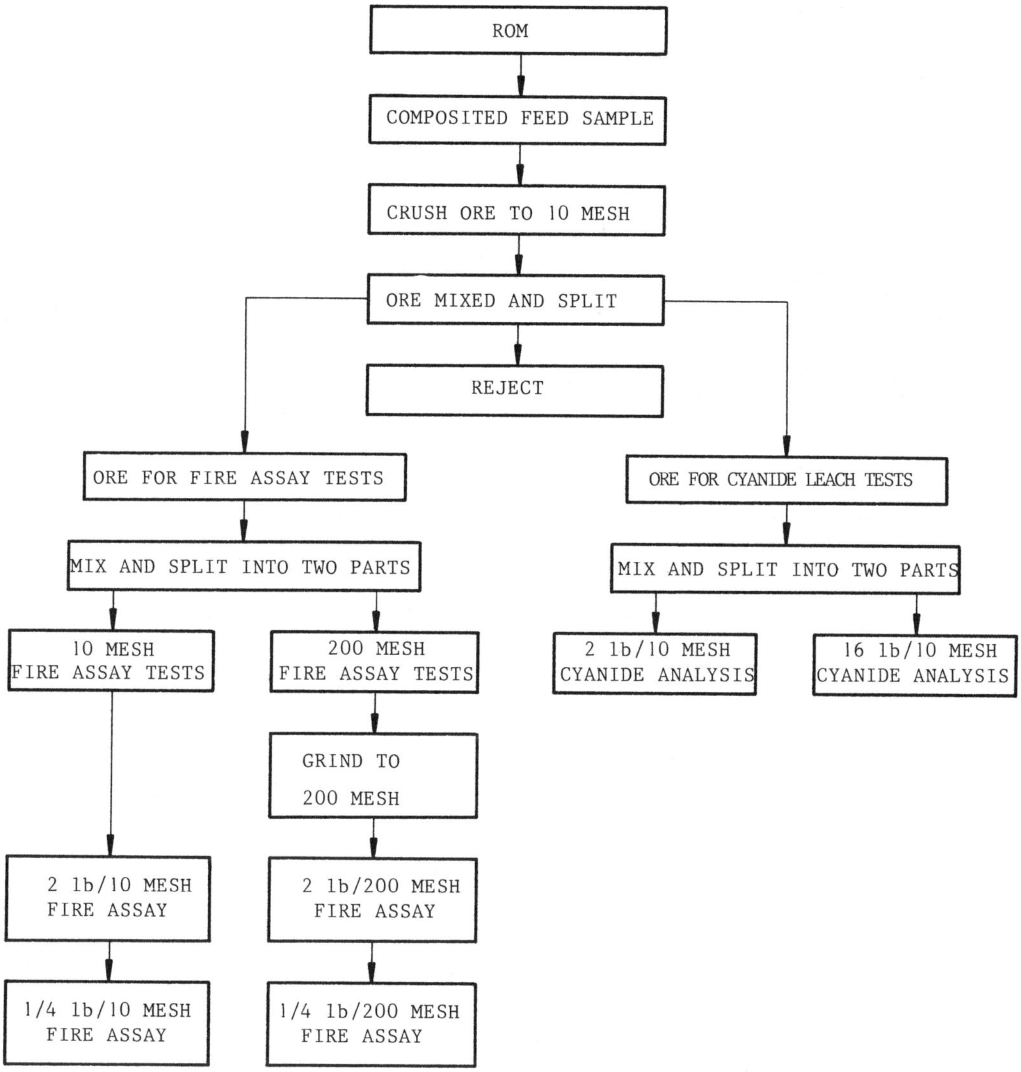

FIGURE 4 : SAMPLE PREPARATION AND ASSAY METHOD STUDY

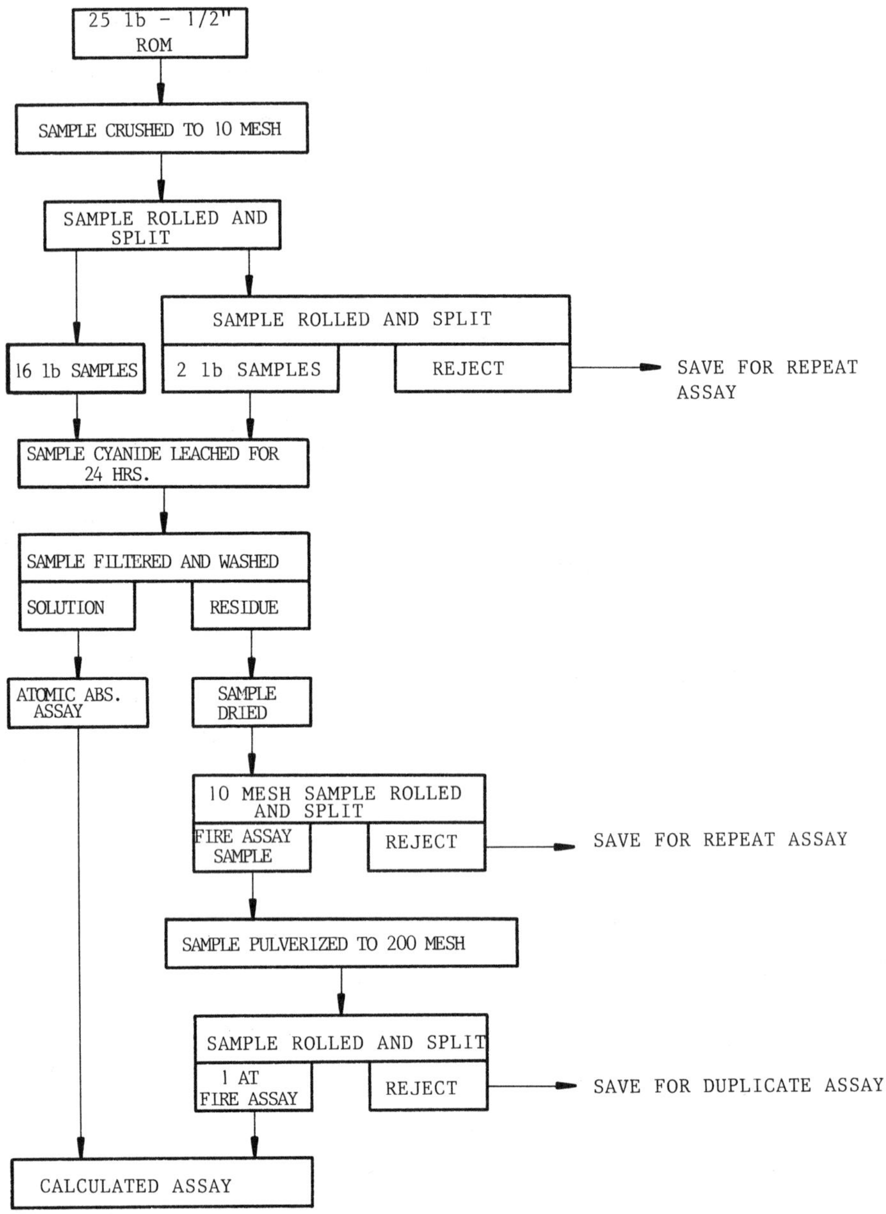

FIGURE 5 : CYANIDE LEACH PROCEDURE FOR CHEMICAL ANALYSIS

to improve the accuracy of the grade estimate :

a) Fire assay 3(1 AT) splits instead of 1 split (3 separate pulverized samples).
b) Flotation test on 1 kg of sample. Tailings are panned to recover any coarse gold.
c) Pilot plant run on composite samples of 4 to 8 tonnes.

The evaluation of the assay is a several step process. Initially a single assay is done and this assay is compared to geology, then two additional assays are run if the sample is ore grade or disagrees with geological information. The mean and standard deviation is calculated for the three assays. The variation is tested by a one-sided F-test through the following inequality.

$$\text{Sample std. dev.} > (F_{1-\alpha}(\nu_1,\nu_2)\ (\text{Manhattan pooled std. dev.})^2)^{1/2}$$

If sample standard deviation is greater than the right hand side of inequality this indicates the sample contains coarse gold and flotation tests are done to obtain a calculated head assay.

The problems posed by the presence of coarse gold was also studied by Ahlrichs. This author mentions that for assaying of suspected coarse gold, the "screen fire assay" technique as described by Fulton and Sherwood may be used. The total sample is crushed and pulverized, so that all but a maximum of 60 grams passes a 100 mesh sieve. All of the plus 100 mesh and 1 AT samples of two splits from the minus 100 mesh fraction are fire assayed.

If more than 25 % of the gold is suggested to be coarse by this method, more detailed studies are recommended. This is especially true for samples assaying less than 0,1 oz/t gold.

However it is pointed out that the amount of gold retained on 100 mesh can be exaggerated if gold is locked with gangue.

Ahlrichs compared two methods for preparing a sample of 200 lb under 10 mesh. Firstly 5 lb were split out, screened on 45 mesh, grinding the oversize, and fire assayed in the fractions - 45/+ 65, - 65/+ 100, - 100/+ 150, - 150/+ 200, - 200/+325 and - 325 mesh. The data is shown in Table 9. Additionally Ahlrichs screened out 10 lb of minus 10 mesh material on 28 mesh and ground the oversize. Then the minus 28 mesh was classified into fractions of - 28/+150, - 150/+325 and - 325 mesh and the first two fractions were passed over a Deister shaking table. All final fractions were fire assayed and the data is shown in Table 10.

TABLE 9 : Gold Assays and Distributions for Screened Fractions (8)

Sieve Ranges (mesh)	Size Ranges (mm)	Fractions Distrib. (% Weight)	Au (oz/t)	Dist(%)
- 45/+ 65	0,350-0,208	6,2	0,075	12,9 } 40,8 } 55,6
- 65/+100	0,208-0,149	21,8	0,046	27,9
-100/+150	0,149-0,105	15,7	0,034	14,8
-150/+200	0,105-0,074	11,9	0,031	10,2
-200/+325	0,074-0,044	11,9	0,024	8,0
-325 sands	0,044-0,015	11,0	0,033	10,1
-325 slimes	< 0,015	21,5	0,027	16,1
Calculated head		100,0	0,036	100,0
Assayed head*		100,0	0,037	100,0

* 5 lb used with the 'screen fire assay method'

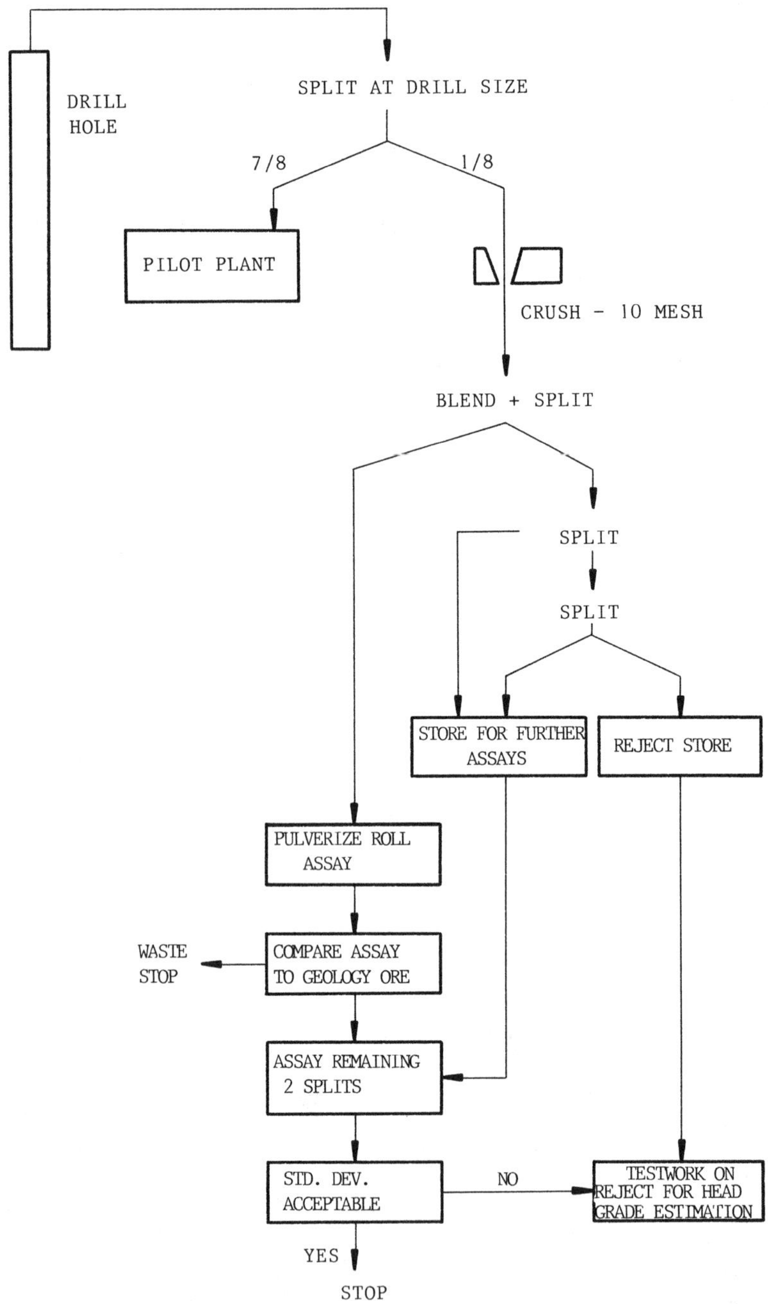

FIGURE 6 : SAMPLING FLOWSHEET USED AT MANHATTAN MINE

TABLE 10 : Gold Assays and Distributions for Tabled Products (8)

Product	Gold Sizes (mm)	Distribution (%)	Au oz/ton	Au Dist (%)
-28/+150 Mesh				
Tabled Concentrate	0,589-0,105	2,5	0,35	24,3
Tabled Tailings	< 0,105	30,3	0,024	20,2
-150/+325 Mesh				
Tabled Concentrate	0,105-0,044	1,1	0,12	3,7
Tabled Tailings	< 0,044	29,1	0,033	26,7
-325 Mesh				
Sands	< 0,044	25,5	0,020	14,2
Slimes	< 0,044	11,5	0,034	10,9
Calculated Head		100,0	0,036	100,0
Assayed Head		100,0	0,037	100,0

From these tables it can be noted that both methods agree well with the assayed head. The percentage of coarse gold is almost the same in both methods : 22,4 % vs 21,8 %, the last value being recalculated considering that part of gold is fine and locked within gangue.

The third aspect afore mentioned is related with the sample size for assay. Concerning with this Ahlrichs has also calculated the minimum sample size requirements for assay in accordance with a method prescribed by Clifton et al who indicate that any sample which contains twenty particles of gold is sufficient for assay. This is based on the particle diameter of the gold and the assay of a particular screened fraction. The effective diameter (De) was calculated from the screened analysis according to Gy's equation :

$$De = \left(\sum_j \frac{Mj}{M} dj^3\right)^{1/3} \qquad (2)$$

Where : Mj = mass of gold in sized fraction (in micrograms)
M = total mass of gold in sample
dj^3 = mid point of the sized fraction

Applying the effective diameter (0.17 mm) and the calculated head of 0,036 oz/t gold for the screened fractions to Clifton's method, a minimum sample size of 100 grams is required for assay of this particular sample.

On the other hand, Mc Lean has studied the effect of gold particle size on sampling and analytical methods and has calculated the weights of single gold particles at various particle sizes and also the number of particles of each size per gram of gold. From this data he estimated the number of gold particles to be expected in the assay sample sizes usually used in commercial assay work (1/4 to 2 AT).

The results, summarized in Table 11, make obvious why assays for coarse gold samples (+ 65 mesh Au) are highly erratic and unreliable, since the chances of a single sample containing one particle of gold are extremely low. According to Mc Lean this emphasizes the need for screening large sample at 100 mesh to remove coarse gold for separate assay before subdividing into conventional assay pulp splits. Only in the minus 200 mesh size range is the number of particles per sample large enough so that the effect of one particle will not effect the gold assay significantly.

One quarter and one half assay ton samples are the most vulnerable to gold sizes greater than 200 mesh.

Mc Lean has also calculated the effect of one gold particle at various sizes on the usual assay sample sizes (see Table 12). This data ilustrates the impact of one gold particle on the assay known as "nugget effect". It is seen that with gold coarser than 150 mesh (104 microns) single particles can induce sensationally high assay even when the actual average gold content in the ore is extremely low. The data also shows why sporadic high assays are to be expected and hence should not be discarded from ore reserve estimations since they are possible indicators of the presence of coarse gold. This, in turn, indicates the possible need for different sampling procedures.

Assuming that the "nugget effect" should not exceed 0.01 to 0.03 oz/ton, Mc Lean derived Table 13 to indicate the required assay sample size vs gold particle size.

TABLE 11 : Gold Particle Content of Various Sample Sizes as a Function of Gold Particle Size (Cubes) for a 0.05 oz/ton Ore

Gold Mesh Size	No.of Au Particles per10 lb Sample	Number of Au Particles per Assay Sample					
		1 A.T.	2 A.T.	5 A.T.	1000 GM	2000 GM	10.000GM
10	0,17	0	0	0	0	0	0
20	0,71	0	0	0	0	0	0
28	2	0	0	0	0	1	
48	16	0	0	0,5	3	6	35
65	46	0,30	0,60	1,5	10	20	101
100	128	0,84	1,68	4,20	28	56	281
150	370	2,4	4,8	12,0	81	163	368
200	1,000	6,6	13,2	33,0	220	440	2,193
325	4,588	30,4	60,8	91,2	1,011	2,022	
400	7,959	52,6	105	263	1,736	3,472	
20u	49,920	330	660	1,650	10,890	21,780	
5u	3,276,000	21,671	43,342	108,355	-	-	
2u	50,000,000	330,000	660,000	-	-	-	

* A.T. : Assay Ton

TABLE : 12 : "Nugget Effect" of Gold Particle Sizes vs Sample Weight

Gold Mesh Size	Wt. of one Gold Part. (mg)	Change in Gold Assay per Particle Gold (oz/t)						
		1/2 AT	1 AT	2 AT	5 AT	32 AT 1000 GM	64 AT 2000 GM	320 AT 10,000 GMS
10	88	176	88	44	17,6	1,75	1,38	0,27
20	11	22	11	5,5	2,2	0,34	0,17	0,03
28	4	8	4	2	0,80	0,125	0,062	0,021
48	0,50	1,0	0,50	0,25	0,10	0,016	0,008	0,002
65	0,17	0,34	0,17	0,18	0,03	0,005	0,002	0,001
100	0,061	0,12	0,06	0,03	0,01	0,002	0,001	0,001
150	0,021	0,04	0,02	0,01	0,004	0,001	0,001	0,001
200	0,0078	0,02	0,01	0,00	0,002	0,001	0,001	0,001
325	0,0017	0,004	0,002	0,001	0,001	0,001	0,001	0,001
400	0,0010	0,001	0,001	<0,001	<0,001	<0,001	<0,001	<0,001
20 u	1,56E-4	0,001	<0,001	<0,001	<0,001	<0,001	<0,001	<0,001
5 u	2,41E-6	<0,001	<0,001	<0,001	<0,001	<0,001	<0,001	<0,001
2 u	1,56E-7	<0,001	<0,001	<0,001	<0,001	<0,001	<0,001	<0,001

NOTE : E-6 = 10^{-6}

TABLE 13 : Required Assay Sample Size vs Gold Particle Size

Gold Particle Size Range (mesh)	Assay Sample Size Required (AT)
- 48	32 to 64
-100	2 to 5
-150	1
-200	1/2
-325	1/4
-325	5 grs (AA)
- 20u	Any sample size

Alluvial placer deposits, which most often have gold much coarser than 10 mesh, need extremely large samples for accurate reserve estimates. Good practice is to use a full scale gold gravity pilot plant.

If large samples cannot be obtained, Mc Lean recommends running replicate assays and calculating average values.

All these results show that whenever coarse gold appear special sample preparation procedures must be implemented. These must include three steps :

a) Sampling and preparing a large amount of ore samples.
b) Screening large samples at 100 mesh and/or 150 mesh and analyzing replicate samples whose size is larger than usual.
c) Calculating head assay by processing large samples in gravity equipments or other methods. This step will be necessary whenever the results from step b) show a high dispersion.

A procedure based on these three steps have been applied at CIMM by the author to different free gold milling ores. For instance for gold ores containing up to a 50 % of gold on the plus 150 mesh a sample preparation method similar to that shown in Figure 7 was used, but instead of one 0,5 kg sample four 1 kg samples were pulverized, screened in the fractions + 100 mesh - 100/+150 mesh and -150 mesh and analized all the material in the first two fractions and 1 AT in the - 150 mesh fraction. Finally the average assay was compared with calculated head from gravity concentration and column leaching.

In these conditions the dispersion of results, in terms of CV, depend on the amount of gold in the + 150 mesh fraction. For instance, when the gold in this fraction reached up to 25 %,coefficient of variation,CV,variated between 5 % and 14 % and when was higher than 25 % the CV variated between 20 % and 30 %.

Concerning the average values, these fitted very well with that calculated from gravity concentration and column leaching.

This procedure is now under optimization and the results will be published somewhere in the future.

CONCLUSIONS

Sampling and sample preparation of gold and silver ores must accomplish several requirements, in procedures and equipments,in order that samples be representative of sampled stream. If these requirements are not met, significant errors, in terms of precision and bias, may be introduced which will depend on the sampled material and the methods used. These errors are even more significant whenever the ore contains coarse gold.

In this case special sampling procedures must be implemented which must include the following steps :

a) Sampling and preparing a large amount of ore.
b) Screening large samples at 100 mesh and/or 150 mesh and analyzing replicate samples whose size is larger than usual.
c) Calculating a head assay by processing large samples in a gravity equipment or other method. This step will be necessary if the results from step b) show a high dispersion.

ACKNOWLEDGEMENTS

The author appreciates the authorization given by the Executive Direction of CIMM to publish this paper.

REFERENCES

Zárate, G. (1984). General considerations in the sampling of precious metal ores. In V. Kudryk, A. Corrigan, W.Liang (Ed.) Precious Metals: Mining Extraction and Processing. AIME, pp.67-78.

Zárate, G. (1987). Consideraciones generales en muestreo de materiales a granel. To be published in Minerales.

JIS M-8101 (1976). Methods for sampling of metal bearing ores of copper, lead, zinc, tin, gold, silver and other. Translated and published by Japanese Standards Association.

Merks, J.W.(1985). Sampling and sample preparation at mineral processing plants. Paper presented to the SME Fall Meeting, Oct.1985, Albuquerque, New Mexico, USA.

Mc Lean, D.C.(1982).Practical considerations in the sampling and estimation of reserves in low grade deposits. Paper presented to IMPI International Seminar, Oct. 1982, London, England.

Tippin, R.B., B. Robertson (-). Sampling test of a Nevada free-gold Milling ore. Paper presented to the SME Fall Meeting, Oct. 1985, Albuquerque, New Mexico, USA.

Dixon, S.N., T.R. Chapman, W.J. Robertson (1984). Sampling of a coarse gold deposit. In V. Kudryk, A. Corrigan, W. Liang (Ed.) Precious Metals : Mining, Extraction and Processing, AIME, pp. 79-91.

Ahlrichs, J.W.(1981)Some mineralogical applications for investigation of gold in geological and metallurgical samples. In D. Hausen,W. Park (Ed.) Process Mineralogy, Extractive Metallurgy, Mineral Explotation Energy Resources, AIME, pp. 381-396.

Technical Sessions

Session No.5

Refractory Gold

DIRECT DETERMINATION OF INVISIBLE GOLD IN REFRACTORY SULPHIDE ORES

S.L. Chryssoulis*, L.J. Cabri** and R.S. Salter***

*Surface Science Western, University of Western Ontario, London, Ontario, N6A 5B8

**Canada Centre for Mineral and Energy Technology, 555 Booth St., Ottawa, Ontario, K1A 0G1

***Lakefield Research, 185 Concession Street, Lakefield, Ontario, K0L2H0

ABSTRACT

Refractory ores can be classified into four types according to mineralogical criteria. One of these types contains "invisible" gold dispersed in the form of a solid solution or sub-micron inclusions in sulphide minerals. Analysis of gold in most sulphide minerals requires detection limits of fractional parts per million such that electron microprobes with detection limits of about 200 ppm for gold cannot determine gold in solid solution in common sulphides. Surface Science Western has developed a unique quantitative method combining an ion microprobe, secondary ion mass spectrometry and ion implantation to determine gold in solid solution in pyrite, arsenopyrite, pyrrhotite, chalcopyrite and sulpho salts to less than 0.5 ppm. In addition, the in-depth profiling capability of the microprobe allows the identification of sub-micron inclusions.

KEYWORDS

Refractory Gold Ore, Invisible Gold, Sub-Micron Inclusion, Ion Microprobe, Secondary Ion Mass Spectrometry, SIMS, Solid Solution, Ion Implantation.

INTRODUCTION

Gold minerals are native gold, gold alloys, and compounds of the element with sulphur, antimony, tellurium and selenium (Gasparrini, 1983). The most common of these are native gold and electrum. The other minerals are rare. In many ores, calculations of the amount of gold present in these forms agree generally with the assay values within the experimental errors inherent in such work. There are, however, a number of ores where the assay values greatly exceed the amount of gold minerals visible at the highest magnification under the microscope. The remainder of the gold is in an 'invisible' form which can only be released by roasting or pressure oxidation of the ores.

Mineral processing transforms mined ore into concentrates suitable for subsequent metallurgical treatment designed to extract the desired metal, in this case gold. Flow sheets developed for any ore are based on knowledge of the gold mineralogy, past practice with similar ores, and on bench and pilot plant test work. The mineral-dressing engineer's principal concern is to liberate the gold mineral from the gangue for gravity separation or flotation, or to expose the gold mineral to make it amenable to cyanidation. Thus, knowledge of mineral associations, grain sizes and textures are important.

If a portion of the gold is dispersed within a sulphide mineral lattice or is in the form of sub-micron inclusions this gold cannot be liberated by grinding and is for the most part not amenable to direct cyanide leaching. Therefore, in the case of refractory gold ores, the additional knowledge of the gold content of the common sulphide minerals present in the ore (i.e. an accurate mineralogical balance for gold) may not only guide the metallurgical engineer in his test work and save valuable time, or save a gold ore from being classified as refractory, but also may improve gold recoveries by introducing a supplementary recovery method. In 1984 gold losses to tailings from Canadian mines, exclusive of placer mining, totalled nearly 14,000 kg (or about 446,000 troy oz.; Cabri, 1987A and 1987B) which is currently worth over U.S. $200 million (@ $450/oz).

Examples where detailed mineralogical investigations combined with bulk chemistry of mineral separates provided a strong case for allocation of the "invisible" or unaccounted gold to different minerals are discussed by Boyle (1979) and Cabri (1987A). However, the purity of the samples always casts a shade of ambiguity on the results, to the extent that experienced mineralogists question the actual possibility of gold in solid solution (Gasparrini, 1983). Accurate data on the gold content of all the gold-bearing minerals, coupled with a data base of the various gold-bearing minerals for different deposits would permit the mineral-dressing engineer to consider the economics of concentrating minerals for separate metallurgical treatment and rejecting other minerals. The engineer would be in a better position to research and develop extractive methods for specific deposit types, especially where roasting, with its usual sulphur and arsenic contaminations, is considered (Cabri, 1987A).

The objective of this paper is to introduce a unique method for the in-situ determination of gold dispersed in the crystal lattice of pyrite, arsenopyrite, pyrrhotite, chalcopyrite, and other common sulphide minerals. This involves Secondary Ion Mass Spectrometry (SIMS) to detect the gold and ion implantation to calibrate the technique.

The ion microprobe (Secondary Ion Mass Spectrometer) has been calibrated to detect gold in sulphides in the ppm concentration range. This technique is presently the most sensitive in-situ microanalytical technique for the determination of gold. The electron microprobe has a detection limit of nearly 200 ppm for gold when employing the special analytical conditions necessary for gold analysis in iron sulphide minerals (Cabri, 1987B). The SIMS detection limit for gold has been determined to be less than 0.5 ppm with margins for further improvement. In addition, the in-depth profiling capability of the ion microprobe offers the advantage of identifying sub-micron inclusions.

THE ION MICROPROBE

Operating Conditions

The instrument used in this study was a CAMECA IMS-3f ion microprobe. Chryssoulis and co-workers (1986) described briefly the instrument and mode of operation.

Instrumental operating conditions are summarized in Table 1. An incident Cs^+ ion beam was found to give high ion yields for gold. A preselected mineral grain was located using the optical microscope. The beam was rastered over an area of 250 by 250 μm or 150 by 150 μm, and the secondary ion image was examined to confirm that the beam was centered over the grain of interest. The area that was analyzed was a circle of 60 μm in diameter in the centre of the rastered area as determined by a field aperture within the secondary optics, and was confined to a single mineral grain. Gold (^{197}Au) was monitored together with sulphur (^{33}S), which was used to monitor instrumental stability.

TABLE 1 Experimental Parameters

ION IMPLANTATION	
Instrument	2 MeV High Voltage Mass Separator, Peletron, (National Electrostatics Corp.,Madison, Wisc., U.S.A.)
Available ion energies	300 - 2000 KeV
Operating ion energy	500 - 900 KeV
Dispersion	$2R \frac{\Delta M}{M} = 3.73$ cm (M=107)
Ion source	Danfysik 910
Implanted species	^{197}Au
ION MICROPROBE ANALYSIS	
Instrument	CAMECA IMS-3f Ion Microprobe
Primary ions	Cs^+
Primary ion energy	16.1 KeV relative to specimen
Primary crater size	100 - 250 μm circular
Sampling area	60 μm diameter circle
Mass resolution	500 - 2000
Kinetic energy offset	50 - 150 eV
DEPTH MEASUREMENT	
Instrument	DEKTAK II stylus-type surface profiler

Standardization

The use of the ion microprobe for analysis of minerals has been limited by several factors inherent in the instrument and to the technique (McIntyre and co-workers, 1984). The use of closely matching standards is the most widely recognized method for obtaining quantitative measurements. McIntyre and co-workers (1984) utilized synthetic minerals which were homogeneously doped with the trace element of interest, while Chryssoulis and co-workers (1986) employed a solid state standard addition method (Leta and Morrison, 1980). The present investigation utilized ion implantation of gold (^{197}Au) in naturally occurring sulphide minerals. This technique reproduces the mineral matrix and produced satisfactory results.

The internal standardization method (Leta and Morrison, 1980) requires the ion implantation of each sample to be analyzed. In the case of gold which has only one isotope it is necessary to profile in depth far beyond the implant region in order to establish the background signal level due to the original trace element content of the specimen.

The background signal can then be extrapolated for the implant region. The accuracy of the technique is affected by sample homogeneity, short term variations in instrumental parameters as well as the implant dose (Leta and Morrison, 1980). Improved accuracy and precision can be obtained by implanting a radioactive isotope of gold with a long half like ^{195}Au. While analyzing, both isotopes of gold are monitored together, the radioactive isotope (^{195}Au) is used for calibration, and the natural isotope (^{197}Au) is used to measure the original gold content of the mineral.

SAMPLE PREPARATION

Large grains of natural pyrite, arsenopyrite and chalcopyrite together with two synthetic pyrrhotite crystals were used in this investigation. Polished sections were prepared and examined with the optical microscope under reflected light (detection limit 0.8 µm) and a scanning electron microscope (detection limit 0.1 µm). Inclusion free areas were selected for ion microprobe analysis. The area of analysis was again examined after sputtering to ensure that a second mineral phase had not become exposed to the primary ion beam.

All specimens were implanted with ^{197}Au for use as an internal calibration standard. Instrumental operating conditions are summarized in Table 1. The insulating nature of the mineral specimens required the application of a conductive coating to minimize charging of the specimen surface during ion microprobe analysis.

ANALYSIS

Quantification

The original concentration (C) of gold in sulphides doped with ^{197}Au can be expressed as (Chryssoulis and co-workers, 1986):

$$C = \frac{FtS}{DI} \quad (1)$$

The original concentration (C) of gold in atoms. cm^{-3} is equal to the product of the implant fluence (F) in atoms cm^{-2}, the analysis time

(t) and the integrated background signal (S); and divided by the product of the depth of analysis (D) in cm as illustrated in Fig. 1, and the integrated signal (I) of the implanted gold. The integrated signal of the implanted gold (I) equals the difference between the integrated signal of ^{197}Au minus the integrated background signal. The implanted gold has a characteristic concentration distribution with depth (Fig. 1). The time of analysis (t) is measured by a computer clock and the integrated gold ion signal (I + S) is corrected for dead time in the ion microprobe's counting system. The implant fluence (F) is measured by the current delivered to the specimen for a preset time during the ion implantation. The depth of analysis (D) is measured with a stylus type depth profiler after the analysis is completed.

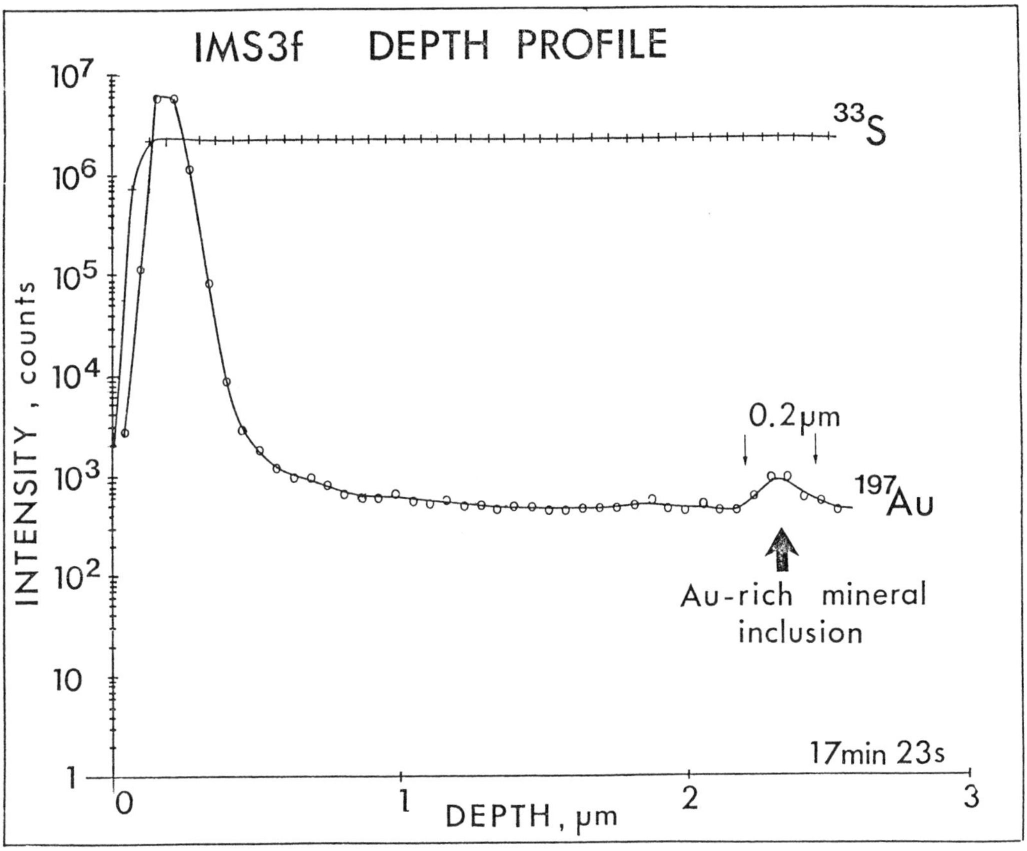

Fig. 1 Ion microprobe analysis by a modified internal standardization method.

Accuracy, Sensitivity and Precision

General considerations on the accuracy and precision of the technique have been discussed by Chryssoulis and co-workers (1986) in the case of silver determinations in sulphide minerals. The Cesium beam used in the gold determinations formed smoother crater bottoms therefore resulting in more accurate depth profile measurements. The accuracy of depth measurements greater than 10,000 Å is 5% or less. The accuracy of the implant fluence (F) was determined to be 5% by Rutherford backscattering for bismuth ion implantation in high purity silicon. The time of analysis is considered to be free of error.

The error introduced by extrapolating the background signal for the implant region is a function of the homogeneity of the gold distribution in a given mineral grain. It can only be measured by implanting a duplicate mineral sample with the radioactive gold isotope (^{195}Au).

The most common source of error in ion microprobe analysis is interference from molecular ions with the same nominal mass as the secondary ions of interest, in this case ^{197}Au. Four molecular interferences at mass 197 were recognized during the analysis of pyrite, arsenopyrite, pyrrhotite and chalcopyrite, which were eliminated either by using high mass resolution or by applying a kinetic energy offset. The sensitivity of the ion microprobe is significantly reduced as a result of decreased transmission of secondary ions, but detection limits in the parts per billion range can nevertheless be achieved.

The accuracy of the ion probe microanalysis was estimated to be better than 15% by comparing the gold analyses of the synthetic pyrrhotite with results determined by instrumental neutron activation analysis (INAA) in Table 2. In the case of the natural arsenopyrite the larger difference between the SIMS and INAA data is attributed to micro-inhomogeneity and the limited number of SIMS determinations.

The precision of the modified internal standardization method is estimated to be worse than that for silver (18%) in samples implanted with ^{197}Au. However it should be comparable to that of silver in samples implanted with ^{195}Au.

TABLE 2 Gold Background Content of Doped Sulphides (ppm)

	Method of analysis INAA*	SIMS+	Percent
Mineral			
Synthetic pyrrhotite 641	0.43 - 0.49	0.57, 0.47	13.5
Synthetic pyrrhotite 644	1.42	1.40, 1.36	
		1.42, 1.51	0.2
Arsenopyrite	1.12 - 1.21	1.66, 1.08**	42.7

* Instrumental Neutron Activation Analyses (INAA), McMaster University
\+ Secondary Ion Mass Spectrometry, Surface Science Western
** Inhomogeneous gold distribution

APPLICATIONS

Common sulphide minerals including pyrite, arsenopyrite, chalcopyrite, pyrrhotite, galena and sphalerite contain trace amounts of precious or semi-precious metals. The ion microprobe was used for the quantitative determination of gold and silver, which with few exceptions (Aubert and co-workers, 1964; Ag in galena) are below the detection limit of the electron microprobe. Therefore it can provide an important supplement to the established techniques employed by the geologist, applied mineralogist and metallurgical engineer. The capability of mapping elemental distribution in three dimensions is also of interest and can be used to differentiate between nano-inclusions and metals in solid solution or dispersed in the crystal lattice of the host sulphide mineral.

Recovery of Gold from Sulphide Ores

Gold is found in an increasing number of sulphide ores containing iron, copper, lead or zinc. The gold content of the deposits varies widely but is in general sufficient to be of economic importance in the exploitation of the ore. The ores are typically treated by flotation to produce a number of concentrates.

The gold mineralogy of the deposits also varies and includes native gold or electrum and to a lesser extent other gold minerals and alloys, as well as the several major sulphide minerals (Badalov, 1966; Badalov and Badalova, 1968; Mironov, and co-workers, 1986). Gold like silver can be distributed in a complex manner between several mineral phases (Vaughan & Craig, 1978; Chryssoulis, 1986) and is therefore distributed between the various concentrates and the plant tailings. As a result gold recovery losses represent an important loss of potential revenue.

This study has confirmed (Table 3) what was suspected based on indirect evidence (Boyle, 1979): that chalcopyrite, arsenopyrite and pyrite may contribute to the observed gold content of massive base metal sulphide ores. Arsenopyrite, pyrite and pyrrhotite are of particular concern due to their predominance in some deposits and because gold in these minerals generally reports to the tailings. Having identified the gold-bearing sulphide mineral in the tailings it may be possible to consider supplementary metallurgical treatment resulting in improved gold recoveries.

Preliminary results of the present study for chalcopyrite, arsenopyrite and pyrite from Canadian sulphide deposits indicate gold contents from less than 1 ppm up to 20 ppm (Table 3). Additional data is required in order to establish the range of gold concentration in the various sulphide minerals.

TABLE 3 Gold Content of Sulphide Minerals

Mineral	Gold content (ppm)	No. of grains analyzed
Pyrite	< 0.5 - 4.5	17
Arsenopyrite	< 0.5 - 4.8	8
Chalcopyrite	< 0.5 - 20.0	6

Gold may be recovered from sulphide ores, concentrates and tailings by a variety of metallurgical processes. Gold which is in solid solution, or dispersed in a mineral matrix may only be recovered by chemical processing, while finely disseminated gold values which cannot be efficiently separated by physical means may also be recovered.

The nature of the contained gold is a material consideration in the design of any metallurgical process to treat gold containing feeds. The ion microprobe can be used to investigate the mineralogy of feeds, calcines and leach residues in order to select an appropriate recovery strategy.

CONCLUSIONS

Ion implantation has been successfully used to produce internal calibration standards for the in-situ analysis of the 'invisible' gold in sulphide minerals. The method was demonstrated for silicates by Zinner and co-workers (1976) and silver in sulphides (Chryssoulis, 1986).

Ion microprobe analysis is a sensitive technique for the quantitative determination of trace amounts of precious metals in minerals, and can be used to complement the established techniques of applied mineralogy such as electron probe microanalysis and image analysis in order to produce accurate material balances for trace elements in ores and in mineral processing or metallurgical plants.

The developed technique has direct application to:

(i) the characterization of refractory gold ores. Some ores which are classified as refractory may not be refractory to a specific recovery method which can be identified based on the SIMS results.

(ii) improving the recovery of gold by introducing supplementary ore processing based on the identification of additional gold-carrier minerals.

(iii) providing guidance to metallurgical tests for gold by means of a more accurate gold mineralogical balance.

(iv) the reduction in the environmental contamination due to processing of gold concentrates. Sulphide minerals which carry insignificant amounts of gold can be identified and perhaps rejected prior to pyrometallurgical or hydrometallurgical extractive processing.

(v) lowering of processing costs for gold recovery from concentrate due to rejection of insignificant gold carriers resulting in higher concentrate grade, lower reagent and fuel costs, smaller processing facilities, and reduced residue handling and storage.

ACKNOWLEDGEMENTS

This study was made possible with an industrial fellowship from

Surface Science Western to the principal method and with the financial support kindly provided by CANMET. Dr. W. Lennard and Mr. G. Sims of Atomic Energy Canada Ltd. assisted with the ion implantation and Dr. N.S. McIntyre and Mrs. D. Johnston of the University of Western Ontario contributed to the success of the study. Dr. J.H. Crocket of McMaster University is greatfully acknowledged for performing the INAA determinations.

REFERENCES

Aubert, G.; P.K. Duong, and J. Geffroy (1964). Sur la localisation de l'or dans le mineral du Châtelet (Creuse). Bull. Soc. Fr. Miner. Crist., 87, 623-624.

Badalov, S.T. (1966). The role of major components in the geochemistry of minor and rare elements of ore deposits. Geoch. Intern. 857-860.

Badalov, S.T. and R.P. Badalova (1968). Some regularities of distribution of gold and silver in principal ore minerals of hypogene deposits of Karamazar and Western Kazakhstam. Geoch. Intern., 660-668.

Boyle, R.W. (1979). The Geochemistry of Gold and Its Deposits. Geological Survey of Canada, Bulletin 280.

Cabri, L.J. (1987A). The mineralogy of precious metals: new developments and metallurgical implications. Can. Min., 25, 1 - 7.

Cabri, L.J. (1987B). New developments in determination of the distribution of precious metals in ore deposits. Proceedings of the VII IAGOD, Lulea, Sweden. in press.

Chryssoulis, S.L.; W.J. Chauvin, and L.J. Surges (1986). Trace element analysis by secondary ion mass spectrometry with particular reference to silver in the Brunswick sphalerite. Can. Met Quart., 25, 233-239.

Chryssoulis, S.L. (1986). Determination of silver in the Brunswick sphalerite, pyrite and chalcopyrite by secondary ion mass spectrometry. RPC MDP/86/03, unpublished report.

Gasparrini, C. (1983). The mineralogy of gold and its significance in metal extraction. CIM Bull., 76, 144-153.

Leta, D.P. and G.H. Morrison (1980). Ion implantation for in-situ quantitative ion microprobe analysis. Anal. Chem., 52, 277-280.

McIntyre, N.S.; L.J. Cabri; W.J. Chauvin, and J.H.G. LaFlamme (1984). Secondary ion mass spectrometric study of dissolved silver and indium in sulphide minerals. Scanning Electron Microscopy, 3, 1139-1146.

Mironov, A.G.; V.F. Geletiy; I.N. Nesterova and A.J. Tsepin (1986). A Radioisotope Study of the Traces of Gold in Sulfide and Magnetite. Geochemistry International, 23, 117-128.

Vaughan, D.J. and J.R. Craig (1978), Mineral Chemistry of Metal Sulphides. Cambridge University Press, Cambridge.

Zinner, E., R.M. Walker, J. Chaumont and J.C. Draw (1976). Ion probe analysis or artificially implanted ions by terrestrial samples and surface enhanced ions in lunar sample 762-577. Proceedings the Lunar Science Conference, 953-984.

THE PROCESS DEVELOPMENT FOR DIFFICULT TO TREAT GOLD BEARING ORE IN A SMALL MINE

K. Heiskanen, H. Laapas, T. Julken, Helsinki University Technology
Helsinki, Finland and E. Barrantes, Cia Minera Trinidad S.A., Lima, Peru

Cia Minera Trinidad is a small Peruvian polymetallic sulphide ore producer. It has a 170 tpd flotation plant near Trujillo about 600 km north of Lima. The main flotation products are silver-lead-copper bulk concentrate and zinc concentrate. The ore is in a hydrothermal vein system. It contains galena, sphalerite and pyrite as the most abundant sulphides. Other important sulphides are tetrahedrite and arsenopyrite. The ore contains on average 2.6 g/t Au. The geological sampling shows a tendency of increasing gold values at lower mining levels. During the initial flowsheet development the flotation of other products was readily obtained, but gold could not be recovered. Since the gold content is an important source of revenue for the mine, a test program was started. The program was restricted by the financial resources of the small company both in terms of research cost and possible investment cost of the treating plant. The paper describes the details of process development from completely unsuccessful direct cyanide leaching to the final flowsheet which consists of flotation and acid leaching. The main problems were high arsenic and antimony contents and the presence of soluble copper minerals together with finely disseminated gold. The paper will also discuss the low cost treatment plant which is under design according to the developed treatment flowsheet.

AQUEOUS PRESSURE OXIDATION OF REFRACTORY FEED STOCKS

D.R. Weir and R.M.Berezowsky
Sherritt Research Centre
Fort Saskatchewan, Alberta
Canada, T8L 2P2

The first commercial plant to incorporate pressure oxidation of ore as a pretreatment for liberating refractory gold started up in early 1985, while the first to use pressure oxidation on concentrate started up in late 1986. Other projects are at the detailed engineering or pre-construction stage. Over the past seven years a variety of ores and concentrates have been subjected to integrated miniplant testing and process development by Sherritt. Data derived from these studies have been used in the design and construction of the two commercial plants, and have led to the design or process flowsheets for many other projects. Some factors affecting the selection of pressure oxidation in gold recovery process are discussed.

REFRACTORY GOLD ORE OXIDATION -- SIMULATION OF CONTINUOUS FLOW

by

A.S. Hayden*, P.G. Mason** and W.T. Yen***

*EH Associates, Toronto
**Wright Engineers Limited, Vancouver
***Lakefield Research
Lakefield, Ontario

ABSTRACT

Pressure oxidation is an effective method of liberating gold values from many refractory sulfide ores. Pressure oxidation testwork is most conveniently and economically conducted in batch laboratory autoclaves; however, the chemistry and kinetics may not represent plant scale continuous flow conditions. A simple laboratory technique has been developed which allows simulation of continuous flow in a single autoclave and provides much of the metallurgical information required for process design.

KEYWORDS

Refractory gold ore; pressure oxidation; Getchell ore; batch test; semi-continuous flow; reaction rate; residence time; acid consumption; gold liberation; sulphide oxidation.

INTRODUCTION

In recent years, the treatment of the refractory gold ores has attracted increasing interest. Pre-treatment of such materials is usually required to allow satisfactory gold extraction during cyanidation. One of the proven pre-treatment methods is the pressure oxidation in an autoclave.

Laboratory scale autoclave testwork is most conveniently and economically conducted as batch tests. However, batch pressure oxidation testwork usually produces conditions that are not representative of plant scale continuous flow systems, and the results may therefore be unreliable or difficult to interpret. Process development on pilot scale is very expensive and conditions often cannot be precisely controlled.

This paper will describe a semi-continuous testing technique which allows simulation of continuous flow conditions in a laboratory scale autoclave. The technique can be used to bridge the gap between batch and pilot testwork, and can generate useful design information at moderate cost. Testwork results on refractory gold ore from FRM Minerals' Getchell property in Nevada have been consistent and predictable, and data from this work are presented for purposes of illustration.

EXPERIMENTAL

Ore Samples

Three gold ore samples were used in this study. The chemical analyses of the samples are shown in Table 1.

TABLE 1 Chemical Analyses of Samples

	Comp. I	Comp. II	Comp. III
Au (g/t)	3.9	3.98	3.85
Ag (g/t)	3.7	4.00	4.4
As (%)	2.31	2.15	2.31
Fe (%)	2.9	2.85	3.15
Hg (%)	0.0072	0.0071	0.007
CaO (%)	5.94	5.67	6.51
Al_2O_3 (%)	9.66	9.53	9.56
MgO (%)	1.03	1.04	1.01
Na_2O (%)	0.065	0.069	0.078
K_2O (%)	2.71	2.39	2.44
SiO_2 (%)	61.7	63.6	58.9
S (total, %)	3.51	3.54	4.27
S (sulfide, %)	2.16	2.16	2.38
SO_4 (%)	1.20	1.16	1.26
C (total, %)	1.61	1.64	1.83
C (graphite, %)	0.54	0.64	0.49
CO_2 (%)	4.06	4.07	4.59

Gold and silver are invisible under normal microscopy. The sulphide minerals are primarily pyrite, realgar and orpiment, with minor amounts of arsenopyrite,chalcopyrite, pyrrhotite, sphalerite, galena, stibnite and cinnabar. Other opaque minerals include hematite, magnetite and graphitic carbon. The non-opaque minerals are quartz, calcite, kaolinite, illite, apatite, montmorillonite, gypsum, fluorite and barite.

The significant differences among the three composite samples are the relative amounts of realgar, orpiment, carbonate minerals and graphitic carbon. Due to the fine particle size, the gold, sulphide and carbonate minerals could not be efficiently separated by a simple flotation method.

Apparatus

A two-litre Parr autoclave assembled with a mixer, thermowell, gas

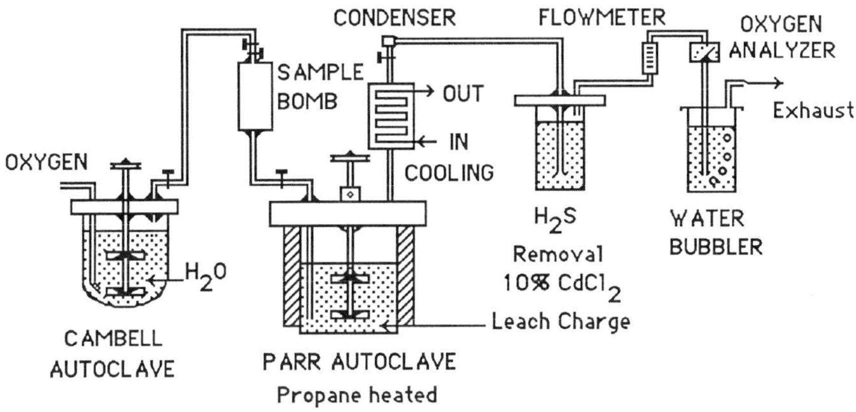

Fig. 1 Pressure Oxidation Leaching Equipment

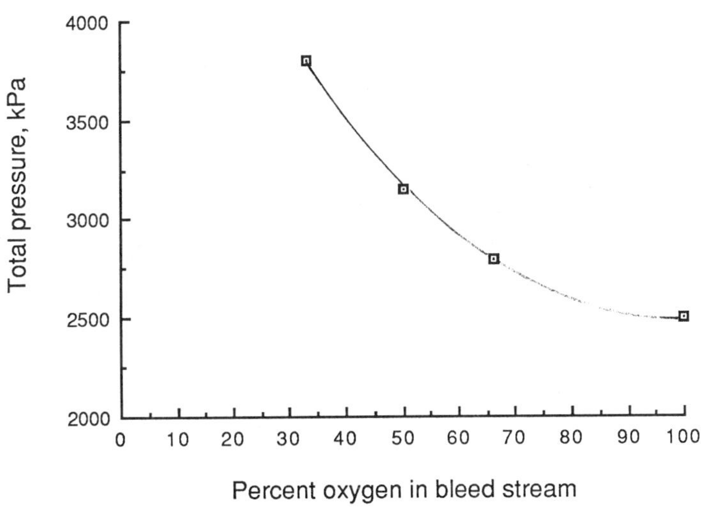

Fig.2 Oxygen pressure adjustment chart

inlet and outlet pipes as shown in Fig. 1 was used for pressure oxidation. The gas inlet pipe was connected with a sample bomb, which allowed injecting or removing sample slurry during the leaching period. The gas outlet pipe was connected with a cooling condenser, which prevented moisture loss to bleed gas.

The autoclave was heated with a propane burner, which was capable of raising the slurry from room temperature to 210 degrees C in less than 15 minutes.

Procedure

Samples were ground to 85% minus 74 µm or finer and pulped with fresh water and depending on test requirements, varying amounts of sulphuric acid. Oxidation was conducted at 195 degrees C to 210 degrees C under an oxygen enriched atmosphere. Oxygen was admitted into the bottom of the autoclave to facilitate dispersion. A controlled gas bleed from the unit was fed to an oxygen analyser to monitor oxygen content.

Batch testing. During the initial heat-up period, oxygen was not admitted to the vessel. After the leach slurry reached the designated temperature, oxygen gas was introduced into the system to raise total pressure to 3170 kPa. The initial flowrate of bleed gas was set at 800 mL/min to maintain a minimum of 50% oxygen in the exhausted gas mixture. As oxygen concentration increased, the autoclave pressure and the bleed rate were reduced; a minimum of 50% oxygen was always maintained. The relationship between total operating pressure and percent oxygen in the bleed gas is shown in Fig. 2. Once oxygen content reached 95 % or higher, the gas flow was adjusted to 300 mL/min and the total applied pressure was reduced to maintain 690 kPa oxygen partial pressure. The test duration was 90 minutes. The oxidized slurry was then quenched with cool water and cyanided at pH 10.5 with 0.5 g/L NaCN.

Semi-Continuous testing. Simulation of a single stage continuous leaching operation was conducted by removing a small amount of slurry at fixed time intervals and immediately replacing it with a similar amount of fresh slurry.

A continuous test was always preceded by an one hour batch leach conducted as described above. Then, 75 mL or 150 mL of leached slurry were removed using a sample bomb and immediately replaced by 75 mL or 150 mL of fresh slurry. The sample replacement interval was normally 9 or 10 minutes. Measurements of pH, emf and free acid were conducted on each slurry sample removed from the autoclave. The test was terminated when the pH, emf and free acid had stabilized; usually after two complete volume replacements. The final slurry and occasionally some of the intermediate slurry samples were subjected to standard cyanidation tests.

CHEMICAL REACTIONS

High temperature oxidation of the major sulphide ores can be summarized as follows:

$$2\ FeS_2 + 7O_2 + 2\ H_2O = 2\ FeSO_4 + 2\ H_2SO_4 \quad (1)$$

$$2\ As_2S_2 + 9O_2 + 6\ H_2O = 4\ HAsO_2 + 4\ H_2\ SO_4 \quad (2)$$

$$As_2S_3 + 6\ O_2 + 4\ H_2O = 2\ HAsO_2 + 3H_2SO_4 \quad (3)$$

Formation of elemental sulphur is not favoured under the range of conditions tested, although minor amounts of sulphur have been occasionally observed. Ferrous iron and trivalent arsenic are oxidized further to their ferric and arsenate states:

$$4\ FeSO_4 + 2\ H_2SO_4 + O_2 = 2Fe_2(SO_4)_3 + 2\ H_2O \quad (4)$$

$$2\ HAsO_2 + 2\ H_2O + O_2 = 2\ H_3AsO_4 \quad (5)$$

Ferric iron may then participate in further sulphide oxidation.

Some of the iron and arsenic in solution will precipitate as ferric arsenate, and additional iron may be precipitated as a basic iron sulphate or as hematite.

$$Fe_2(SO_4)_3 + 2\ H_3AsO_4 = 2\ FeAsO_4 + 3\ H_2SO_4 \quad (6)$$

$$Fe_2(SO_4)_3 + 2\ H_2O = 2\ Fe(OH)SO_4 + H_2SO_4 \quad (7)$$

$$Fe_2(SO_4)_3 + 3\ H_2O = Fe_2O_3 + 3\ H_2SO_4 \quad (8)$$

Hydronium jarosite may also form.

Acid produced during oxidation of sulphides or added during the test will attack gangue minerals, solubilizing some aluminum, magnesium and silica, and precipitating calcium sulphate. Potassium and sodium present in solution may precipitate additional iron as jarosite.

REACTION RATES AND RESIDENCE TIME ANALYSIS

Material processed continuously in a single stage, well mixed reactor is subjected to an infinite and continuous distribution of residence times. The periodic nature of slurry withdrawal using the described "semi-continuous" test technique represents a major deviation from this ideal condition, and must be accounted for in any analysis of extraction data.

Interpretation of data also requires kinetic information for the reaction of interest. A common form for expressing reaction rate is shown in the following equation:

$$dC/dt = -kC^n \quad (9)$$

where: C = Concentration
k = Rate Constant
n = Order of Reaction
t = Time

The reaction order is normally expected to be a small integer. The rate constant is a function of the test conditions. Semi-continuous testing at more than one nominal retention time will allow calculation of n.

Equation (9) can be integrated to yield the following:

For first order reactions,

$$C/C_o = e^{-kt} \tag{10}$$

where: C_o = Initial concentration at zero time

For all other reaction orders,

$$C/C_o = \left(C_o^{(n-1)}(n-1)kt + 1\right)^{1/(1-n)} \tag{11}$$

Each withdrawn sample represents material that has been in the reaction vessel for a finite range of retention times, and the average value of C/C_o can be obtained by summing the products of the increments and their corresponding extraction:

For a first order reaction, and substituting $N\partial t$ for t, where N is the number of increments taken and ∂t is the sample period,

$$C/C_o = e^{-kN\partial t}(1-F)^{(N-1)} + F\{e^{-k(N-1)\partial t}(1-F)^{(N-2)} + e^{-k(N-2)\partial t}(1-F)^{(N-3)} + \dots\dots\dots + e^{-k\partial t}\} \tag{12}$$

where: F = Fraction removed per sample

Where N is large, the contribution of remaining original material is small and the first term can be dropped. The remaining expression is a geometric progression with a common ratio of

$$e^{-k\partial t}(1-F) \tag{13}$$

which then summed yields:

$$C/C_o = F\left\{\frac{1}{e^{k\partial t} - (1-F)}\right\} \tag{14}$$

A solution for reaction order other than 1 may be obtained in a similar manner.

GOLD LIBERATION

The degree of gold liberation from refractory ore by pressure oxidation treatment is measured by subjecting oxidized material to a standardized cyanidation test. A portion of the gold may not be refractory, and cyanidation test results may be adjusted accordingly. In the case of the study refractory ore 10% - 15% of the contained gold is cyanide recoverable without oxidation pre-treatment.

The "rate of gold liberation" for the ore under the tested conditions has been found to be approximately second order with respect to gold content.

$$dC/dt = -kC^2 \tag{15}$$

However, limited semi-continuous laboratory work indicates a rate that is closer to first order and further work is required to develop the most appropriate rate expression.

With the assumption that Equation (15) holds for all tests, final gold extractions can be projected for any oxidation time and tested set of conditions. Further, oxidation performance can be evaluated in terms of gold liberation rate constants derived from the tests, which allows direct comparison of tests having different effective retention times. Table 2 records the results of selected tests on the study ore.

TABLE 2 The Semi-continuous Oxidation Performance

Acid, kg/t		Free Acid	emf	Cyanide Residue		Rate Constant	Projected**	
Added	Consumed*	g/L	mv	S, %	Au,g/t	k	Residue Au,g/t	% Au Extraction
35	24	8.6	580	0.19	0.46	0.034	0.27	92.5
45	30	10.6	590	0.20	0.42	0.038	0.21	93.2
55	35	14.2	620	0.17	0.24	0.075	0.15	96.4
55	34	14.1	545	0.30	0.37	0.068	0.15	96.0

*Defined as added acid less residual free acid.

**Predicted for an oxidation time of 90 minutes at 210 degrees C and 690 kPa of oxygen.

Table 3 records oxidation terminal solution conditions for a number of batch tests, and Table 4 presents similar information for semi-continuous tests.

TABLE 3 Gold Extraction Result from Batch Test on Composite I

Acid Added kg/t	Gold Extraction %	Residue		Autoclave Leach Solution				
		Au g/t	$S^=$ %	pH	Terminal emf mv	Terminal Free Acid g/L	Total Fe mg/L	Fe^{+++} mg/L
11	66.9	1.43	1.29	6.8	140	0	-	-
36	80.7	0.90	0.09	1.0	560	12.9	234	220
36	86.0	0.57	0.05	1.2	560	12.5	282	270
116	91.7	0.31	0.03	0.8	620	28.9	320	310
116	95.3	0.18	0.05	0.7	640	26.6	503	500

TABLE 4 Gold Extraction Result from Semi-Continuous Flow Test

Comp. Sample	Acid Added kg/t	Gold Extraction %	Residue Au g/t	Residue $S^=$ %	Autoclave Leach Solution pH	Terminal emf,mv	Terminal Free Acid g/L	Total Fe mg/L	Fe^{+++} mg/L
I	40	94.0	0.29	0.05	1.0	595	10.8	121	117
II	25	85.5	0.70	0.32	1.7	510	2.8	38.3	31.1
II	35	89.1	0.46	0.19	1.1	580	8.6	511	499
II	45	90.5	0.42	0.20	1.0	590	10.6	185	169
II	55	94.0	0.24	0.17	1.0	620	14.2	228	221
III	55	92.0	0.31	0.27	1.1	530	10.8	147	143

ACID CONSUMPTION

The ore sample used in this investigation consumes substantial amounts of acid through acid attack of carbonate and silicate minerals, and acid consumption increases with increasing free acid level. Moreover, a direct relationship exists between free acid level during oxidation and subsequent gold extraction. In general, more acid is required for satisfactory results than can be supplied through oxidation of sulphides. It is necessary for economic reasons to minimize the use of purchased acid.

The relationship between acid consumption (defined as added acid less residual free acid) is illustrated in Fig. 3 for two ore composites. Increasing consumption in accompanied by increasing amounts of aluminum and magnesium in solution.

The semi-continuous testing technique allows fairly precise control of solution conditions, including free acid level. For Getchell material, it has been possible with this method to demonstrate significantly improved gold recovery at moderate free acid levels. Fig. 4 compares gold extraction data vs terminal free acid for batch tests and semi-continuous tests, and Fig. 5 re-states these data in terms of (second order) rate constants.

CONCLUSION

The primary advantages of the semi-continuous testing technique is the ability to maintain a stable solution environment throughout a test. Reproducibility is improved compared with batch testing, and it is possible to measure the effects of process variables such as acid concentration with some degree of confidence. Information not available from batch tests may be obtained which will allow better planning of any pilot plant studies contemplated.

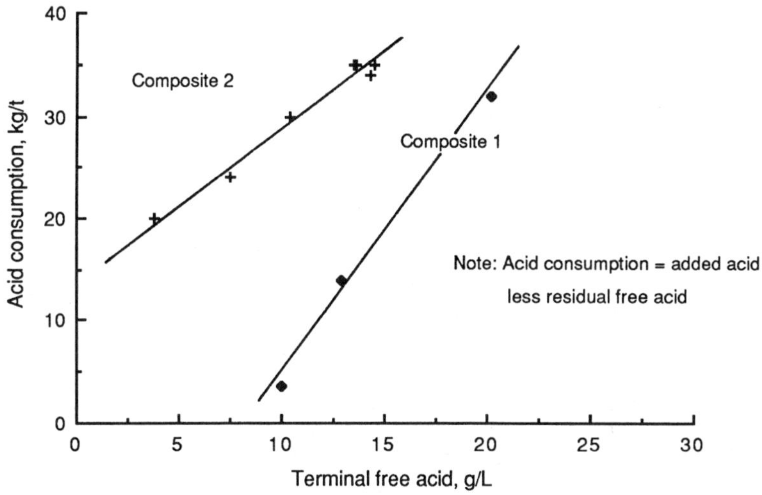

Fig. 3 Acid consumption vs. free acid level for two composite samples

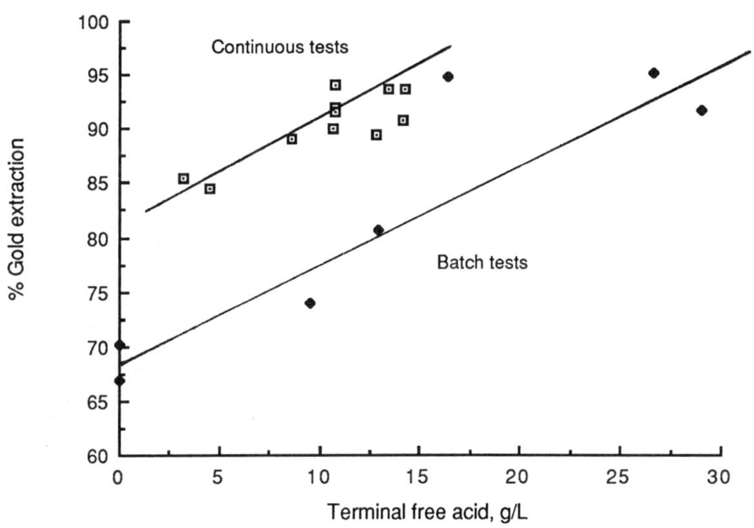

Fig. 4 Gold extraction vs. free acid level during oxidation

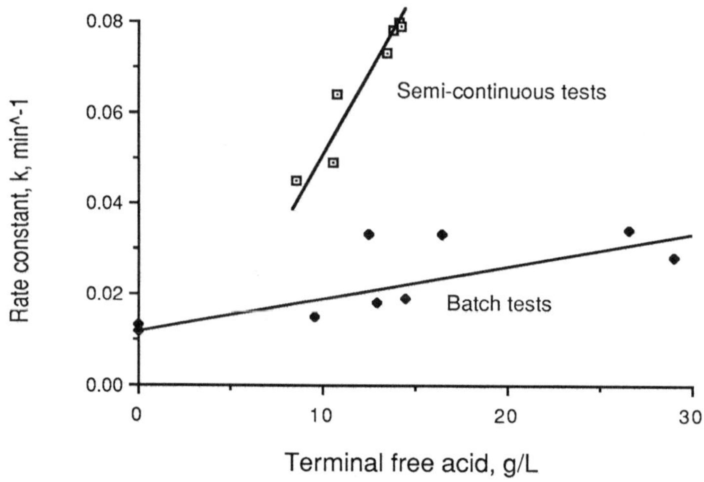

Fig. 5 Gold extraction vs. free acid level

One limitation of the technique is that a good direct measure of gold liberation obtainable in a commercial continuous operation cannot be obtained, and it is necessary to extrapolate results based on a knowledge of kinetic behaviour. A second limitation is that conditions found in continuous multi-stage operation cannot be exactly duplicated (although the range of conditions can be investigated in a series of tests). Consequently, semi-continuous single stage laboratory scale testwork is unlikely to eliminate the need for pilot scale testing. It should, however, substantially reduce the amount of pilot testing required.

ACKNOWLEDGEMENT

The authors wish to thank FRM Minerals Inc. and Lakefield Research for permission to publish the informations contained in this paper.

THE RECOVERY OF GOLD FROM THIOUREA LEACH LIQUORS WITH ACTIVATED CARBON

C.A. Fleming

Council for Mineral Technology, Private Bag X3015, Randburg, South Africa

ABSTRACT

The extraction of the gold-thiourea complex from aqueous leach liquors with activated carbon is examined, and the important factors that influence the rate of loading and the loading capacity are identified. It is shown that, under most conditions that would be experienced in a practical leaching situation, the gold-thiourea complex is loaded onto the carbon without undergoing any chemical changes. In addition, the loading capacity of the carbon is suppressed by the chemicals that are required for effective leaching (excess thiourea and ferric ions) and, in this respect, the leaching process and the gold-extraction process are antagonistic. Other parameters that influence the loading characteristics are the pH of the leach liquor, the concentration of anions in solution, and the temperature.

Various methods for the elution of the gold-thiourea complex from activated carbon are examined and optimized.

KEYWORDS

Gold thiourea complex; activated carbon; extraction; elution.

INTRODUCTION

There has been a rapid escalation, over the last few years, of research effort into novel methods for the leaching of gold from its ores. The main reasons for this have been the environmental hazards attendant on the traditional cyanidation route, and the inability of cyanide to leach certain refractory ores efficiently. Most of the research effort has involved the thiourea ligand, which, in acid solution with ferric ion as the oxidizing agent, dissolves gold as a cationic complex $Au(Th)_2^+$ (Th = $CS(NH_2)_2$). At least one plant, in Australia (Hisshion and Waller, 1984), has already converted from cyanide to thiourea leaching for the recovery of gold from an arsenopyrite concentrate. Despite this activity and widespread interest, little attention has been paid to the downstream recovery of gold from alternative leach liquors or pulps. The carbon-in-pulp (CIP) process offers many technical and economic advantages over the conventional filtration and cementation processes, and is generally the preferred route on new cyanide gold plants. It would, therefore, probably be the preferred route on any gold plant employing an alternative lixiviant to cyanide.

The loading and stripping characteristics of the gold-thiourea complex on carbon have been examined, and are reported in this paper. The most important factors influencing the kinetics and thermodynamics of the absorption reaction have been identified, and a tentative reaction mechanism is proposed.

EXPERIMENTAL

Synthetic leach liquors were prepared by the dissolution of gold powder in an aqueous solution containing 1 g/ℓ ferric ions and 5 g/ℓ thiourea. It was possible to dissolve 5g of gold quite readily in 1 litre of this solution, and this stock solution, containing 5000 mg of gold per litre, was then diluted 20 times before the carbon-loading experiments were carried out. The iron and thiourea concentrations in the final solution were, therefore, 50 mg/ℓ and 250 mg/ℓ (0.003M) respectively - values that are sufficiently low for their influence on the efficiency of gold extraction by activated carbon to be ignored.

The rates of extraction and elution were determined in batch experiments in which a small mass of carbon was mixed, with an overhead stirrer, in a large volume of either the synthetic leach liquor or an eluate solution. Samples of solution were withdrawn at regular intervals for analysis to determine the rate of change in the gold concentration. The sample solutions were analysed for gold by atomic absorption spectrophotometry, and samples of carbon were also analysed periodically by X-ray fluorescence as a check on mass balances.

For the equilibration studies, 200 mℓ aliquot portions of the synthetic leach liquor were treated with different masses, in the range 0.05 to 0.5g, of powdered activated carbon in a rolling bottle for a period of 72 hours.

The carbon used in all the experiments was the coconut-shell product, Le Carbone G210 AS, supplied by the French company, Pica.

RESULTS

Extraction

Theoretical considerations. The standard reduction potentials and the reduction potentials under typical leaching conditions are given in Table 1 for the reduction of various gold complexes to the metal. On the basis of these values, it would be predicted that activated carbon, with a reduction potential of between 0.1 and 0.4V, would reduce the gold-chloride and -thiocyanate complexes to the metallic state on the carbon surface, but not the gold-cyanide or -

thiosulphate complexes. The gold-thiourea complex is a border-line case, but should be reduced to the metallic state at thiourea concentrations below about 0.1M. (This concentration is, in fact, fairly typical of a thiourea leach liquor.)

However, while the reduction mechanism has been confirmed for the gold-chloride complex (by the observation of a coating of gold metal on the superficial surface of the carbon), this was not possible with the thiourea complex, and was possible only under certain conditions with the gold-thiocyanate complex. The evidence suggests, in fact, that the gold-thiourea complex is loaded onto activated carbon without chemical change. For example, the data in Fig. 1 show the equilibrium loadings of gold and sulphur on carbon that had been used to extract the gold from thiourea leach liquors under a wide variety of chemical conditions. The ratio of sulphur to gold is, in all cases, close to the theoretical value of 2 that would be expected for the absorption of $Au(NH_2CSNH_2)_2^+$ without chemical change.

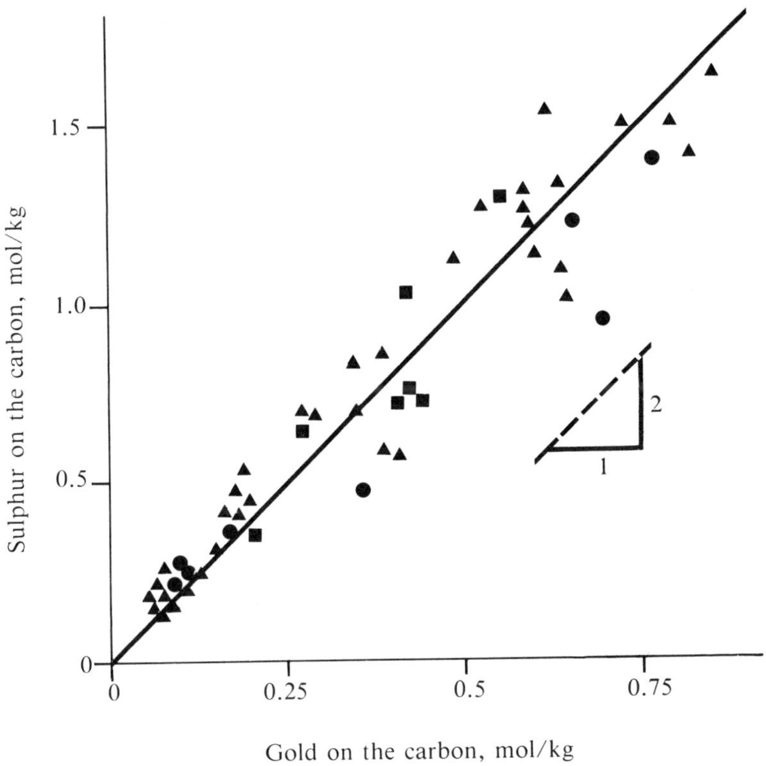

Fig. 1. The absorption of sulphur (as thiourea) and gold onto carbon from a solution containing the gold thiourea complex
● Excess acid (HCl, H_2SO_4) added to solution
■ Excess oxidant (Fe^{3+}) added to solution
▲ Excess cations (Ca^{2+}, Mg^{2+}, Na^+, K^+) or anions (SO_4^{2-}, ClO_4^-, NO_3^-, Cl^-) added to solution

The mechanism by which gold is loaded onto carbon from cyanide solution has not been firmly established, but the weight of evidence suggests that aurocyanide also retains its chemical integrity during absorption, being loaded as an ion pair of the type $M^{n+}[Au(CN)_2^-]_n$ ($M = Ca^{2+}$, Mg^{2+}, Na^+, K^+, etc.) in alkaline solution, and an ion pair of the type $HAu(CN)_2$ in acidic solution (McDougall and Fleming, 1987). Since the aurocyanide complex is more stable than either the thiourea or thiocyanate complexes (Table 1), this is not an unexpected result, and earlier mechanistic schemes (Green, 1913; Dixon, *et al.*, 1976), which proposed that the absorption of gold onto carbon from cyanide solution involved reduction of the gold, can probably now be rejected.

TABLE 1. Potentials for the Reduction of Gold Complexes to Gold Metal (Gold in solution = 10^{-5}M)

Ligand	Ligand concentration (M)	E^0 (V)	E(V)
Cyanide	10^{-2}	-0.57	-0.63
	10^{-3}		-0.51
Thiosulphate	0.5	0.03	-0.23
	10^{-2}		-0.03
Thiourea	10^{-1}	0.35	0.17
	10^{-2}		0.29
Thiocyanate	10^{-1}	0.71	0.54
	10^{-2}		0.66
Chloride	1.0	1.00	0.88
	10^{-2}		1.53

The rate of extraction. The rate of loading of the gold-cyanide complex onto carbon has been studied fairly extensively in recent years, and has been shown to be reasonably fast in the initial stages but extremely slow as equilibrium is approached. Experiments conducted in the laboratory have indicated that the loading of aurocyanide onto carbon may take longer than 6 months to reach equilibrium, although the loading after 24 hours could be as much as 90 per cent of this value. This is not attributed to any property of the aurocyanide complex, but to the heterogeneous nature of the internal pore structure of the carbon (McDougall and Fleming, 1987). The pore sizes of the carbon vary from between 10 and 100 Å for the micropores (and these comprise some 90 per cent of the total surface area) to between 500 and 20 000 Å for the macropores. It is reasonable to expect that gold will diffuse rapidly into the macropores and slowly into the micropores and, as the cross-sectional area of the micropores approaches that of the aurocyanide ion, the resistance to mass transfer will become infinite, and the rate of loading infinitely slow. Since this is a property of the carbon and not the gold complex, it is likely that the gold-thiourea complex will behave similarly. The results presented in Table 2 indicate that this is indeed so, the absorption process continuing for up to 5 months after the initial contact between the leach liquor and the carbon. In this regard, it is important to note that gold loadings that are referred to as 'equilibrium loadings' elsewhere in this paper are not true equilibria, since the equilibration period in these experiments was always only 3 days (a period that was chosen for its practical relevance).

TABLE 2. The Rate of Loading of the Gold Thiourea Complex onto Activated Carbon
Conditions: Solution volume = 1000 mℓ, carbon mass = 1g

Time	Gold in solution (mg/ℓ)	Calculated gold on carbon (g/t)
0	250	
6h	175	75 000
4 day	60	190 000
7 day	50	200 000
11 day	39	211 000
14 day	28	222 000
28 day	8.3	241 700
150 day	0.34	249 660

The effect of leaching chemicals on extraction efficiency. Experimental work published previously (Groenewald, 1976; Gabra, 1984; Schulze, 1984; Yen and Wyslouzil, 1985; Fleming, 1986) has shown that the ideal environment for the leaching of gold with thiourea is acidic (pH 1.5 to 3.0), contains an excess of thiourea (5 to 15 g/ℓ), and contains ferric ions as the oxidant for gold metal. Since the development of an integrated gold process requires that the leaching chemicals should be compatible with the carbon-extraction process, the effect of these chemicals on the efficiency of absorption of the gold complex onto carbon was examined.

The equilibrium absorption isotherms plotted in Fig. 2 show the effect of various leaching chemicals on the loading of the gold-thiourea complex onto carbon. It can be seen that the presence of excess sulphuric acid results in a slight improvement in the absorption characteristics of the gold-thiourea complex, whereas the presence of either thiourea or ferric ions in excess concentration results in a significant suppression of the gold loading. The most important result, however, is the 90 per cent reduction in gold loading that occurs when excess amounts of both thiourea and ferric ions are present in the same leach liquor, and this synergistic effect would be expected to have a major influence on the efficiency of gold extraction in this system. Besides being analysed for gold, the loaded carbons were also analysed for iron and sulphur. These results are included in Fig. 2, and indicate that the poor gold absorption is associated with abnormally high levels of sulphur on the carbon. Moreover, the concentration of sulphur on the carbon, under these conditions, is about 100 times higher than the concentrations of gold or iron, so sulphur cannot be loading onto the carbon as a metal-thiourea complex. It could be loading simply as the thiourea ligand, but what is more likely is that the thiourea undergoes oxidation to formamidine disulphide, which would be loaded strongly onto carbon, or to elemental sulphur, which would be physically deposited in the carbon macropores, thereby impeding the diffusion of the gold complex. This would explain the poor gold-absorption characteristics and would also explain the synergistic role of the ferric ion in this reaction, since the carbon would be expected to catalyse the oxidation of thiourea by ferric ions.

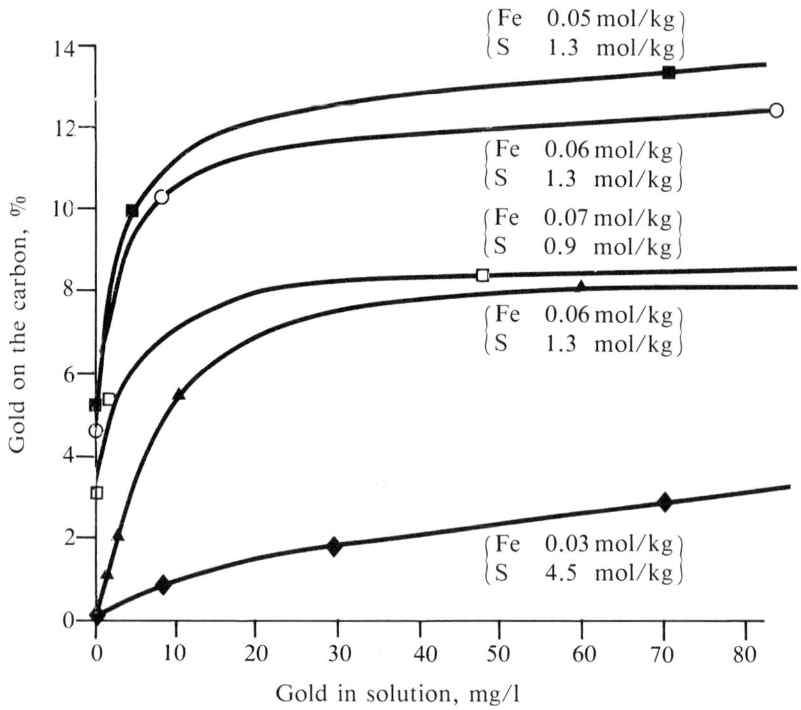

Fig. 2. The equilibrium loading of gold from thiourea leach liquor onto carbon in the presence of typical leaching chemicals
(○) Control (■) 20 g/l H_2SO_4 (□) 50 g/l ferric sulphate (▲) 15 g/l thiourea (◆) 20 g/l H_2SO_4 + 15 g/l thiourea + 50 g/l ferric sulphate

<u>The effect of anions in solution on extraction efficiency</u>. Cations in solution, particularly divalent ions such as calcium and magnesium, enhance the absorption of gold from cyanide solution onto activated carbon, and this has been attributed to the formation of ion pairs $M^{n+}[Au(CN)_2^-]_n$ ($M = H^+$, Na^+, K^+, Ca^{2+}, Mg^{2+} etc.) that are

absorbed more strongly than the constituent ions (Gross and Scott, 1927, McDougall and Fleming, 1987). It is possible that a similar mechanism is operative in the extraction of the cationic gold-thiourea complex by carbon, in which case the loading of this complex should be enhanced by the presence of anions in solution, particularly by large, highly polarizable anions that are able to form strong ion pairs.

The effects of anions in solution on the loading of the gold-thiourea complex are shown by the absorption isotherms in Fig. 3. It is difficult to discern any systematic trends in these results, although it is perhaps significant that the highly polarizable perchlorate anion enhances the loading of the gold-thiourea complex fairly substantially. An anion such as bisulphate also enhances gold loading slightly (as seen earlier, Fig. 2), as does the nitrate anion, but chloride and sulphate have very little effect. The cyanide and sulphide anions suppress gold loading, but it seems likely that this is associated with a change in the gold chemistry in solution rather than with specific reactions at the carbon sur-

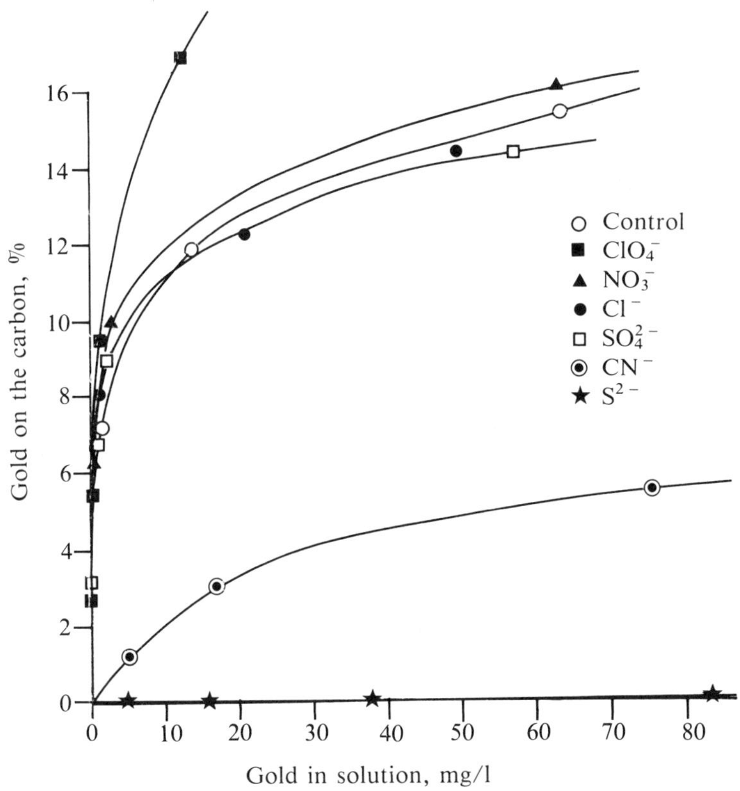

Fig. 3. The effect of different anions in solution on the equilibrium absorption of the gold-thiourea complex on carbon. Anion concentration = 0,5 M (all the anions are present as sodium salts)

face (particularly since the gold-thiourea complex is unstable at the high pH values of 0.5M sodium sulphide or cyanide solutions). The very poor loading of gold in the presence of high concentrations of these anions, particularly sulphide, is a feature of the mechanism that is utilized in the achievement of effective elution, as shown below.

The effect of cations in solution on extraction efficiency. Absorption isotherms for the loading of the gold-thiourea complex onto activated carbon in the presence of various alkaline and alkaline earth cations are shown in Fig. 4. The different cations

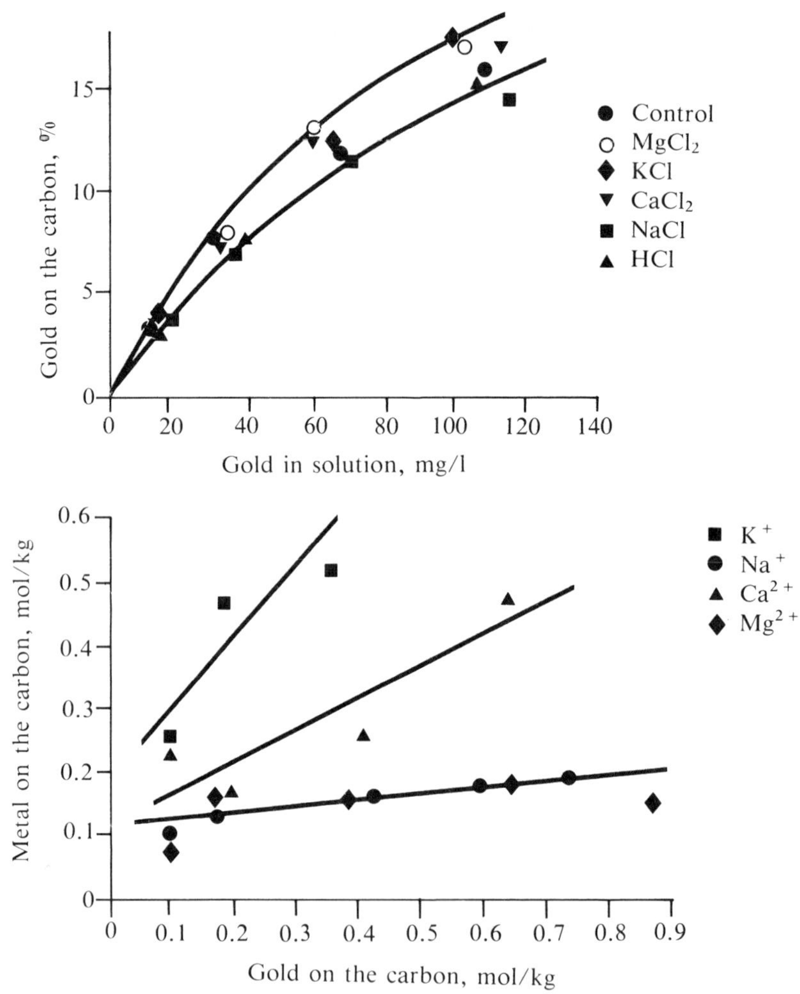

Fig. 4. The equilibrium loading of the gold-thiourea complex and various metal cations on carbon as a function of the presence of the metal cations in the leach liquor $[M]_S = 0.5\,M$

were introduced to the leach liquor as the chloride salts, since the chloride anion was shown previously (Fig. 3) to have little effect on gold loading. Also shown in Fig. 4 are results for absorption of the cations themselves onto carbon, and it is interesting to note that, while these cations have very little effect on the equilibrium gold loading, as was expected, they are themselves, particularly potassium and calcium, loaded onto activated carbon to fairly high concentrations. The mechanism by which these cations are loaded onto carbon was not established, but experiments were carried out which indicated that they are loaded independently of the gold complex.

The effect of competing metal complexes on extraction efficiency. To determine the effect of competing base-metal ions that could be co-leached with gold in a practical situation, synthetic thiourea leach liquors were spiked with cobalt, copper, nickel, zinc, and silver. The effect was examined of each individual metal ion, as well as a composite of all five, on the loading of the gold-thiourea complex onto carbon. The concentration of each metal ion in the synthetic leach liquors (including gold) was between 200 and 220 mg/ℓ. The gold absorption isotherms shown in Fig. 5 indicate that cobalt, zinc, and nickel suppress gold loading slightly, while copper and silver suppress it fairly substantially. A composite of

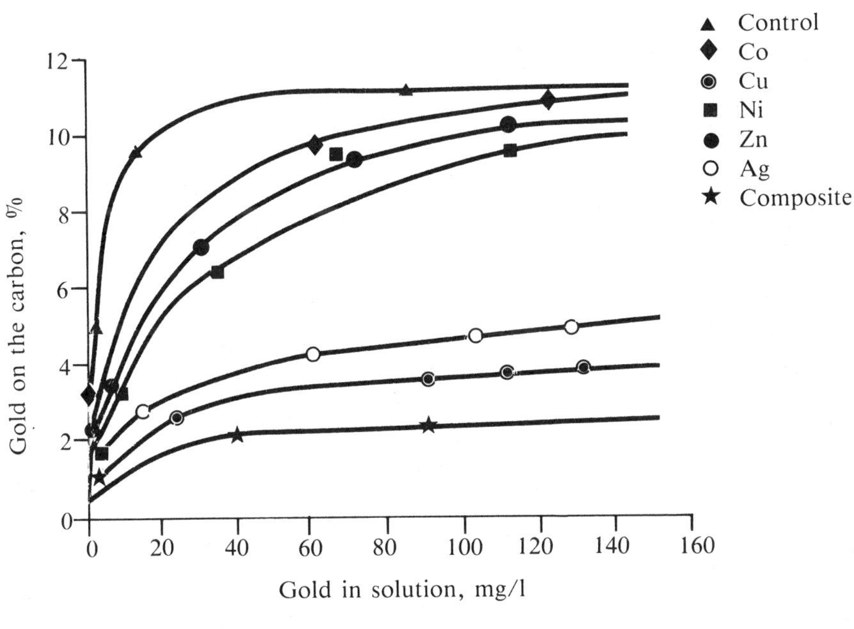

Fig. 5. The effect of various metal ions on the equilibrium absorption of the gold-thiourea complex on activated carbon. $[Au]_S = [Ag]_S = [Cu]_S = [Co]_S = [Ni]_S = [Zn]_S = 200-230$ mg/l. $[NH_2CSNH_2] = 5000$ mg/l

all five metals reduces the gold-loading capacity of the carbon by as much as 80 per cent. In the experiment with the composite solution, the samples of loaded carbon were also analysed for each of the other metals. These results are summarized in Table 3, and indicate that silver, copper, and iron are loaded onto carbon from thiourea solution, whereas zinc, nickel, and cobalt are not. Silver, in fact, is apparently extracted in preference to gold, which could have important implications in the treatment of silver-rich gold ores.

TABLE 3. Equilibrium Loading of Various Metals onto Activated Carbon from a Synthetic Thiourea Leach Liquor
Conditions: Concentration of metal in solution before extraction = 200 to 220 mg/ℓ, thiourea in solution before extraction = 5000 mg/ℓ

Mass of carbon (g)	Volume of solution (mℓ)	Concentration on the carbon (g/t)						
		Au	Ag	Cu	Fe	Zn	Ni	Co
0.05	100	26000	55000	16000	7000	<1000	<1000	<1000
0.10	100	33000	60000	19000	3500	<1000	<1000	<1000
0.20	100	22000	53000	16000	2300	<1000	<1000	<1000
0.50	100	19000	33000	15000	500	<1000	<1000	<1000
1.00	100	14000	19000	13000	300	<1000	<1000	<1000

The effect of temperature on extraction efficiency. The rate of loading and the equilibrium loading of the gold-thiourea complex onto carbon at various temperatures are shown in Table 4. The rate data refer to batch reactions in which 0.5g of carbon was mechanically agitated in 300 mℓ of solution containing, initially, 320 mg/ℓ of gold. The solution was sampled every 10 minutes for the first hour, and the rate data refer to the first-order constant for the rate of extraction over that period. The results indicate that the rate of loading increases with temperature, and an Arrhenius plot yields a value of 3.3 kcal/mol for the enthalpy of reaction, a value that is consistent with a diffusion-controlled process. In contrast, the equilibrium loading of gold decreases with increasing temperature. The trend is the same as that found for the absorption of the aurocyanide complex onto carbon (McDougall and Fleming, 1987) but, in that instance, the effect of temperature is much greater (i.e. an increase in the solution temperature from 25 to 95°C results in a reduction of 90 to 95 per cent in the loading capacity for aurocyanide).

The reversibility of absorption. An important aspect of the absorption mechanism, particularly in continuous countercurrent applications, is the reversibility of loading. If the reaction is reversible, operational factors such as back-mixing of the carbon in the absorption stages or the recycling of poorly eluted carbon will

TABLE 4. The Effect of Temperature on the Kinetics and Thermodynamics of Absorption of the Gold-Thiourea Complex on Activated Carbon

Temp. (°C)	k (h^{-1})	Loading (g/t)
24	1.22	10 100
40	1.85	9 800
60	2.09	10 000
80	3.00	8 050
97	4.10	6 900

have a significant influence on the overall extraction efficiency. To determine whether the absorption of the gold-thiourea complex onto carbon is reversible, the following experiment was carried out. Large particles (larger than 2 mm) of activated carbon were loaded with the gold complex to a gold concentration of 35 000 g/t. The large particles were then contacted in a rolling bottle, in aqueous solution, with small particles (smaller than 1 mm) of activated carbon that had not been pre-loaded with gold. After a period, the large and small particles were separated by screening, and the gold concentration on each fraction was determined. In one experiment, the large and small particles were separated by a glass frit while aqueous solution was recirculated continuously between the two compartments. The results are presented in Table 5, and indicate that the absorption reaction is reversible, although the rate of transfer from the large to the small particles is somewhat slower than in the cyanide system (Fleming and Nicol, 1984). This is a surprising result in view of the fact that the gold-cyanide complex is loaded more strongly onto carbon from typical leach liquors than

TABLE 5. The Reversibility of Absorption of the Gold-Thiourea Complex onto Carbon as Determined by the Rate of Transfer of Gold from Large to Small Particles of Carbon

Method of contact of particles	Aqueous medium#	Contact time	Gold in solution	Gold on carbon (g/t)		Transfer efficiency
		(h)	(mg/ℓ)	>2mm	<1mm	(%)
Mixed	water	24	0.09	33 000	1 700	5
Mixed	water	168	0.00	29 300	7 600	21
Mixed	Fe^{3+}/Th	24	1.39	26 000	8 900	26
Separate	Fe^{3+}/Th	24	1.89	35 400	500	2

10 g/ℓ ferric sulphate, 10 g/ℓ thiourea (Th).

the thiourea complex (see below). The results also indicate that the reverse reaction is enhanced by the presence of leaching chemicals (Fe^{3+}, thiourea) in solution, as might have been expected from the trends displayed in Fig. 2, and that a rapid transfer of gold from one particle of carbon to another is dependent on intimate contact between the particles. This effect, known as contact ion-exchange, has been observed with ion-exchange resins (Bunzyl and Schultz, 1982).

<u>Comparison with absorption of gold onto carbon from cyanide solution</u>. In a comparison of the absorption characteristics of the thiourea and cyanide complexes of gold onto carbon under practical leaching conditions, ore from a South African gold mine was first leached under optimum conditions with each of the ligands. At the end of the leaching period, activated carbon was added to the pulp, and the rate of uptake of gold was monitored in batch reactions, under identical hydrodynamic conditions, in the two reactors. The results, which are presented in Table 6, indicate that the cyanide system is more efficient than the thiourea system, in both the leaching and carbon-extraction areas. The poor extraction of the thiourea complex is due to the high concentration of thiourea in the leach liquor.

TABLE 6. <u>Relative Rates of Absorption of the Thiourea and Cyanide Complexes of Gold onto Activated Carbon from Leached Pulp from the Western Areas Gold Mine</u>
Conditions: Mass of ore = 500g, volume of leach liquor = 500 mℓ, mass of carbon = 1g; thiourea leach: 10 g/ℓ $CS(NH_2)_2$, pH 1,5, potential = 420 mV (NHE), t = 24h; cyanide leach: 1 g/ℓ NaCN, pH 10,5 (lime) t = 24h

Time	Gold concentration (mg/ℓ)					
	Thiourea leach			Cyanide leach		
(h)	Ore	Solution	Carbon	Ore	Solution	Carbon
0	0.74	3.46	0	0.25	3.39	0
1		2.85			2.17	
2		2.48			1.39	
3		2.23			0.93	
5		1.69			0.40	
7	0.64	1.32	1145	0.22	0.25	1560

An analysis of the rate data in Table 6 indicates that the kinetics are first-order over the 7-hour period, with rate constants of 0.15 and 0.42 h^{-1} for the thiourea and cyanide systems respectively. The economic implication of the slow kinetics in the thiourea system is that, for a given extraction performance, a higher carbon concentrations will be required than in the cyanide system, and this means that more gold and carbon will be locked up in the plant as a fixed capital cost to the process.

Elution

Since no work on the elution of the gold-thiourea complex had been published previously, a number of quick screening tests were initially carried out to show whether it would be possible to recover the gold with inexpensive bulk chemicals. Solutions of sulphuric acid, hydrochloric acid, sodium hydroxide, ammonium thiocyanate, sodium cyanide, and sodium sulphide, plus various mixtures of these, were tested with carbon that was loaded with the gold-thiourea complex. It was found that only sodium cyanide and sodium sulphide were effective, and that the rate of elution with cyanide was enhanced significantly by the addition of ethanol, or other polar organic solvents, to the eluate.

Elution with sodium cyanide. A series of batch experiments was carried out in an attempt to determine the optimum cyanide concentration for elution, and to establish whether the presence of sodium hydroxide has any beneficial effect on elution. In these tests 100 mℓ of eluate solution was used to treat 1 mℓ of carbon; this volume ratio is far higher than would be used in practice, but the conditions were designed to allow the elution reaction to go almost to completion. The results presented in Table 7 all refer to elution at the boiling temperature of the eluate; experiments were also carried out at room temperature, but the elution efficiency after 6 hours was always less than 10 per cent.

TABLE 7. The Efficiency of Elution of the Gold-Thiourea Complex from Carbon with an Aqueous Solution of Sodium Cyanide and Sodium Hydroxide.
Conditions: Temperature = 90 to 95°C, time = 6h

Eluate composition		Gold in solution	Gold on the carbon	Elution efficiency
NaOH (M)	NaCN (M)	(mg/ℓ)	(g/t)	(%)
Loaded carbon			18 200	
0.0	0.5	92	2 700	85
0.1	0.5	79	2 900	84
0.5	0.5	91	3 600	80
2.0	0.5	85	4 000	78
0.2	0.1	95	5 700	69
0.2	0.2	93	3 800	79
0.2	0.5	118	1 900	90
0.2	2.0	118	2 100	89

The results indicate that sodium hydroxide has no beneficial effect at concentrations above about 0.2M, and that the optimum sodium cyanide concentration is 0.5M. The elution is fairly fast, 80 to 90 per cent of the gold being recovered in 6 hours under atmospheric conditions.

Elution with sodium cyanide in mixed water-ethanol solvent. Batch elution tests were carried out under the same conditions with eluate solutions containing 0.5M sodium cyanide in solvent mixtures of water and ethanol in various proportions. In the mixed solvents, reasonable elution efficiencies were obtained even at room temperature, so these results are included in Table 8. A solution containing equal quantities of ethanol and water seems to be the optimum, and it should be possible to obtain good elution efficiency, even under ambient conditions, with this method.

TABLE 8. The Efficiency of Elution of the Gold-Thiourea Complex from Carbon with Sodium Cyanide in Ethanol/Water Solvents
Condition: Elution time = 6h

Solvent composition		At 25°C		At boiling point of solvent	
Water (%)	Ethanol (%)	Gold on the carbon (g/t)	Elution effic. (%)	Gold on the carbon (g/t)	Elution effic. (%)
100	0	18900	0	3360	82
87	13	11900	35	180	99
65	35	9000	51	120	99
50	50	3700	80	60	99.6
30	70	5900	68	150	99
20	80	5500	72	270	98

Elution with sodium sulphide. Sodium sulphide is also a very effective eluate for the gold-thiourea complex, and reasonable elution efficiencies are obtained even at ambient temperatures, although the rate of elution is somewhat faster at higher temperatures. The effect of the sodium sulphide concentration on the elution efficiency is shown by the results in Table 9.

Elution with continuous recovery of gold by electrolysis. This is the method adopted in the Zadra technique for the stripping of gold cyanide from activated carbon, in which the hot eluate solution is recirculated continuously between the elution vessel and the electrowinning cell. In the electrowinning cell, gold is continuously extracted from the eluate solution, and the equilibrium of the stripping reaction is shifted in favour of the solution phase. In this way it is possible to effect almost complete elution with relatively small volumes of eluate solution. This technique was tested with each of the three eluates mentioned earlier. The results, which are summarized in Table 10, show that the most efficient elution was obtained with 0.5M sodium cyanide in an ethanol-water mixture at a temperature of 55°C. The elution was so fast, in fact (98 per cent recovery in 1 hour), that the same solvent was also tested at 22°C, but the conductivity of 0.5M sodium cyanide in the mixed solvent was too low for effective electrolysis

TABLE 9. The Effect of Sodium Sulphide Concentration on the Efficiency of Elution of the Gold-Thiourea Complex from Carbon
Conditions: Carbon volume = 1 mℓ, eluate volume = 100 mℓ, elution time = 6h

Sodium sulphide (M)	At 25°C Gold on the carbon (g/t)	At 25°C Elution efficiency, (%)	At 94 to 100°C Gold on the carbon (g/t)	At 94 to 100°C Elution efficiency (%)
0.1	17 600	3	12 800	30
0.2	13 500	26	9 700	47
0.5	9 700	47	3 400	81
1.0	4 300	76	1 700	91
2.0	3 200	86	1 200	94

at the lower temperature.

Electroelution with sodium sulphide was also extremely fast under ambient conditions, although the residual gold in solution after 24 hours was fairly high (15 mg/ℓ). As a result, it was not possible to reduce the gold concentration on the carbon to less than 200 g/t. This problem has been observed previously (Zadra, 1950), and is attributed to oxidation of the sulphide to polysulphide at the anode of the electrowinning cell; the polysulphide, which is a complexing ligand for gold, leaches the gold deposited at the cathode at the same rate at which the gold is deposited, thus limiting the

TABLE 10. The Efficiency of Stripping of the Gold-Thiourea Complex from Activated Carbon with Various Eluates by the Electro-elution Method.

Time	Concentrations of gold in solution and on the carbon (g/t) for the different eluates							
h	0.5M NaCN, 70% ethanol, 22°C		0.5M NaCN, 70% ethanol, 55°C		2.0M Na_2S, 100% water, 22°C		2.0M NaOH, 0.5M NaCN, 55°C	
	Soln.	Carb.	Soln.	Carb.	Soln.	Carb.	Soln.	Carb
0	0	6400	0	10100	0	7990	0	5950
0.5	60	3730	142	622	70	1320	7	5550
1	80	2600	84	203	70	740	4	5510
2	210	1250	35	71	31	630	3	5040
4	230	660	6	<50	19	490	3	4150
6	240	490	4	<50	18	360	2	3840
24	300	400	4	<50	15	230	2	3540

electrowinning efficiency. Zadra (1950) reported that the problem can be minimized by adding sulphite to the eluate to reduce the polysulphide back to sulphide. Another problem with sulphide elution is that silver and base metals such as copper, nickel, and iron are poorly eluted because of the low solubility of their sulphide complexes.

Finally, the rate of elution with aqueous caustic cyanide solution was very slow, but this was because it was not possible to increase the temperature above 55°C in the apparatus that was constructed for these experiments. This method of elution would undoubtedly work well if higher temperatures were used, particularly temperatures above 100°C, in pressurized elution vessels.

The effect of temperature on elution efficiency. Results showing rates of elution with the three eluates at different temperatures are plotted in Fig. 6. In each case the rate of elution increases with temperature, but the sensitivity to temperature is different for the three eluates. In the case of the aqueous caustic cyanide eluate, satisfactory elution can be achieved only at high temperatures (higher than 80°C). In the case of sodium cyanide in the mixed water-ethanol solvent, the extent of elution after 6 hours is almost independent of temperature, and an ambient elution will probably be most cost-effective. Finally, elution with sodium sulphide is strongly temperature-dependent over the whole temperature range, and the temperature should be as high as possible (80 to 90°C) for rapid elution.

DISCUSSION

The results presented here indicate that, under most conditions that would be encountered in a practical situation, the gold-thiourea complex is loaded onto activated carbon without undergoing chemical change, i.e. as the $Au(CS(NH_2)_2)_2^+$ species. Moreover, there is some evidence to suggest that the mechanism of loading is similar to that in the cyanide system, absorption occurring as an ion pair, $[Au(CS(NH_2)_2)_2]^+X^-$ (where X is ClO_4^-, Cl^-, or NO_3^-). The rate of loading is slow, with equilibration times of at least several months. This is due to the extremely slow diffusion of the gold-thiourea complex into the carbon micropores, which comprise a large percentage of the total pore volume of typical coconut-shell activated carbons. However, the initial rate of loading into the carbon macropores and mesopores is relatively fast, and this is the loading regime that would be utilized in most practical applications.

In the absence of any competition from other species in solution, the gold-thiourea complex is loaded to very high concentrations onto activated carbon. However, excess thiourea in solution plus an oxidant such as ferric ions, both of which are necessary for effective leaching kinetics, have a deleterious effect on the absorption of the gold-thiourea complex onto carbon, and can reduce the capacity by up to 90 per cent. This is believed to be due to oxidation of the excess thiourea by the ferric ions in solution, producing elemental sulphur (as one of the oxidation products), which is deposited in the pores of the carbon and impedes the diffusion of the gold complex. Other species that reduce the capacity of activated carbon for the gold-thiourea complex are the

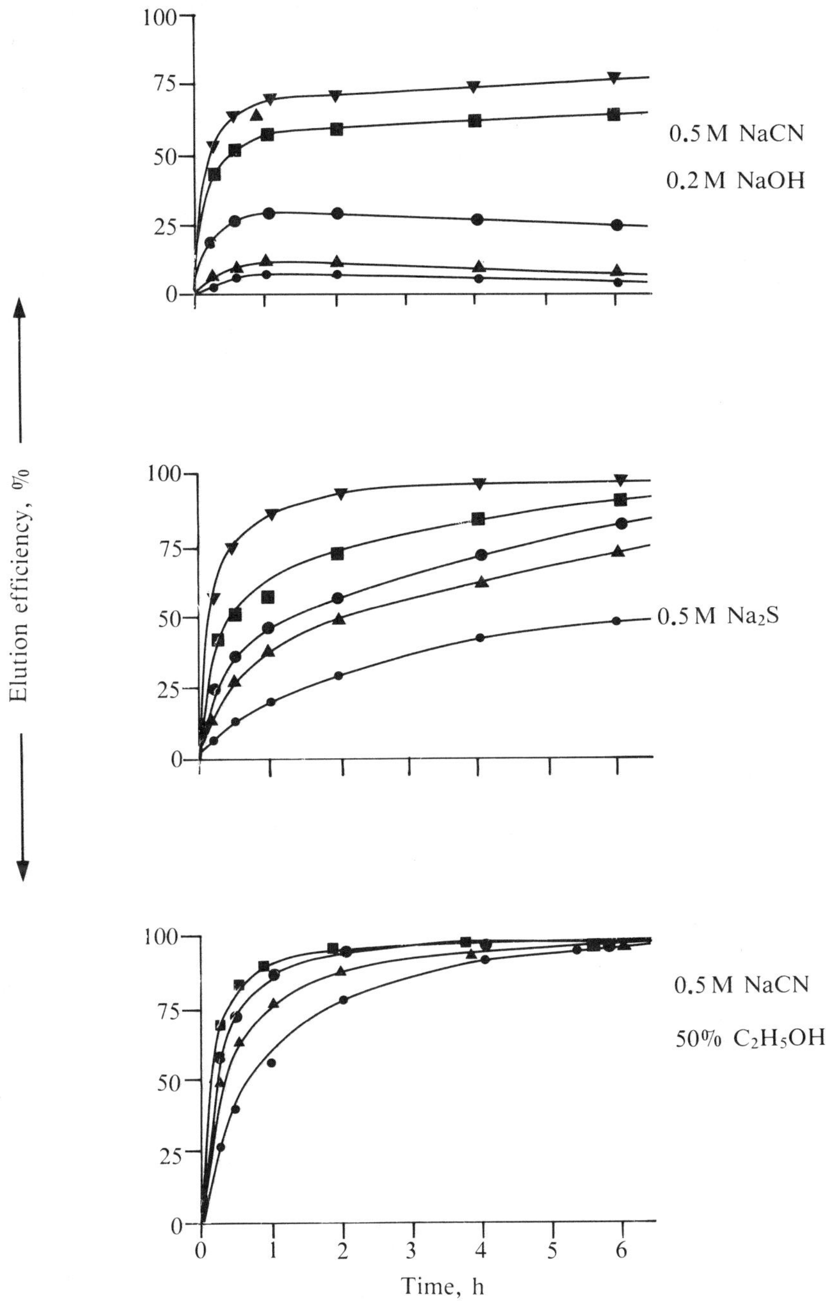

FIGURE 6. The effect of temperature on the rate of elution of the gold-thiourea complex from activated carbon with various eluates. (●) 22 °C (▲) 40 °C (●) 60 °C (■) 80 °C (▼) 94 °C. Carbon mass = 1 g. Eluate volume = 100 ml

silver- and copper-thiourea complexes, both of which are loaded fairly strongly onto carbon in competition with gold.

It is possible to strip the gold-thiourea complex from activated carbon quite readily, and to recover gold from the strip solution by electrolysis. The most effective eluates are sodium cyanide in a mixed water-ethanol solvent and sodium sulphide in water. In both cases, gold-elution efficiencies of over 90 per cent can be achieved in 1 hour under fairly mild conditions (less than 60°C). The elution techniques currently practised on most CIP plants use aqueous caustic cyanide solutions at elevated temperatures and pressures for the elution of the gold cyanide-complex from carbon; these will also work for the gold-thiourea complex.

ACKNOWLEDGEMENTS

The author is indebted to Mrs O.L. Wellington and Mr M.U. Conway for their excellent technical assistance. This paper is published by permission of the Council for Mineral Technology (Mintek).

REFERENCES

Bunzyl, K., and Schultz, W. (1982). Kinetics of differentially small conversions in contact isotopic and contact ion-exchange. Anal. Chem. 54, 272-277.

Dixon, S., Cho, E.J., and Pitt, C.H. (1976). The interaction between gold cyanide, silver cyanide and high surface area charcoal. A.I.Ch.E. Symposium, Chicago.

Fleming, C.A., and Nicol, M.J. (1984). The absorption of gold cyanide onto activated carbon. III Factors influencing the rate of loading and the equilibrium capacity. J.S. Afr. Inst. Min. Metall. 84(4), 85-93.

Fleming, C.A. (1986). A process for the simultaneous recovery of gold and uranium from South African ores. Extractive Metallurgy of Gold, Proceedings of GOLD 100, 2 (C.E. Fivaz and R.P. King eds.), S.Afr. Inst. Min. Metall. Symp. Ser. S8, Johannesburg, 301-321.

Gabra, G. (1984). A kinetic study of the leaching of gold from pyrite concentrates using acidified thiourea. Precious Metals: Mining, Extraction and Processing, (Y. Kudryk, D.A. Corrigan and W.W. Liang eds.), Warrendale, Pa.: AIME, Metallurgical Society, 145-173.

Green, M. (1913/1914). The effect of charcoal in gold bearing cyanide solutions with reference to the precipitation of gold. Trans. Inst. Min. Metall., 23, 65-84.

Groenewald, T. (1976). The dissolution of gold in acidic solutions of thiourea. Hydromet. 1, 277-290.

Gross, J., and Scott, J.W. (1927). Precipitation of gold and silver from cyanide solution on charcoal. United States Bureau of Mines, (Washington D.C.), Tech Pap. 379.

Hisshion, R.J., and Waller, C.G. (1984). Recovering gold with thiourea. Mining Mag. (Lond.), Sept., 237-243.

McDougall, G.J., and Fleming, C.A. (1987). The extraction of precious metals on activated carbon. Critical Reports on Topics on Ion Exchange and Adsorption in Hydrometallurgy, Society of Chemical Industry, London. (In Press).

Schulze, R.G. (1984). New aspects in thiourea leaching of precious metals. J. Metals, 62-65.

Yen, W.T., and Wyslouzil, D.M. (1985). Gold extraction from refractory ore by pressure oxidation and thiourea leach. 17th Canadian Mineral Processors Operators Conf., Ottawa, Ontario.

Zadra, J.B., (1950). A process for the recovery of gold from activated carbon by leaching and electrolysis. United States Bureau of Mines, (Washington, D.C.), Report of Investigations No. 4672.

OPTIONS IN THE NITROX PROCESS®

K. J. Fair, J. C. Schneider, and G. Van Weert

Hydrochem Developments Ltd., Brampton, Ontario L6W 3X3

ABSTRACT

The NITROX PROCESS utilizes nitric acid to render refractory gold and silver ores and concentrates amenable to the recovery of the valuable constituents, usually by cyanidation. The oxidation of the sulphides is accompanied by the reduction of HNO_3 to NOx, generally NO. An integral part of the process is the recovery of the NOx as nitric acid for reuse. Precipitation of the solubilized metals and the subsequent recycle of the calcium nitrate is important in terms of the water balance and nitrate recovery. The discussion will highlight recent findings on the recovery of gold and silver in various NITROX PROCESS options, in particular on the interaction between the nitrate and cyanide segments of the process.

KEYWORDS

Gold and silver; nitric acid; refractory gold; cyanidation; NOx scrubbing; nitrate.

INTRODUCTION

The NITROX PROCESS for gold and silver concentrates is an oxidation pretreatment using nitric acid which converts sulphides into oxides, thereby making the gold and silver values amenable to extraction by a process such as cyanidation.

The following discussion will first present an overview of nitric acid metallurgy. The reactions of arsenopyrite and pyrite in nitric acid media are of the greatest importance to the NITROX PROCESS as applied to gold and silver concentrates and will be reviewed. Solubilizing iron, sulphur and arsenic necessitates removal of these elements in a form that is easily and safely disposed of. The removal of these elements and the subsequent gold and silver recovery result in various options in the flowsheet which will be discussed in detail. Finally, the recovery of the nitric acid will be reviewed, as efficient recovery of the reagent is important economically and environmentally.

NITRIC ACID METALLURGY

A large amount of work has been performed using nitric acid to extract a variety of metals such as copper (Prater, Queneau and Hudson, 1973; Habashi, 1973a; Brennecke and co-workers, 1981), uranium (Schisler, 1986), molybdenum (Vizsolyi and Peters, 1980), nickel (Habashi, 1973b), cobalt (Ritcey and Ashbrook, 1981), and silver (Kunda, 1984; Ackerman and Bucans, 1986). All of these involve solubilizing the element of interest and subsequently separating it from the other elements that were solubilized. Nitric acid treatment of gold bearing ores and concentrates differs in this regard as the gold is not solubilized but remains with the other acid insolubles.

Gold in refractory ores and concentrates is often intimately tied up with pyrite, arsenopyrite and pyrrhotite (Weir, Robinson and King, 1986; Weir and Berezowsky, 1984). The NITROX PROCESS involves oxidizing these sulphide minerals to sulphates and arsenates which are removed from solution by precipitation, generally by using limestone for pH adjustment. Each mineral behaves differently in nitric acid. Prater, Queneau and Hudson (1973) determined that minerals have different reactivities in nitric acid and yield varying amounts of elemental sulphur. It should be noted that much of the previous work attempted to maximize the elemental sulphur yield because it meant lower reagent costs for neutralization, a lower load on the nitric acid recovery system and is easier to dispose of. Bjorling and Kolta (1964) worked on various sulphides and found that lower order sulphides such as pyrrhotite, and sphalerite have high yields of elemental sulphur while higher order sulphides such as pyrite have a low sulphur yield. Our work indicates that arsenopyrite behaves as a lower order sulphide with an elemental sulphur yield of about 70%. Moderate strength nitric acid will react quickly with arsenopyrite and pyrite (Fig. 1). A 750 mV solution corresponds to approximately 12 wt% HNO_3. It can be seen that the arsenopyrite is oxidized very rapidly, partially due to its finer size, but at the conditions chosen of 80°C, 750 mV and 10% solids even the fairly coarse pyrite is completely reacted in 1 hour.

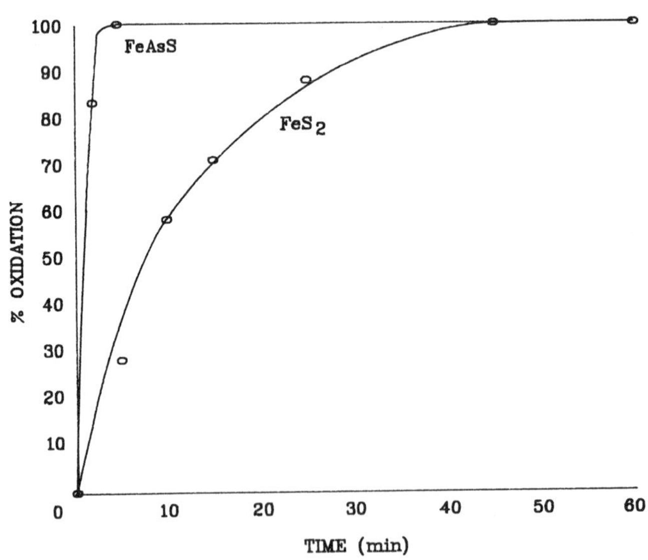

Fig. 1. Rates of pyrite and arsenopyrite oxidation in nitric acid at 80°C, 750 mV (SCE), 10% solids. FeAsS 48% - 75 microns, FeS_2 15% -75 microns.

Under the usual conditions for the NITROX PROCESS the arsenic remains in solution. This allows for various options in gold recovery, as well as giving much greater freedom in removing the arsenic, iron and sulphur from solution without the possibility of forming precipitated species that tie up the precious metal values.

GOLD AND SILVER RECOVERY

Once the sulphides have been oxidized and the major elements such as iron, arsenic and sulphur are in solution (along with minor elements such as copper, zinc, cobalt, nickel and magnesium) there are generally two flowsheet options. All of the elements taken into solution eventually have to be removed, preferably in a form that does not interfere with subsequent processing steps. Option 1 precipitates the major elements in the oxidation slurry.

Option 1

The sulphur exists in the oxidation liquor as sulphate which can be removed in acidic environments by the addition of calcium units:

$$H_2SO_4 + CaCO_3 + H_2O \rightarrow CaSO_4.2H_2O + CO_2\uparrow \quad (1)$$

$$H_2SO_4 + Ca(NO_3)_2 + 2H_2O \rightarrow CaSO_4.2H_2O + 2HNO_3 \quad (2)$$

$$H_2SO_4 + Ca(NO_2)_2 + H_2O \rightarrow CaSO_4.2H_2O + NO\uparrow + NO_2\uparrow \quad (3)$$

Vizsolyi and Peters (1980) describe the selective precipitation of gypsum from nitric acid liquors containing other elements in solution such as Mo, Fe and Cu. The gypsum precipitated from NITROX solutions filters well, even when the liquor is 1 to 2 molar HNO_3. Consistent rates of 400 $kg.h^{-1}.m^{-2}$ have been achieved with the cake thickness about 15mm. The gypsum precipitate is free of arsenic.

Removal of iron and arsenic starts at about pH 1 (Ritcey and Ashbrook, 1981). Most of the present iron and arsenic precipitation work was conducted at pH 3 to pH 4, usually at 80°C. Within this pH range $CaCO_3$ is efficiently and quickly utilized and the precipitation products are iron arsenates, iron hydroxides, and gypsum.

Previous workers (Krause and Ettel, 1985; Ritcey and Ashbrook, 1981) have shown the importance of the iron to arsenic ratio for effective removal of arsenic from solution. Krause and Ettel (1985) concluded that an Fe/As ratio of 4 or greater results in extremely insoluble basic ferric arsenates. Current work has operated with ratios ranging from roughly 4:1 to greater than 25:1 and has found that essentially all of the iron and arsenic is precipitated at pH 4. The precipitate filters extremely well - rates of 1800 $kg.h^{-1}.m^{-2}$ with a 13mm cake thickness are routinely achieved.

Trace elements such as copper, zinc, nickel and cobalt have been found to also precipitate at pH 4. Table 1 shows how some of the trace elements deport to the precipitate.

Gold recoveries from precipitates of oxidation slurry made at pH 4 have been very high, as seen in Table 2. This option involves precipitating the dissolved materials in the presence of the gold, having a single train of filtration and cyaniding all the solids. Fig. 2 shows the flowsheet for this option, which has been previously described (Van Weert, Fair, and Schneider, 1986).

TABLE 1 Minor Metals Removal by Precipitation of Oxidation Solution with $CaCO_3$, at pH 3 to pH 4, 80°C

Test #	% Removed From Solution Cu	Zn	Co	Ni
9	94	38	ND	ND
57E	71	ND	<.1	4
Ritcey & Ashbrook (1981) 60°C	ND	ND	3	3

ND - Not Determined

TABLE 2 Flowsheet 1 Gold Recoveries

Material Source	Feed Grade g/t	Gold Recoveries Without Oxidation	With Oxidation
N.W. Ontario	34	5	99
Utah	12	31	97
New Guinea	18	19	98

Very little if any silver jarosite is formed, which results in high silver recoveries when the precipitated material is cyanided. This is in contrast to pressure oxidation which operates such that iron precipitation is performed in the autoclave. This leads to silver being bound as an argentojarosite which must be broken down by a further treatment. The problem of silver precipitation in pressure oxidation is discussed concerning treating Porgera concentrates where, without a treatment the silver recovery is less than 10% (Weir, Robinson and King, 1986). The process consumes lime as $Ca(OH)_2$ according to:

$$2KFe_3(SO_4)_2(OH)_6 + 3Ca(OH)_2 + 6H_2O \rightarrow 6Fe(OH)_3 + K_2SO_4 + 3CaSO_4.2H_2O \quad (4)$$

$$Fe(OH)SO_4 + Ca(OH)_2 + 2H_2O \rightarrow Fe(OH)_3 + CaSO_4.2H_2O \quad (5)$$

A hot lime step before cyanidation of NITROX residues has been found to be beneficial at times for improving gold extraction (Van Weert, Fair and Schneider, 1987) but has had inconclusive results on silver recovery (Table 3).

TABLE 3 Flowsheet 1 Silver Recoveries

Material	Without Hot Lime	With Hot Lime
A	92	90
B	91.1	96.7

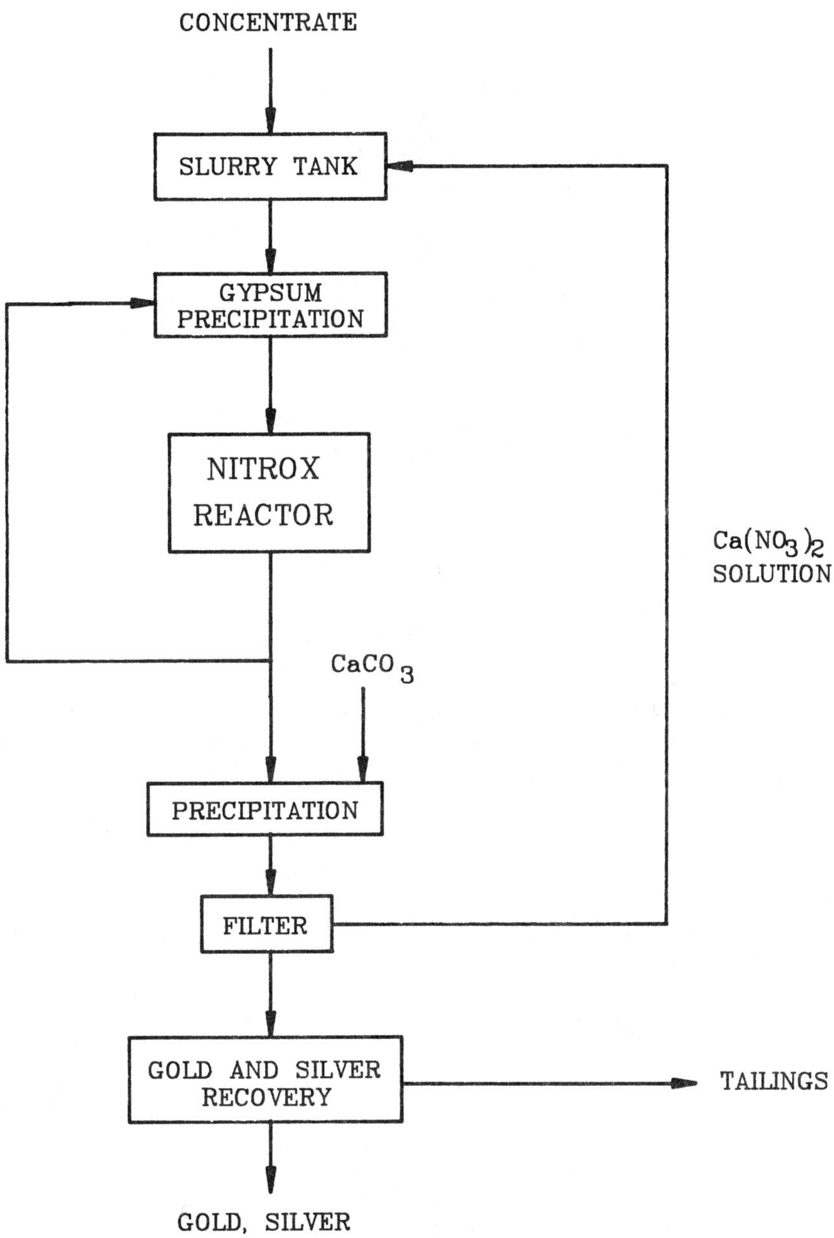

Fig. 2. Flowsheet 1: Precipitation at pH 4 with gold and silver recovery treatment of the combined residue and precipitate.

Precipitation of the iron arsenate and gypsum also brings down varying percentages of elements such as copper, zinc and nickel. Copper especially is troublesome in cyanide circuits as it forms cyanides, resulting in higher cyanide consumptions. Table 4 shows the relatively high extraction of copper in cyanidation.

TABLE 4 Copper Extraction in Cyanidation of Precipitated (pH 4) Cake

Test #	(% Cu) Head Grade	% Extracted
8	0.069	51
9	0.41	45
42	0.026	25
B4	0.36	89

Option 2

Feed materials that are high in elements such as copper are best processed without such elements entering the cyanide circuit. This is achieved by subjecting the oxidation slurry to an acid filtration step, shown in Fig. 3. Option 2 differs from option 1 in having filtration after the oxidation stage and a smaller gold recovery circuit since fewer solids proceed to cyanidation.

The concentrate is slurried with return $Ca(NO_3)_2$ solution and proceeds to the gypsum precipitation. The amount of gypsum produced at this stage is a function of the calcium units returning from iron precipitation. From gypsum precipitation the slurry proceeds to the oxidation reactor where the iron, arsenic, sulphur and copper are solubilized, leaving gangue and gold in the residue. The slurry is filtered after oxidation, with part of the filtrate returned to the gypsum precipitation stage to regenerate nitric acid. The remainder proceeds to precipitation which is operated to optimize reagent addition and produce the most desirable form of precipitate. Should there be enough value in the dissolved metals (such as copper, cobalt or nickel) to justify recovery, then the flowsheet is easily modified to incorporate metal recovery.

The residue from the filtration consists of the gold, gangue, gypsum and any elemental sulphur formed. Table 5 shows that the arsenic concentration is greatly reduced.

TABLE 5 Oxidation Residues Produced From Flowsheet 2 Configuration. Feed: Flotation Concentrate Containing 10.8% As, 33.6 g/t Au.

Test	Mass of oxidation residue (% of feed)	Residue Assay % As	Residue Assay Au (g/t)	Gold Recovery in Cyanidation (%)
42	77	0.08	43.2	96.0
48	21	0.19	159.8	97.0
58	19	0.08	170.8	91.8

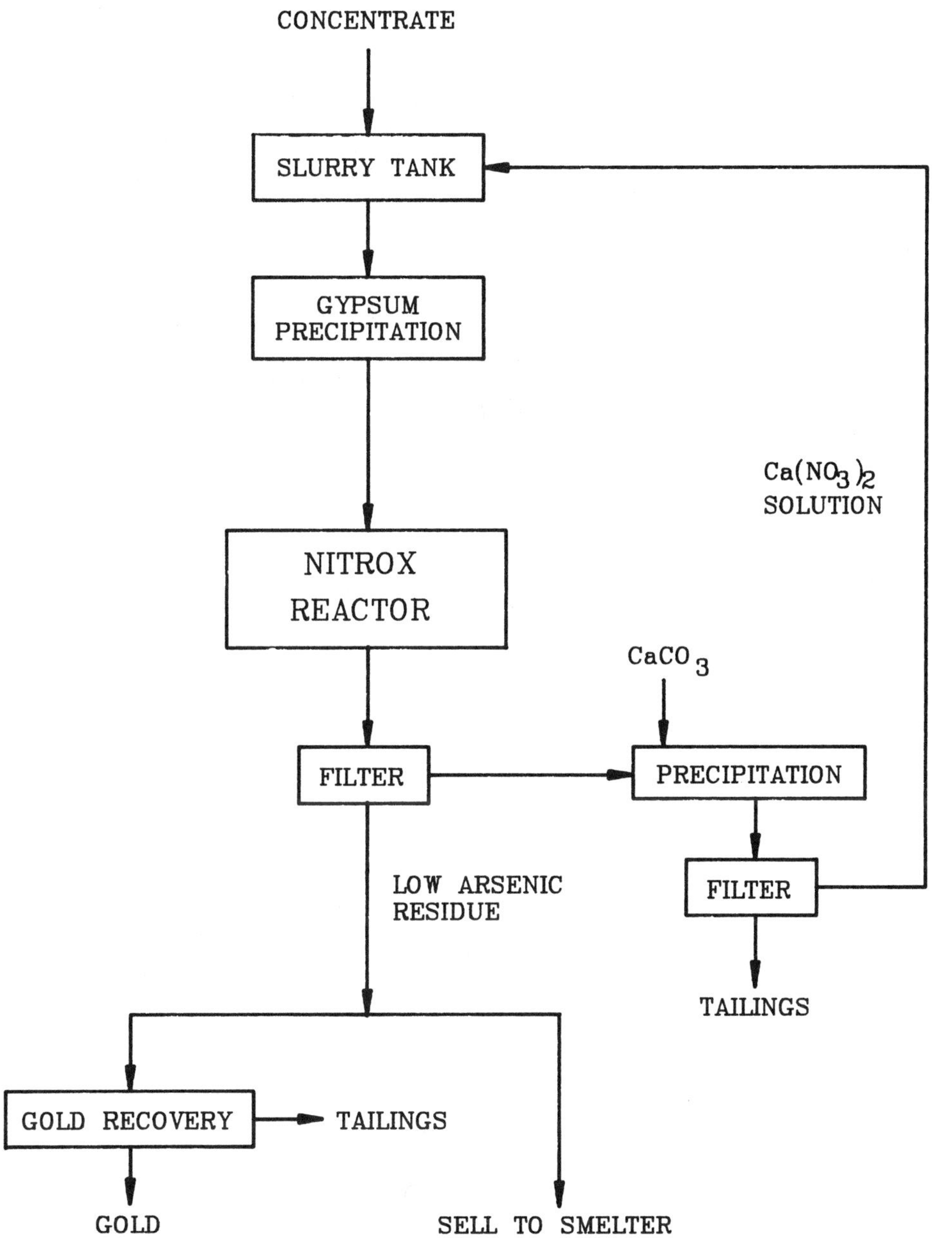

Fig. 3 Flowsheet 2: filtration after oxidation.

With the loss in mass comes a corresponding upgrading in the gold concentration. Table 5 illustrates the option of producing a low arsenic residue that could be either treated in a cyanide circuit or sent to a smelter. Conditions in oxidation for test #58 were such that a large amount of elemental sulphur was produced. This resulted in the filtered residue containing 24.8% S°. The low arsenic, high gold residue could be smelter feed which would yield greater than 98% of the gold.

The presence of significant cyanide consumers in a feed material leads to the use of flowsheet 2 - filtration after oxidation. This results in a low arsenic, higher gold content residue which may either be treated in a gold recovery circuit such as cyanidation or shipped to a smelter. For shippers the advantages to the latter option are the reduced mass and the removal of arsenic which increases the number of smelters that will accept the materal.

NITRIC ACID RECOVERY

To oxidize arsenopyrite and pyrite requires oxygen and in the NITROX PROCESS this is supplied by nitric acid. The oxygen demand of the sulphide minerals of interest is given in Table 6.

TABLE 6 Oxygen Demand of Sulphide Oxidation

Mineral	% S Oxidation	t O_2/t mineral	t HNO_3/t mineral	t air/t mineral
FeS_2	100	1.00	2.63	4.35
FeAsS	100	0.69	1.81	3.00
FeS	100	0.82	2.16	3.57
FeS	50	0.55	1.45	2.39

The amount of nitric acid required to supply this oxygen is given, and, as the nitric acid is regenerated using air, the air requirement is given. The reduction of the nitric acid to supply oxygen is:

$$4HNO_3 \rightarrow 2H_2O + 4NO + 3O_2 \quad (6)$$

The reduced species is NO with the nitrogen having a valence change from +5 to +2; for every mole of NO formed 3 electrons are transferred. Reactions such as the oxidation of pyrite in nitric acid proceed equally well in the presence or absence of oxygen or air:

$$2FeS_2 + 10\ HNO_3 \rightarrow Fe_2(SO_4)_3 + H_2SO_4 + 10\ NO + 4H_2O \quad (7)$$

The kinetics thus do <u>not</u> involve the diffusion of gaseous oxygen across the gas liquid interface and through the liquid to the mineral surface. The nitric oxide formed from the breakdown of the sulphides must be recovered for both environmental and economic reasons; the same is true of nitrate remaining in solution after oxidation.

Gas Phase

The regeneration of nitric acid involves the oxidation of NO to NO_2 and absorption of the NO_2. The absorption of NO_2 causes 33% of the NO_2 to form NO instead of HNO_3. This NO then has to be oxidized and absorbed and so on. The overall reaction of 100% NO recovery is Eq. (10).

$$6NO + 3O_2 \rightarrow 6NO_2 \qquad (8)$$
$$6NO_2 + 2H_2O \rightarrow 4HNO_3 + 2NO \qquad (9)$$
$$4NO + 3O_2 + 2H_2O \rightarrow 4HNO_3 \qquad (10)$$

The formation of NO upon absorption of NO_2 means that several stages are required for recovery of the HNO_3. Commercial nitric acid plants have over 20 stages to recover this nitric acid. Due to the slower kinetics of oxidizing the NO to NO_2 as both the NO and O_2 become more dilute, many more stages are required to recover the last portion of nitric acid than the first: the first 7 stages of an absorption column will recover 90% of the nitric acid while the remaining 13 stages will be required to recover the next 8% (Chilton, 1960). To recover the final 2% would involve a tremendous number of stages. Commercial plants do not recover all the NO - the balance is either vented, or catalytically reduced to N_2 and O_2 to meet environmental regulations.

An alternative to extended absorption is to capture the final NO by scrubbing. Absorption in an alkaline media has a different chemistry than that of nitric acid production. Only 50% of the NO has to be oxidized to NO_2 and this species, N_2O_3, is absorbed as nitrite. The following equations show the absorption of N O in three alkaline media.

$$4NO + O_2 \rightarrow 2N_2O_3 \qquad (11)$$
$$2N_2O_3 + 4NaOH \rightarrow 4Na(NO_2) + 2H_2O \qquad (12)$$
$$2N_2O_3 + 2Ca(OH)_2 \rightarrow 2Ca(NO_2)_2 + 2H_2O \qquad (13)$$
$$2N_2O_3 + 2CaCO_3 \rightarrow 2Ca(NO_2)_2 + 2CO_2\uparrow \qquad (14)$$

It has been found that the best scrubbing occurred with NaOH and that $Ca(OH)_2$ is a more efficient scrubbing media than $CaCO_3$, as shown in Table 7 which compares the two media at different initial NOx concentrations and temperatures.

TABLE 7 NOx Removal by 2 Stage Scrubbing Comparing $Ca(OH)_2$ vs $CaCO_3$ and Effect of Temperature and Initial NOx

Media	Temp. (°C)	Residence Time (s)	NOx initial (ppm)	NOx final (ppm)	Recovery (%)
$Ca(OH)_2$	25	57	46,000	1,000	98
$Ca(OH)_2$	75	50	46,000	9,000	80
$Ca(OH)_2$	25	50	159,000	1,000	99
$Ca(OH)_2$	75	50	159,000	15,000	91
$CaCO_3$	25	57	46,000	6,500	86
$CaCO_3$	75	57	46,000	12,000	74
$CaCO_3$	25	50	159,000	8,000	95
$CaCO_3$	75	50	159,000	15,000	91

Even though using $CaCO_3$ as a scrubbing media is not as efficient as NaOH or $Ca(OH)_2$, it works remarkably well as seen from Table 7. There is a large cost advantage to using limestone instead of lime, especially when limestone units must be added to the system for precipitation. Consistent with process requirements, the scrubbing system can be divided into two: the initial scrubbing by $CaCO_3$, to remove the bulk of the NOx proceeding to scrubbing, and a final stage of $Ca(OH)_2$ or NaOH scrubbing to reduce the NOx in the tail gas to environmentally acceptable levels. This scrubbing takes place counter-currently in low pressure drop agitated reactors, similar to the oxidation reactor, which allows for a slurry of limestone or lime to be used.

Reducing final NO levels to below 1000 ppm means almost total recovery of the NO formed from oxidizing the sulphides. Table 8 shows the relationship between the tail gas NO and the overall NO recovery for a concentrate feed containing 50% pyrite.

TABLE 8 Relationship Between Final NOx Emissions and Overall NO Recovery for a Concentrate Feed of 50% FeS_2 with 100% Oxidation and Stoichiometric Air

Tail gas NO Concentration (ppm)	% NO Recovery
100	99.95
200	99.91
500	99.77
1000	99.53
2000	99.06
5000	97.65

Although the NITROX PROCESS will recover almost all the NO formed and recycles it back into the system as HNO_3 and $Ca(NO_3)_2$ or $Ca(NO_2)_2$, NO is not the only reduced species of HNO_3. Prater, Queneau and Hudson (1973) found that N_2O is also formed: 1% N_2O was evolved at 95°C and up to 3% N_2O at 75°C. It is anticipated that these numbers are similar but lower for the NITROX PROCESS. The gas side recovery of the reduction of HNO_3 is thus expected to be about 99%, with 90% or greater occuring as absorption and making HNO_3 and the final 10% or less from scrubbing.

The advantages of making elemental sulphur as an oxidation product is that the system is considerably reduced in size. It was shown earlier (Table 6) that if pyrrhotite has 50% of its sulphur report as elemental sulphur the HNO_3 demand, and thus the NO production and subsequent recovery, is reduced by 33%.

Liquid Phase

The recovery of the nitrate in solution is a function of the nitrate level in the slurry after oxidation, and the washing efficiency which in turn depends on the products to be washed and the amount of wash water that can enter the circuit.

Earlier workers (Prater, Queneau and Hudson, 1973) found that increasing amounts of residual free H_2SO_4 after oxidation resulted in a decrease of residual HNO_3: from 6 gpl HNO_3 at 0.02M H_2SO_4 to 1 gpl at 1.00M H_2SO_4 (after 120 minutes leach time, 90°C, 20% solids). Current work has found that to maintain +750 mV @ 85°C the liquor can range from 2M HNO_3 and no H_2SO_4 to 0.7M HNO_3 and 1M H_2SO_4 plus $Fe_2(SO_4)_3$. Further reduction of the nitrate level to 0.15M was made by adding an excess of FeS concentrate. This concentrate addition also consumed free acid, raising the pH above 1 at which point precipitation occured. This is similar to the work by Bjorling and Kolta (1964) on pyrrhotite. The advantage of this is that if Flowsheet 2 is used, barren highly reactive sulphides such as pyrrhotite, or ferrous scrap such as iron filings could be used to reduce the acidity and the nitrate level.

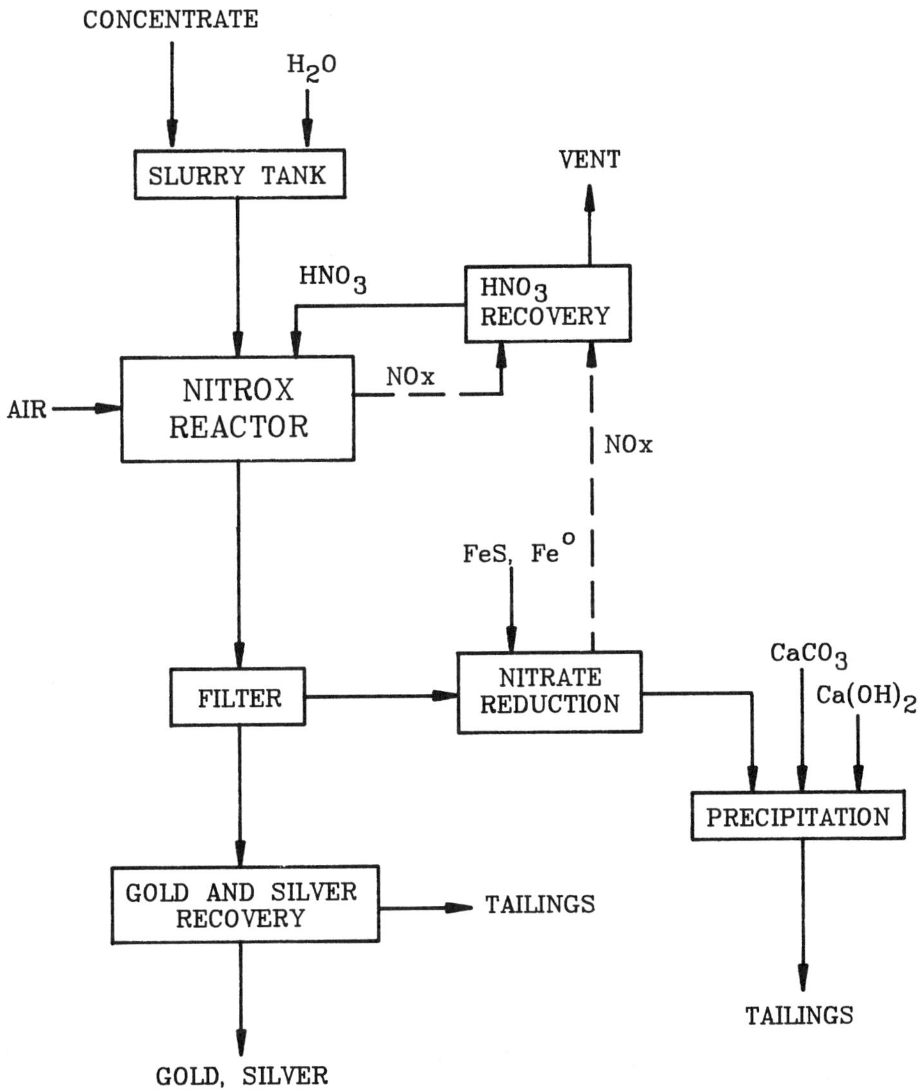

Fig. 4. Effect of nitrate reduction on the NITROX PROCESS flowsheet.

The entire stream could be disposed of which would remove most constraints of water balance. Using this flowsheet variation could represent a loss of nitrate as low as 0.3% of the HNO_3 required for oxidation (50% pyrite concentrate).

Previously reported (Van Weert, Fair and Schneider, 1987) filtration tests of the precipitated cake, pH 4 with $CaCO_3$ at 80°C, has shown that the material is easily washed and has a high recovery of the solute (99.7%). The nitrate recovered is recycled into the circuit and contacted with free H_2SO_4 formed during oxidation to produce gypsum and nitric acid.

Summary

The efficient recovery of the NO formed during oxidation of sulphides, and of residual nitrate remaining in the oxidation liquor is important both economically and environmentally. The NO recovery on the gas side can exceed 99% but a loss of approximately 1% is anticipated in the production of inert N_2O. Efficient washing is possible for the oxidation and precipitated products and at most a further 1% loss is anticipated. Overall the recovery is expected to exceed 98%. Recovering the nitric acid efficiently and for low cost is obviously very important to the NITROX PROCESS which optimizes the recovery of the nitric acid by innovative use of process liquors for absorption, and maximizes the efficiency of gas recovery by agitation.

ACKNOWLEDGEMENTS

Hydrochem gratefully acknowledges the support for the NITROX PROCESS by the COSCH Group, the IRAP office of the National Research Council, and by Dickenson Mines Limited. Thanks are due to N.S. Wood and G. Popenko who assisted in the laboratory studies.

REFERENCES

Ackerman, J.B., and C.S.Bucans (1986). Plant and process startup of the Sunshine refinery. Minerals and Metallurgical Processing, vol.3, No.1, 20-32.

Bjorling, G., and G.A. Kolta (1964). Oxidizing leach of sulfide concentrates and other materials catalyzed by nitric acid. Proceedings, 7th International Mineral Processing Congress, New York, 127-138.

Brennecke, H.M. and co-workers (1981). Nitric-sulfuric leach process for recovery of copper from concentrate. Mining Engineering, 1259-1266.

Chilton, T.H. (1960). The manufacture of nitric acid by the oxidation of ammonia. Chemical Engineering Progress Monograph Series, No. 3, Vol. 56.

Habashi, F. (1973a). Action of nitric acid on chalocopyrite. AIME Transactions, Vol. 254, 224-228.

Habashi, F. (1973b). Treatment of a low-grade nickel-copper sulfide concentrate by nitric acid. AIME Transactions, vol. 254, 228-230.

Krause, E. and V.A.Ettel (1985). Ferric arsenate compounds: are they environmentally safe? Solubilities of basic ferric arsenates. 15th Annual CIM Hydrometallurgical Meeting. Vancouver, B.C.

Kunda, W. (1984). Hydrometallurgical processing of silver concentrate. In Kudryk, V. Corrigan, D.A., and Lang, W.W. (Ed), Precious Metals: Mining, Extraction and Processing. Metallurgical Society of AIME, p.p. 397-423.

Prater, J.D., Queneau, P.B., and T.J. Hudson (1973). Nitric acid route to processing copper concentrates. AIME Transactions, vol. 254, 117-122.

Ritcey, G.M., and A.W. Ashbrook (1981). Treatment of cobalt arsenide feeds for the subsequent recovery of cobalt. CIM Bulletin, vol. 74, No. 834, 103-112.

Schisler, J.M., and Staff of Eldorado Resources Ltd. (1986). Commissioning and startup of the Blind River uranium refinery. Minerals and Metallurgical Processing, vol. 3, No. 1, 14-19.

Van Weert, G., Fair, K.J., and J.C. Schneider (1986). Prochem's NITROX process. CIM Bulletin, vol. 79, No. 895, 84-85

Van Weert, G., Fair, K.J., and J.C. Schneider (1987). The NITROX PROCESS for treating gold bearing arsenopyrites. 116th Annual TMS/AIME Meeting, Denver, Colorado.

Vizsolyi, A., and E. Peters (1980). Nitric acid leaching of molybdenite concentrates. Hydrometallurgy, 6, 103-119.

Weir, D.R., and R.M.G.S. Berezowsky (1984). Gold extraction from refractory concentrates. 14th Annual CIM Hydrometallurgy Meeting, Timmins, Ontario.

Weir, D.R., Robinson, P.C., and J.A. King (1986). Precious metals recovery from pressure oxidized Porgera concentrates. Mineral and Metallurgical Processing, vol. 3, No. 4, 208-215.

Technical Sessions

Session No.6

New Developments

NEW TECHNOLOGY FOR MORE EFFICIENT CYANIDATION OF GOLD ORES

C.E. Elmore*, R.J. Brison**, C.W. Kenney**
*Kamyr, Inc., Ridge Center, Glens Falls, N.Y.
**Hazen Research, 4601 Indiana Ave., Golden, CO

ABSTRACT

This paper describes three recent developments in the utilization of oxygen to increase the rate of leaching gold ores by cyanide.

A practical carbon-in-leach process utilizing oxygen (CILO) has been developed. This process permits reduction of the size of a conventional C-I-L section by 75-80%. Alternatively, it can eliminate the leaching section of a conventional CIP plant, with both leaching and adsorption being accomplished in a modified C-I-P section. Oxygen usage and costs are expected to be minor. Cyanide consumption, contrary to expectations, is typically substantially reduced compared to conventional cyanidation.

In addition, processes have been developed for the utilization of oxygen to increase the rate of leaching and/or gold recovery in cyanide heap leaching and in the previously reported Kamyr countercurrent column leach process. The latter is now entering semi-commercial scale development.

KEYWORDS

Cyanidation, Oxygen, CIL, Heap, Kamyr, Countercurrent, Kinetics.

INTRODUCTION

In 1982, Kamyr, Inc. with the assistance of Hazen Research, Inc. initiated a project to develop improved technology for recovery of gold and silver from ores and other materials. One of the objectives was to adapt some of the technology that Kamyr had successfully developed for the pulp and paper industry. The initial focus was on development of a countercurrent tower cyanidation process, which has been reported previously (Elmore, 1984; Sherman, 1985). This paper reports on further developments both on the countercurrent tower process and on the effective utilization of oxygen to improve cyanidation efficiency in this and other cyanide processes.

KAMYR COUNTERCURRENT PROCESS

This process involves simultaneous leaching and solid-liquid separation in a tower with continuous downflow of ore and upflow of solution. The ore is stabilized by addition of fiber and flocculant as seen in sketch on Fig. 1.

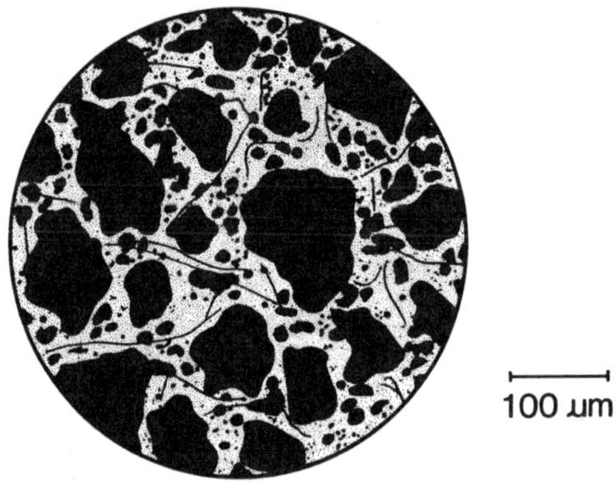

Fig. 1 Sketch of solids-fiber-flocculant mixture as observed under a microscope.

The Kamyr tower for mineral leaching is seen in Fig. 2.

Laboratory tests on various ore and tailing samples indicated that, under equivalent conditions, Kamyr leaching resulted in the same gold and silver extraction as conventional agitation leaching. However, the Kamyr process offered the possibility of carrying out the entire leaching and solid-liquid separation step in one tower rather than in about 10 separate tanks normally required for leach-CCD systems. In addition, the Kamyr process may be used to treat ores ground as coarse as -3.5mm to as fine as -40 µm . This could result in great savings in grinding costs over agitated leaching for ores which show good gold extraction with -3.5mm grind.

Because the tower overflow solution is clear, it may be treated directly by either carbon adsorption or zinc precipitation for metal recovery, thus eliminating the need for the relatively expensive CIP or CCD system treatment with associated precious metal loss.

As compared to heap leaching, the Kamyr process permits higher metal recovery, allows treatment of ore at finer sizes, is better suited to rugged terrain, and permits year-round operation in cold climates.

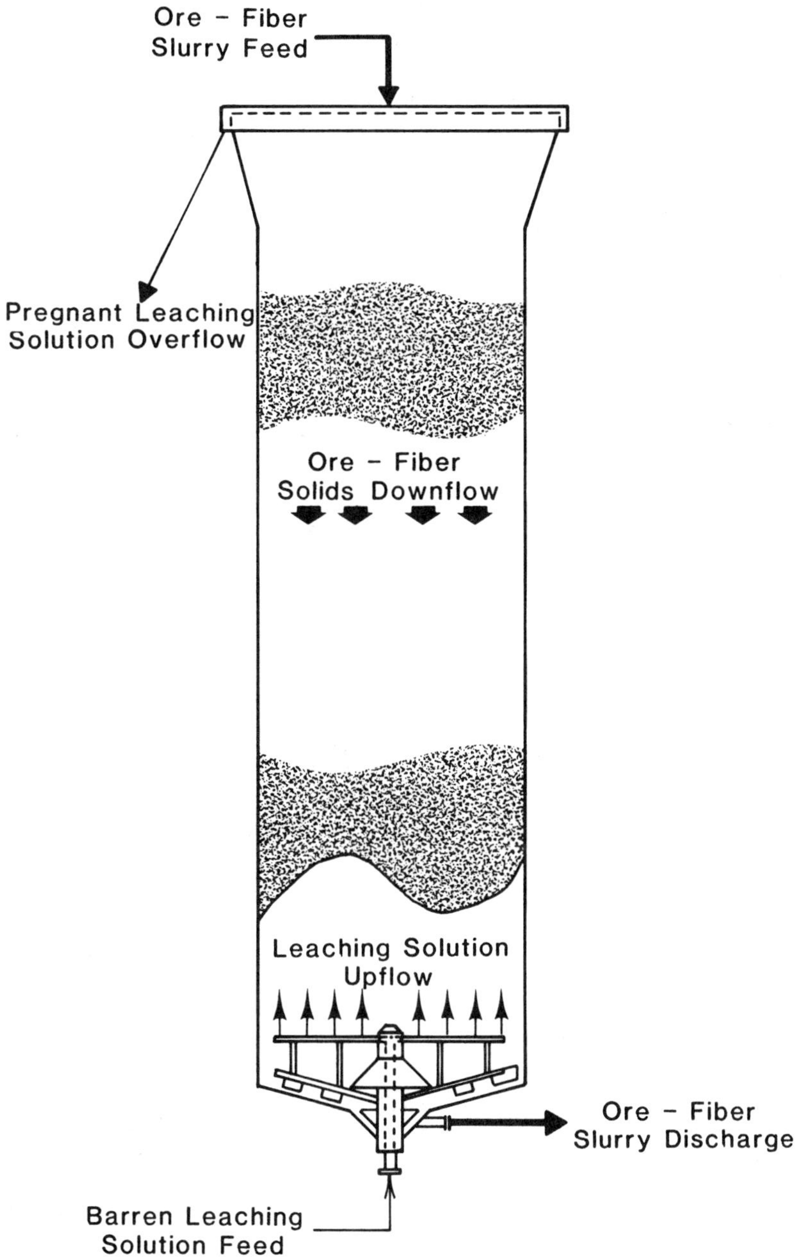

Fig. 2 Kamyr tower for mineral leaching.

Over the last four years laboratory tests were conducted at Hazen to develop and successfully prove out the Kamyr process on a large number of gold and silver ores.

A 200 tonne per day Kamyr demonstration plant is now under construction at a major mining company and will be starting up early next year. The Kamyr tower at this plant is 5 meters in diameter and 30 meters high.

EFFECT OF INCREASED OXYGEN CONCENTRATION

To enhance the Kamyr process further, the possibility of improving the process kinetics by increasing the concentration of oxygen in solution was explored. A literature search provided data on oxygen solubility and on prior studies of the effect of oxygen on the leaching rate of gold.

Cyanidation is normally carried out with the solution in contact with air at atmospheric pressure. Table 1 shows that at 20°C and 1 atmosphere the measured dissolved oxygen (Linke, 1958) is about 4.8 times higher for water in equilbrium with oxygen than for water in equilibrium with air. As predicted by Henry's Law this ratio is nearly the same as the ratio of gas phase oxygen partial pressure.

TABLE 1: Oxygen Solubility at 20°C

Gas Phase Oxygen Partial Press.		Liquid Phase Dissolved Oxygen ml O_2 gas/liter
Air	0.21	6.36
Oxygen	1.00	31.0
Ratio	4.76	4.87

In 1937, USSR researchers (Tronev) reported increased dissolution rates at increased temperatures and air pressures in KCN solutions.

Subsequent work, however, has indicated increasing temperature is not necessary and has the effect of increasing cyanide consumption (Pietsch, 1983).

More recent work reported by American Cyanamid Company (Hedley, 1968) indicated that the rate of gold dissolution in 10 g/l NaCN solution is directly proportional to the partial pressure of oxygen, as shown in the following:

TABLE 2: Au Dissolution Rate Per Oxygen in Gas Phase

% O_2 in Gas	Rate of AU Dissolution mg/cm^2/hr	Ratio
(1)	(2)	(2):(1)
0	0.04	-
9.6	1.03	0.11
20.9	2.36	0.11
60.1	7.62	0.13
99.5	12.62	0.13

Rate control in cyanidation of gold was discussed by Finkelstein (1972). He stated that, when the (CN-)/O_2 ratio exceeds a critical value, the rate is porportional to the O_2 concentration and independent of cyanide concentration. However, he pointed out that this relation is complicated by other considerations such as (1) the questions of rate control by diffusion or by surface reaction and (2) passivation of the gold surface.

A German patent (Pietsch, 1983) claims a process for leaching gold and silver from ores in a pipe reactor under any oxygen pressure of 25 to 130 atmospheres. Retention times of 10 to 25 minutes were reported to result in gold extractions in the range from 94 to 97%.

Two of the authors of that patent (H.B. Pietsch and W.M. Turhe, 1983) presented a paper on pressure leaching of precious metal ores. As illustrated in Fig. 3, their data showed rapid extraction of gold under oxygen pressure. At 20 atmospheres, optimum extraction in 30 minutes was reported.

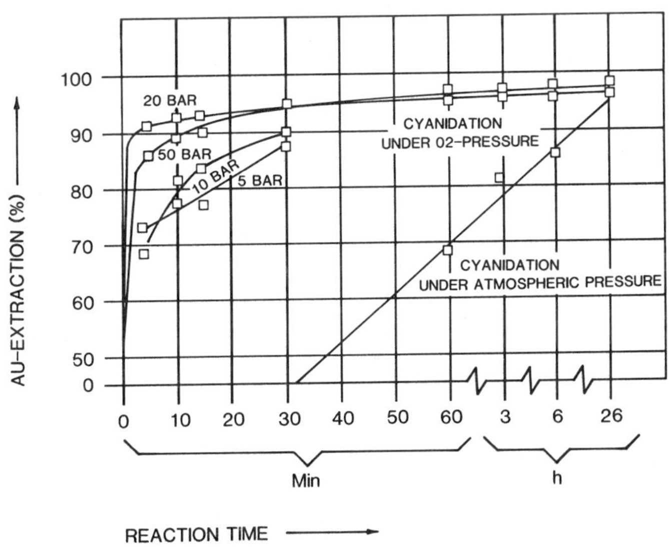

Fig. 3 Gold dissolution at elevated oxygen pressures.

A concern in using oxygen at increased pressures was that decomposition of cyanide may be increased. However, data reported in the same paper (Pietsch, 1983) indicated that cyanide decomposition is not a major factor even at 50 atmospheres O_2 pressure, unless the temperature is increased to above 75°C.

Starting in 1984 a number of laboratory tests were run at Hazen with oxygen at 1 Atm and higher pressures to simulate the Kamyr leaching tower at several levels of slurry hydrostatic head. The dissolution rate with the resulting high oxygen concentrations was five to ten times the rate with air. This means a 24 hour normal leach could be accomplished in 2 to 5 hours without any loss in gold extraction.

With oxygen addition, a 100,000 Tonne Per Month production scale system would appear as illustrated in Fig. 4.

The demonstration plant, currently under construction, has a potential capacity of 1000 Tonnes Per Day with the addition of oxygen.

CARBON-IN-LEACH WITH OXYGEN (CILO)

The laboratory work at Hazen with oxygen also led unexpectedly to the development of a new process for cyanidation of gold ores in early 1985. In a series of tests on agitation leaching with oxygen followed by CIP, the oxygen atmosphere was inadvertently maintained during the CIP step. It was expected that the test would show a rather high cyanide consumption due to carbon-catalyzed oxidation of cyanide. However, not only was the gold recovery high, but the cyanide consumption was lower than in conventional leaching with air.

This result immediately suggested the possibility of carrying out leaching and adsorption simultaneously in an oxygen-saturated slurry. Because pure oxygen at atmospheric pressure can increase the leach rate by up to four or five times, the entire leaching and adsorption steps in a typical plant might now be performed in a series of tanks approximately equal in number and size to those formerly required for the CIP section alone, and the much larger tanks formerly required for leaching might be eliminated. This process has been named CILO (carbon-in-leach with oxygen).

To explore the applicability of such a process, comparative tests were made on a wide range of gold-silver ores from the United States, Canada and South Africa. Standard conditions for the tests were: 45% solids, 0.3 g/l initial NaCN concentration, and 11.0 initial pH. The grind varied from sample to sample, but was the same for each pair of tests on a sample. These tests, as summarized in Table 3, showed that, on the average, the same recovery was obtained in five hours of CILO treatment as in 24 hours of conventional CIL treatment, or 20 hours vs. 96 hours in the case of a tailing sample. Furthermore, the cyanide consumption was, in most cases, reduced significantly. The lower cyanide consumption is presumably due to the shorter treatment time and to the reduction of carbon dioxide in solution. To minimize cyanide consumption in this process, it is desirable to

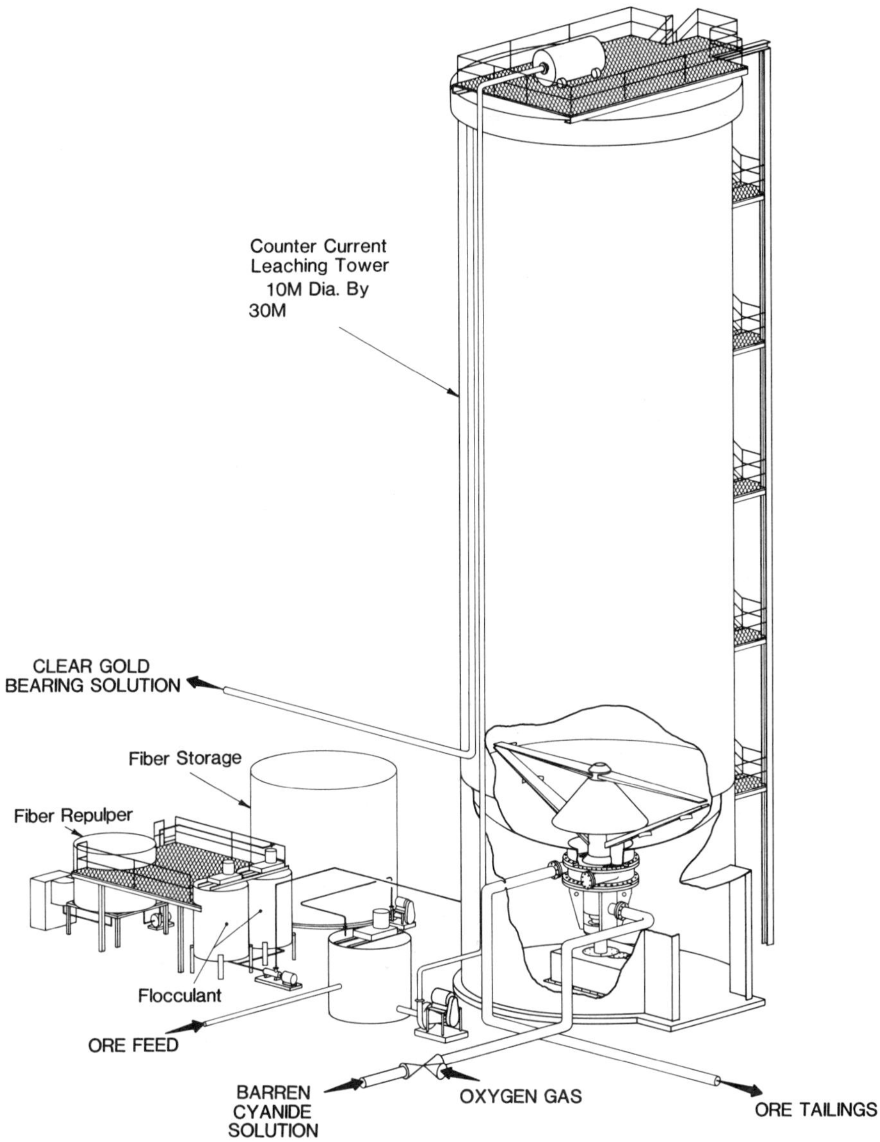

Fig. 4 100,000 Tonne Per Month Kamyr leaching system.

maintain a pH of about 11.0 or higher because carbon-catalyzed oxidation of cyanide is reported to increase sharply as the pH is decreased below this value.

Table 3: Comparison of CIL and CILO Treatment with a Relative Time Factor of 4.8

	Leach Time,hr		Au Extraction, %		NaCN Consumed,g/tonne	
Sample	CIL	CILO	CIL	CILO	CIL	CILO
Ore A	24	5	90.2	89.5	65	15
B	24	5	95.4	95.7	60	10
C	24	5	90.1	89.6	100	25
D	24	5	79.4	76.8	145	80
E	24	5	83.1	78.4	240	160
F	24	5	79.0	84.2	260	190
G	24	5	59.1	61.9	90	20
H	24	5	98.3	97.2	205	10
TailingA	96	20	68.1	68.3	260	130
Average, all tests			82.5	82.5	160	70

To obtain maximum benefit from CILO, it is desirable to maintain near-saturation of the slurry with oxygen. Techniques have been developed for initial saturation and for maintaining saturation during the process, with minimal usage of oxygen. Because this requires essentially complete enclosure of the process, new techniques have also been developed for screening and advancing the carbon. Although the CILO process could be adapted to operate at increased pressure with a further increase in leach rate, it is anticipated that most CILO systems will be designed for operation at or near local atmospheric pressure for economic reasons. Because a typical CIP plant requires about four to six hours retention in the CIP tanks, little would be gained in most cases by increasing the pressure to reduce the leach time below this period.

In brief, the expected advantages of CILO over a typical CIP system are elimination of the entire separate leach system, lower cyanide consumption, reduced in-process metal inventory, and reduced gold loss due to preg. robbing. As compared to a typical CIL system, anticipated advantages are reduction in tank size by a factor of four to five, reduction of gold loss due to abraded carbon by a similar factor, reduced tieup of gold and reduced carbon inventory.

Disadvantages of the CILO process are somewhat added complexity as compared to a CIP system and the need for an oxygen supply. However, the cost of oxygen has been estimated to be not more than a few cents per ton of ore in an efficiently designed system for a typical ore.

The CILO process has been piloted since April of 1986 by a major gold mining organization, and results to date have been very favorable.

HEAP LEACHING WITH OXYGEN

Tests are now underway at Hazen to explore the possibility of increasing heap leaching efficiency by the addition of oxygen, rather than air, to the leach solution and/or the atmosphere within the heap. The tests to date have indicated that a significant improvement is obtainable by the use of oxygen.

Six tests were run on agglomerated minus 25mm ore (1.4-1.7 g/tonne) in 140mm diameter by 1800mm tall leach columns. Oxygen was injected into four of the columns, two of which were also covered. The two remaining "conventional" columns were not covered and no oxygen was injected. Cyanide solutions feeding the four columns with oxygen injection were also presaturated by bubbling oxygen at atmospheric pressure; solutions to the two remaining columns were not presaturated with oxygen. Cyanide solutions were percolated through the columns for 34 days, followed by a 2-day water wash. Column test results are summarized in Table 4.

Table 4: Heap Leaching With Oxygen Versus Air

Time	Average Gold Extraction, %	
	Oxygen	Air
7	79	68
15	86	76
28	90	81
36	91	82

Supporting data are given in Tables 5 and 6.

The average cummulative extraction after 1, 2, 3 and 4 weeks was 9 to 10% higher on tests run with pure oxygen than the standard tests run in air.

There were no significant differences in gold extractions between the oxygenated columns that were covered and uncovered. However, later analyses of the gas inside the columns indicated air infiltration into the covered columns. Cyanide consumptions were less than 0.25 Kg NaCN/t for all tests; slightly lower consumptions were observed in the columns with oxygen compared with the conventional columns.

While confirmatory tests and further studies will be needed to establish the effects on other ores and the economics of oxygen addition for specific heap leaching operations, the trends demonstrated in work to date are very encouraging.

Table 5 Heap Leach Simulation Test Results, California Gold Ore

Test (Column) Number	Total Time Days	Total Volume Liters	Calculated Head Analysis g Au/Tonne Ore	Residue Analysis g Au/Tonne Ore	NaCN Consumed kg/t Ore	Final Gold Extraction %	Description
1	36	128	1.54	.16	0.15	89.44	Oxygen addition, covered
2	36	126	1.44	.11	0.15	92.34	Oxygen addition, covered
3	36	125	1.41	.13	0.16	91.02	Oxygen addition, uncovered
4	36	126	1.47	.12	0.18	91.57	Oxygen addition, uncovered
5	36	118	1.61	.35	0.21	78.22	Conventional, no oxygen, uncovered
6	36	123	1.41	.21	0.24	85.38	Conventional, no oxygen, uncovered

Table 6 Effect of Oxygen Injection on Simulated Heap Leaching - Statistical Comparison

Test Number	Oxygen Source	Column Configuration	Time, Days 7	15	28	36
			Extraction Percent			
1	Oxygen	Covered	75.43	83.64	88.27	89.44
2	Oxygen	Covered	81.18	88.25	91.39	92.34
3	Oxygen	Uncovered	81.33	87.08	90.10	91.02
4	Oxygen	Uncovered	79.20	86.44	90.54	91.57
5	Air	Uncovered	66.00	72.70	77.08	78.22
6	Air	Uncovered	71.06	79.41	84.25	85.38
Statistical Evaluations						
1-4	Mean with Oxygen		79.29	86.35	90.08	91.09
	Std. Deviation		2.57	1.96	1.32	1.23
5-6	Mean with Air		68.53	76.06	80.67	81.80
	Std. Deviation		3.58	4.74	5.07	5.06
	Difference of Means		10.76	10.30	9.41	9.29
Comparison of two means:						
	Variance		8.86	8.50	7.73	7.54
	Std. Deviation		2.98	2.92	2.78	2.75
	t value		4.17	4.08	3.91	3.91
	% confidence		98-99	98-99	98-99	98-99

SUMMARY

With the assistance of Hazen, Kamyr is developing and acquiring patents for three new processes for the recovery of gold and silver from ore. The status of these are as follows:

1. The promising results of Kamyr countercurrent tower lab tests warranted the construction of a 200 ton per day demonstration plant due to start up early in '88.

2. The carbon in leach with oxygen (CILO) process has been piloted by a major gold mining organization and is ready for a full scale application.

3. Laboratory study of Heap Leaching with oxygen is continuing.

REFERENCES

Balakrishnan, R., and Skinner, G.F. (1978) Extraction of Gold and Silver S. African Patent Application No. 78/1184.

Davidson, R., Brown, G.A., Schmidt, C.G., et al (1978). The Intensive Cyanidations of Gold-Plant Gravity Concentrates. J.S. Afr. Inst. Min. Metall., 146-165.

Elmore, C.L., and Brison, R.J. (1984) The Kamyr Process for Leaching Gold and Silver Ores. First International Sumposium on Precious Metals Recovery, Reno, Nevada, June.

Finkelstein, N.P. (1972) The Chemistry of the Extraction of Gold from its Ores. Gold Metallurgy in South Africa. Chapter 10, P. 309.

Hedley, N. and Tabachnick, H. (1968) Chemistry of Cyanidation. Mineral Dressing Notes American Cyanamid Co.

Linke, W.F. (1958) Solubilities of Inorganic and Metal-Organic Compounds (Seidel) 4th Ed. V.1, 250 V.2 1128-30.

Muir, C.W.A., Hendricks, L.P., and Gussman, H.W. (1984) The Treatment of Refractory Gold-Bearing Flotation Concentrates using Pressure Leaching Techniques, Precious Metals: Mining Extraction, and Processing 309-322.

Pietsch, H.B., Turke, W., and Rathje, G.H. (1983) Research on Pressure Leaching of Ores Containing Precious Metals Erzmetall, 261-265.

Pietsch, H., Turke, W., and Bareuther, E., et al (1983) Leaching of Gold and Silver. German Patent DE3126234).

Sherman, M.I., Elmore, C.L, and Brison. R.J. (1985) Leaching and Washing a Flocculated Slurry having a Fiber Content U.S. Patent No. 4501721).

Tronev, Von V.G., and Bondin, M. (1937) On the Dissolution of Precious Metals at High Pressure; 11, Dissolution of Gold in Cyanide Under Air Pressure. Comptes Rendus (Doklady de l'Academie des Sciences del l'URSS, Vol. 16 No. 5 281-284.

DYNAMIC SIMULATION OF THE CIP GOLD RECOVERY PROCESS

C. Carrier*, D. Hodouin*, M. Courchesne**

*Groupe de Recherche sur les Applications de l'Informatique à l'Industrie Minérale. Université Laval, Quebec, Canada.
**Les Ressources Aiguebelle (Div. Mine Yvan Vézina), Rouyn-Noranda, Quebec, Canada

ABSTRACT

The carbon in pulp process for gold recovery from cyanide solutions is now the standard technique in new plants. For a better understanding of the operation of this process, a dynamic model has been developped which involves adsorption and desorption rate parameters, and transport and mixing phenomena. The model is used as a basic module of a dynamic simulator where the following elements are flexible : size and number of tanks in series, quantity of carbon in the tanks, carbon transfer strategy from tank to tank, pumping flow rates, gold content of carbon and solution feed. The simulator is calibrated using in-plant measurements collected in the concentrator of Les Ressources Aiguebelle, Québec. Parameter estimation methods are proposed for pseudo-equilibrium data and for transient regime data. Some applications of the simulator are given to illustrate the effect of the operating and design variables on gold recovery.

KEYWORDS
Carbon-in-pulp; modelling; simulation; gold adsorption; activated carbon

INTRODUCTION

The carbon-in-pulp process for recovery of gold dissolved in dilute cyanide solutions, in the presence of solids, has now replaced the conventional processes of pregnant liquid/barren solids separation followed by Merrill-Crowe precipitation. Most plants built over the past ten years incorporate a CIP section. The process is described as cheaper in capital and operating costs, and few major start-up or operational difficulties are reported. For the purpose of the present paper, the gold adsorption onto carbon is performed in a counter-current cascade of tanks.

The slurry stream coming out of the cyanidation circuit is sent into a series of agitated tanks in which it is mixed with activated carbon. The carbon particles size is much coarser than the top size of the ground ore. The tanks are fitted with screens that allows the slurry to overflow continuously into the next tank, while retaining the carbon in the tank. Carbon is transferred from tank to tank in a discontinuous manner. According to a specific schedule, the carbon from tank one is removed, then carbon from tank two is sent to tank one, and so on, till the last tank, which is filled with regenerated carbon coming from the elution-regeneration section. The slurry going out of the last tank is sent to the tailings pond, and the loaded carbon is sent to an elution-regeneration stage. The amount of carbon removed, the transfer from tank to tank, the sequence of transfer and the carbon inventory in the various tanks are parameters which have to be selected for an optimal operation.

In comparison with other processes in mineral or metallurgical processing, the CIP process presents a major difference in flow characteristics for the slurry, which is in continuous flow, and carbon, which is in continuous agitation in a fixed volume, corresponding to a batch behaviour except during the counter-current transfers. Because of these characteristics, the optimal circuit operating characteristics are not easy to obtain due to the limited information available on the mechanisms of adsorption of gold on carbon and due to the transient conditions which are the rule in such circuits.

The paper presents an attempt to derive a dynamic model to simulate the operation of such a CIP circuit. A first order adsorption process in perfectly mixed vessel is assumed. Model calibration is made using the results of a sampling campaign in the Mine Vezina plant of Les Ressources Aiguebelle Inc. The dynamic simulator usefulness is demonstrated by the evaluation of the influence of various operating conditions and design parameters on the process performances.

DYNAMIC MODEL

To built a dynamic simulator of the CIP process, three aspects have to be taken into consideration: the rate of adsorption, the mixing in the tanks and the exchanges between tanks. For the first aspect it is widely accepted that gold adsorption is a chemical surface phenomenon which is controlled by diffusion of the aurocyanide complex $Au(CN_2)^-$, although the detailed surface mechanisms are not well understood (McDougall, Hancock, 1980). In a first step, gold cyanide migrate from the solution to a carbon particle through a film which surrounds the particle. Then it can be directly adsorbed on the external particle surface when adsorption sites are still available, or it can penetrate the particle through macropores and then micropores where free surfaces are available.

The dynamics of the process are controlled by the rate of adsorption r which for given carbon, solution, equipment and operation, is a fonction of the gold content of the solution and of the carbon. There are two main approaches to the modelling of the rate of adsorption. One is based on a detailed description of the diffusion mechanisms that occur in and around carbon particles (Peel, 1981; Skrivanek, 1976; Van Deventer, 1984) and the other one is a lumped description of the rate process considered as a first order reaction. In the first approach the suitable parameters are the film and macropore transfer coefficients, the macropore to micropore transfer coefficient, the fraction of adsorptive capacity available in macropores and the equilibrium parameters. As it is difficult to estimate the values of these parameters, a simplified approaches leading to more empirical models has been proposed by Nicol and co-workers (1984a). In the lumped approach used here the following first order reaction is considered :

$$[\text{active site}] + [\text{free aurocyanide}] \rightleftharpoons [\text{adsorbed aurocyanide}]$$

and the rate of adsorption is written (Dixon, 1978; Menne, 1982; Williams and Glasser, 1985)

$$r = ka\ x\ (\tilde{y} - y) - kd\ y \tag{1}$$

where x, y are respectively the gold contents in solution and on carbon; Ka and Kd are adsorption and desorption rate constants and y the total number of active sites on the carbon.

To model the mixing behaviour for the type of agitated tank studied here, tracer tests using LiCl impulses were performed in the plant. The results (see Fig. 1) show that the tanks are very close to be perfect mixers. The best model found, a Cholette-Cloutier unit with a stagnant zone of 10% of the tank volume having an exchange rate with the active zone of 80% of the feed rate (Bazin, Hodouin, 1987), does not lead to significant simulated differences in the behaviour of the process and, consequently, the tanks were considered as perflectly mixed.

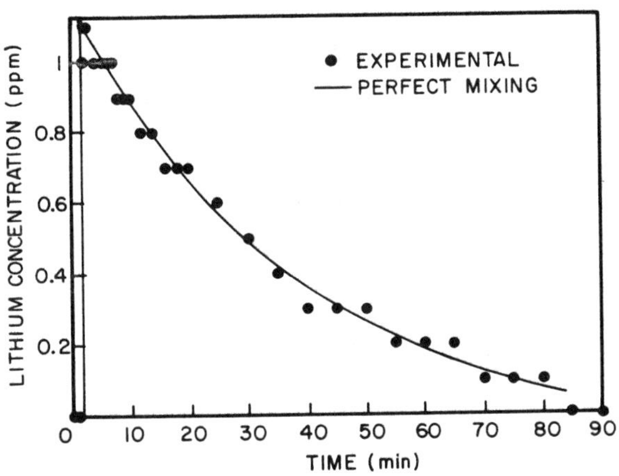

Fig. 1 Tacer test with LiCl impulse

Fig. 2 shows the typical exchange streams that have to be considered in a counter-current cascade of CIP tanks. The mass conservation equations can be written for each species in each tank, and the overall set of equations forms the model of a CIP unit. To express the mass balance we assume that gold dissolution is complete before entering the CIP circuit and that the adsorption rate constants are the sȧme in the n tanks.

The mass balance equation for the adsorbed gold in the ith tank can be written:

$$\frac{dMc_i(t)\ y_i(t)}{dt} = K_a\ x_i(t)\ [\tilde{y} - y_i(t)]\ Mc_i(t) - K_d\ y_i(t)\ Mc_i(t) + C_{i+1}(t)\ y_{i+1}(t) - C_i(t)\ y_i(t) \qquad (2)$$

where the first member is the variation of the adsorbed gold amount in the ith tank and, in the second member, the first term is the quantity of gold adsorbed per unit of time, the second term the quantity desorbed, the third and fourth terms the addition and removal by transfer from tank i+1 to i and from tank i to i-1. These two last terms disappear during periods with no transfer. Generally only one transfer is performed at a time so that during transfer periods one of these two terms cancels.

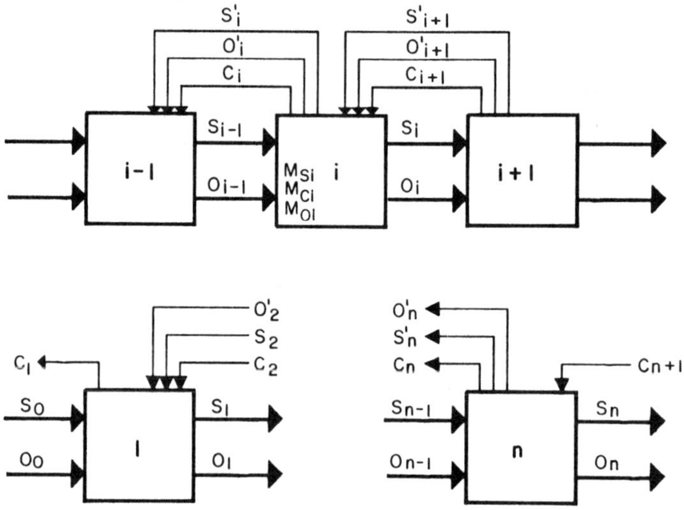

S_i, O_i, C_i: mass flow rate of solution, ore and carbon in the counter-current flow leaving the ith tank.

S_o, O_o, C_{n+1}: Feed-rates of solution, ore and carbon to the circuit.

Ms, Mo, Mc : masses of solution, ore and carbon in the ith tank.

Fig. 2 Exchange streams in a counter-current cascade of CIP tanks.

The mass balance equation for the carbon in the ith tank is the following:

$$\frac{d\,Mc_i(t)}{dt} = C_{i+1}(t) - C_i(t) \tag{3}$$

which have to be written accordingly to the type of transfer made. Outside transfer periods Mc_i is obviously constant.

The mass balance equation for gold in solution is

$$\frac{d\,Ms_i(t)\,x_i(t)}{dt} = S_{i-1}(t)\,x_{i-1}(t) + S'_{i+1}(t)\,x_{i+1}(t) - [S_i(t) + S'_i(t)]\,x_i(t) - Mc_i\,[k_a\,x_i\,(\tilde{y} - y_i) - K_d\,y_i] \tag{4}$$

equation in which the terms depend on the type of transfer and where the prime term are zero outside transfer periods. Accordingly, the variation of the mass of solution in the ith tank is given by:

$$\frac{d\,Ms_i(t)}{dt} = S_{i-1}(t) + S'_{i+1}(t) - S_i(t) - S'_i(t) \qquad (5)$$

and the variation of the mass of ore:

$$\frac{d\,Mo_i(t)}{dt} = O_{i-1}(t) + O'_{i+1}(t) - O_i(t) - O'_i(t) \qquad (6)$$

Equations (2) and (4) contain explicitly the assumption that the tank are perfect mixers for the solution and for the carbon. However the perfect mixing conditions of the three phases with respect to each other must be written . They are:

$$\frac{O'_i(t)}{S'_i(t)} = \frac{O_i(t)}{S_i(t)} = \frac{Mo_i(t)}{Ms_i(t)} \qquad (7)$$

$$\frac{C_i(t)}{S'_i(t)} = \frac{Mc_i(t)}{Ms_i(t)} \qquad (8)$$

Since the ith tank directly overflows in the $(i+1)^{th}$ tank, it can be considered to contain a constant volume V of the 3-phase pulp. The volume V is related to the masses of the 3 phases by:

$$V_i = \frac{Ms_i(t)}{\rho_s} + \frac{Mc_i(t)}{\rho_c} + \frac{Mo_i(t)}{\rho_o} \qquad (9)$$

where ρ_s, ρ_i and ρ_o are respectively the specific gravities of the solution, the carbon and the ore.

The counter-current transfers are assumed to be made by constant volumetric flowrate pump (Q_i). So, during carbon transfer periods, the following equation holds:

$$Q_i = \frac{S'_i(t)}{\rho_s} + \frac{C_i(t)}{\rho_c} + \frac{O'_i(t)}{\rho_o} \qquad (10)$$

For the first and the last tanks only carbon is pumped out or pumped in and the equations 2 to 6 have to be modified accordingly.

THE SIMULATOR

The simulator architecture is depicted in Fig. 3. The simulation of the CIP process is performed by simultaneous integration of the equations (2) to (10) for all tanks. Initial values of carbon loading and solution gold content are selected and a carbon transfer strategy is defined. After a complete cycle simulation, a new calculation is initialized with the values obtained at the end of the previous cycle. The procedure is iterated till carbon and solution have the same gold contents from cycle to cycle. The simulator output gives the gold content of the solution and carbon of each tank and calculates the gold dynamic recovery of the circuit.

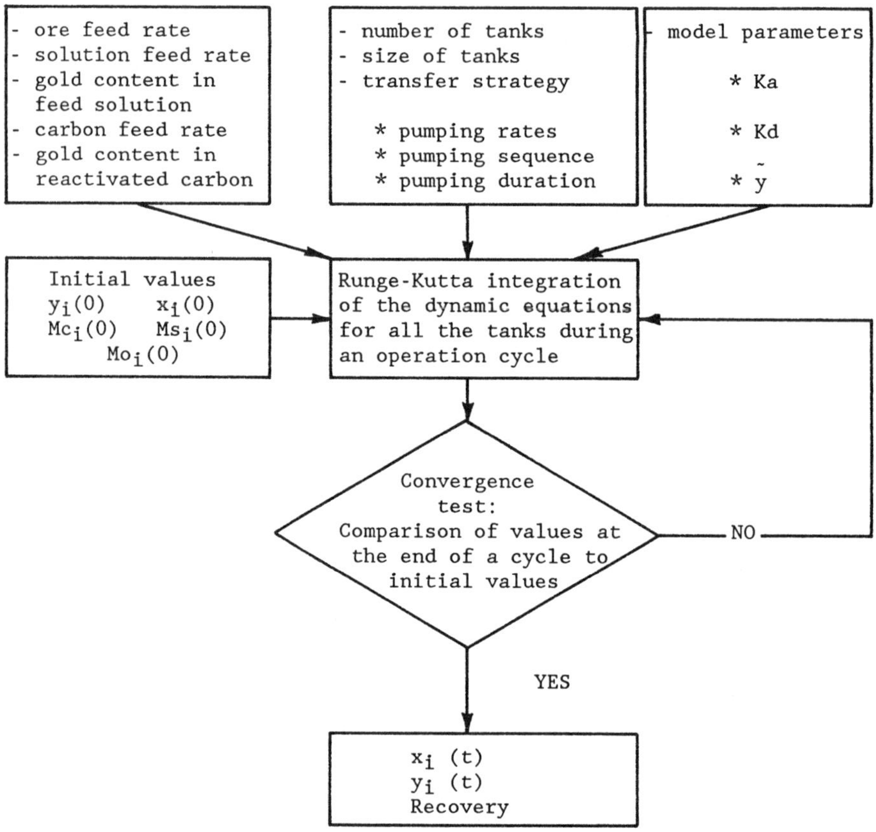

Fig. 3 Simulator architecture

The natural way to evaluate the overall recovery of a CIP process is to evaluate the amount of gold extracted from the feed solution when a steady-state operation is obtained. For a continuous counter-current process, where the carbon is continuously transferred from tank to tank, the recovery of the circuit is given by:

$$R = \frac{x_0 - x_n}{x_0} \tag{11}$$

However for intermittent counter-current operation the expression is more complex since the solution gold content never reaches a constant value. Equation (11) is an approximation which can be applied to the pseudo-steady state values of x between carbon transfer periods. However they give biased estimates of gold recovery. A correct evaluation of the process efficiency requires an integration of the whole story of the solution gold content during a cycle, including the variation of the flow rates which are disturbed during the carbon transfer periods. The mathematical equation which performs these calculations in the simulator is :

$$R = \frac{\int_0^T [\, x_0(t)\; S_0(t) - x_n(t)\; S_n(t) \,]\; dt}{\int_0^T x_0(t)\; S_0(t)\; dt} \qquad (12)$$

Where T is the duration of cycle. In equation (12) it is assumed that the circuit attained a pseudo equilibrium state i.e. that the successive cycles are identical and the feed to the circuit stationnary, i.e. of constant mean.

Fig. 4 gives an example of simulation results for the gold content of the carbon and the solution in the first and last tank of a series of five tanks. The carbon transfer strategy simulated here involves a sequential pulp pumping from the first cell to the last one using a pumping time of 2.25 hr for each tank, and a cycle of 24 hours.

It clearly appears from this simulation that the gold contents are strongly disturbed during the carbon transfers. This has an important implication for the gold recovery, since, in the last tank, there is a very significant increase of the solution gold content during about 6 hours. In the present case this transient behaviour is responsible for half the gold losses to the tailings.

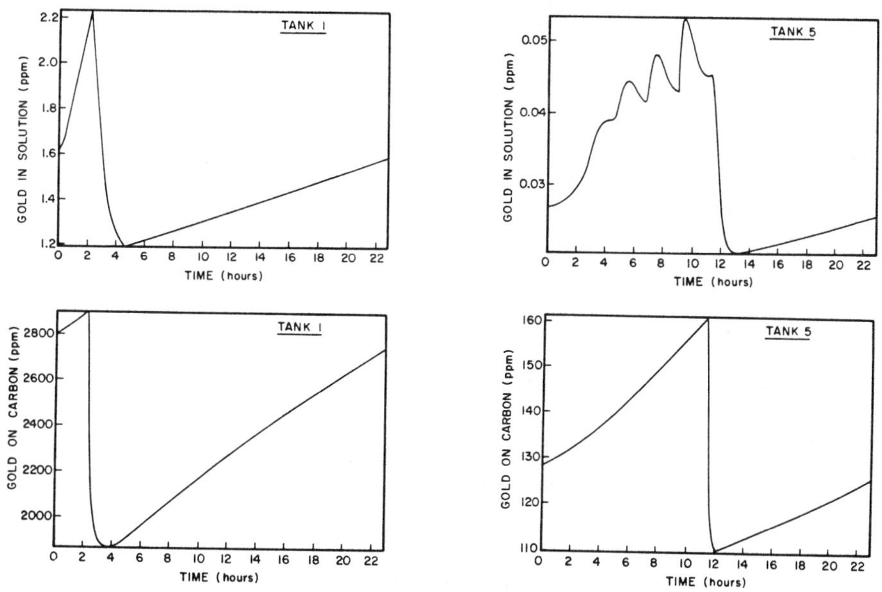

Fig. 4 Dynamic simulation results for the first and last stage.

CALIBRATION OF THE SIMULATOR

The three parameters ($\tilde{y}$, Ka, Kd) of the model must be estimated for each CIP operation since they depend upon the equipment design, the carbon properties and the solution chemistry. For this purpose two kinds of experiments can be performed : pseudo-steady state tests or dynamic tests. In the first method the sampling of the continuous laboratory or plant operation is made during the batch operation with respect to the carbon ; this procedure gives pseudo-steady state values of carbon and solution gold contents.

The experimental procedure of the first method consists in sampling carbon hold-up and feed solution of each tank and the solution going to tailings. For the ith tank, equation (4) can be written:

$$\frac{Ms_i \, dx_i}{dt} = S_0 (x_{i-1} - x_i) - Mc_i [Ka \, x_i(\tilde{y} - y_i) - Kd \, y_i] \tag{13}$$

where the derivative of x_i is small and can be neglected, without significant modification of the estimated parameters as it has been illustrated by simulation. The best parameters are those which minimize the following criterion:

$$J (Ka,Kd,\tilde{y},\hat{x}_{i-1},\hat{x}_i) = \sum_i [(x_i - \hat{x}_i)^2 \frac{1}{\sigma_{xi}^2} + (y_i - \hat{y}_i)^2 \frac{1}{\sigma_{yi}^2}] \tag{14}$$

where $\sigma_{x_i}^2$, $\sigma_{y_i}^2$ are the variances of the gold analyses and where $\hat{y}_i$ is calculated by equation (15):

$$\hat{y}_i = \frac{- [(S_0/Mc_i)(\hat{x}_{i-1} - \hat{x}_i) - Ka x_i \tilde{y}]}{Ka \, \hat{x}_i + Kd} \tag{15}$$

For parameter estimation, data acquisition was performed on the CIP plant of Les Ressources Aiguebelle. This circuit is built to treat 1000 tons of solution per day, i.e. 2000 tons of pulp. The adsorption unit consists in five 70 m^3 adsorption tanks mechanically agitated. The pulp is pumped from the cyanidation plant to the first stage of adsorption. The amount of carbon per stage is about 15 g/l, and is retained in the stage by the mean of interstage screens in the bottom of two launders immersed just below the pulp level. The pulp flows by gravity from the first to the last stage, and the carbon is pumped counter-currently to the pulp by the mean of diaphragm pumps. The carbon transfer is on a daily basis and the amount transferred is about 1 ton. Fig. 5 shows the flowsheet of the adsorption unit. Air jets are used for cleaning the interstage and avoiding sedimentation in the launder. Three sets of pseudo-steady state data (Table 1) were obtained on this CIP plant by sampling the tanks at the end of a cycle just before the carbon transfer period. The parameters estimated for the campaign of july 86 (Table 1) are very different of the two others campaign. Although a different carbon has been used the higher value of the desorption constant for july 86 seems to be unrealistic.

In the second method, laboratory batch tests at constant (Hussey, Salisbury, Potter, 1979) or variable (Hodouin and co-workers, 1987) solution gold content give good estimates of the model parameters when the carbon properties, solution chemistry and hydrodynamics are the same than they are, or will be, in the plant, a situation difficult to obtain. In the technique of immersed baskets (Courchesne, Barbery, 1987), the tests are made under representative conditions of the solution chemistry, however it is difficult to obtain the same hydrodynamics in the basket as in the tank. A better technique consists in using the transient regimes which occur naturally during the carbon transfer periods. Using x_{ij} and y_{ij}, the measured values of the gold contents in the ith tank at time j, the parameters can be estimated by the minimization of the following least-squares criterion :

$$J\ (Ka,Kd,\tilde{y}) = \sum_i \sum_j \left[(x_{ij} - \hat{x}_{ij})^2 \frac{1}{\sigma^2_{x_{ij}}} + (y_{ij} - \hat{y}_{ij})^2 \frac{1}{\sigma^2_{y_{ij}}} \right] \qquad (16)$$

where $\hat{x}_{ij}$, $\hat{y}_{ij}$ are the values calculated by the simulator when a stabilized cycle is obtained.

Table 1 Three Pseudo-steady State Data Acquisitions

	PARAMETERS							
	Ka	Kd	$\tilde{y}$	STAGE	x	$\hat{x}$	y	$\hat{y}$
TEST 1 (JULY 85)	0.021	0.003	4200	FEED	2.120	-	-	-
				1	0.970	1.005	2408	2390
				2	0.430	0.419	1200	1211
				3	0.170	0.165	683	682
				4	0.069	0.073	498	494
				5	0.034	0.034	389	391
TEST 2 (JULY 85)	0.021	0.001	4400	FEED	2.260	-	-	-
				1	1.030	1.056	2501	2486
				2	0.470	0.476	1264	1309
				3	0.160	0.162	582	632
				4	0.060	0.058	361	399
				5	0.021	0.021	240	260
TEST 3 (JULY 86)	0.052	0.025	7119	FEED	2.700	-	-	-
				1	0.400	0.429	2605	2552
				2	0.100	0.088	705	781
				3	0.018	0.022	263	240
				4	0.010	0.010	136	136
				5	0.010	0.010	145	141

As the type of dynamic sampling campaign involved in this estimation method requires a large number of samples, tests were carried out only in the first and last stages of the adsorption unit. Fig. 6 shows the results obtained during the transfer period in the first tank, immediatly following the pseudo-steady state data acquisition. The first hour corresponds to the carbon transfer to the elution process and the following to the carbon transfer from tank 2 to tank 1. The parameters estimated from this test and the corresponding simulated curve are shown in Fig. 6. The simulated curve fits properly the experimental data, with a desorption constant lower that the one estimated with the pseudo-steady state data and a very close number of active sites (y).

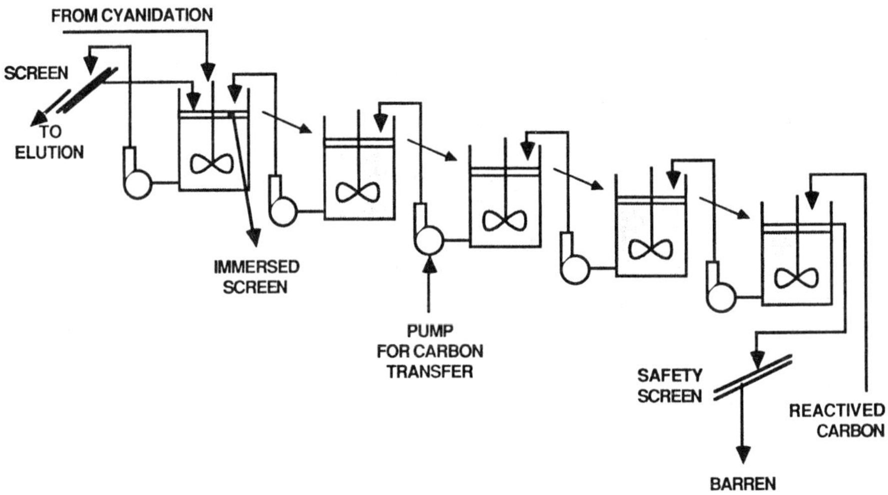

Fig. 5 Flowsheet of the adsorption unit at Les Ressources Aiguebelle (Mine Yvan Vezina).

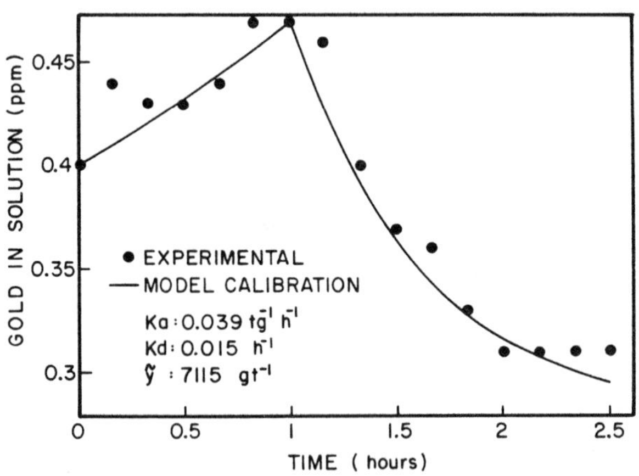

Fig. 6 Dynamic results for the transient regime during the transfer in stage 1.

Fig. 7 shows the results obtained during and after the transfer period in the last stage. The increase in the gold content corresponding to the addition of reactivated carbon is due to the fact that the carbon which comes from the elution was more loaded than the one initially present in the fifth tank. This unusual situation is useful for the estimation of the desorption constant kd. The desorption parameters is much lower that the one estimated in the previous test (Fig. 7).

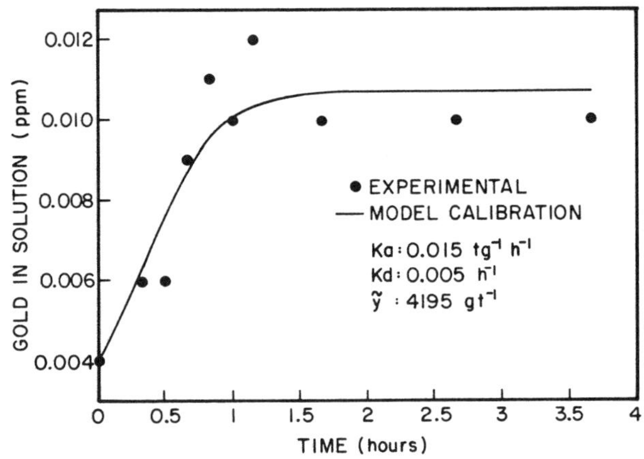

Fig. 7 Dynamic results for stage 5 during reactivated carbon addition.

The three sets of parameters values obtained in July 86 (pseudo steady-state, stage 1, last stage) show that the estimation method can lead to different parameters depending upon the data obtained. This sensivity of the parameters estimation technique can be related to various factors: the data inaccuracies, the interactions between the parameters which create large covariances between their estimates, the influence of the data structure on the variance of the estimates and finally the possibility that the parameters varies along the CIP circuit. Other estimation techniques are being tested, however at this stage it is not posible to propose a completly satisfying estimation scheme.

To test the influence of the parameters inaccuracies on the simulation, they have been varied one at a time around a nominal value and the corresponding gold recovery calculated. It can be seen (Fig. 8) that y is the parameter which has the greatest effect on the recovery. An increase of the parameters y or Ka leads to an increase in the amount of gold sent to the elution process, and a corresponding decrease in the gold of the barren solution. For the parameters Kd the opposite effect is observed, leading to a lower overall recovery. An error of 10% in one of the parameters induce an error of about 20% on the gold losses, an indication of the sensivity of the process to the parameters.

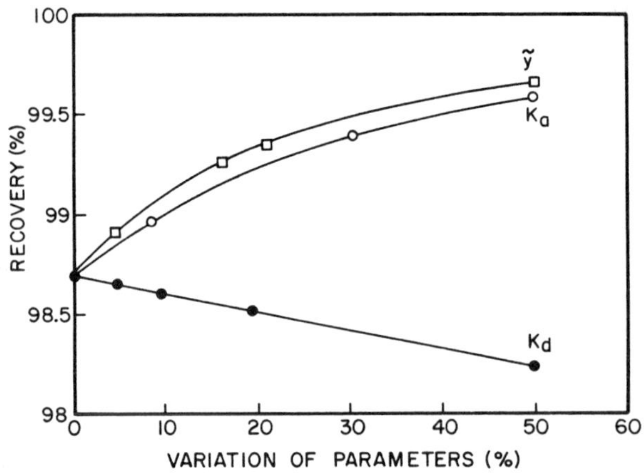

Fig. 8 Sensitivity of the simulated recovery to the variation of the three model parameters.

APPLICATIONS OF THE SIMULATOR

The simulator can be used for the optimization of the process efficiency with respect to various factors as the type of carbon, the circuit design and the operation strategy. The following examples illustrate the potential of the simulator to maximize the gold recovery in an intermittent counter-current CIP circuit. For all simulation it is assumed that the gold dissolution is complete that gold losses by carbon attrition is negligible, and that poisoning of the carbon is not present in the circuit.

Carbon inventory and transferred fraction

For a given amount of gold to be recovered in a selected number of adsorption stages, the carbon inventory in the tanks is a major operating variable. In the case where only a fraction of the carbon hold-up is transferred from stage to stage, the amount of carbon wich is moved is a secondary tuning variable which should be tested. Fig. 9 shows the results of simulations performed for carbon inventories ranging from 1 to 2 tons per tank and for various percentage of the inventory transferred at each operation cycle, all the other design or operating conditions, including the pump volumetric flow rate, being the same. The higher the carbon inventory, the higher the gold recovery. However for high carbon inventory the gold lock-up in the CIP circuit is higher and a larger amount of a less loaded carbon has to be processed at the elution stage. The optimal fraction to be transferred depends upon the inventory. For low inventory a higher fraction must be transferred and the tuning is more critical. This optimum value is the result of the competition between the advantage to transfer a maximum amount of carbon and the drawback of a long pumping rate which increases the losses related to the transfer disturbances.

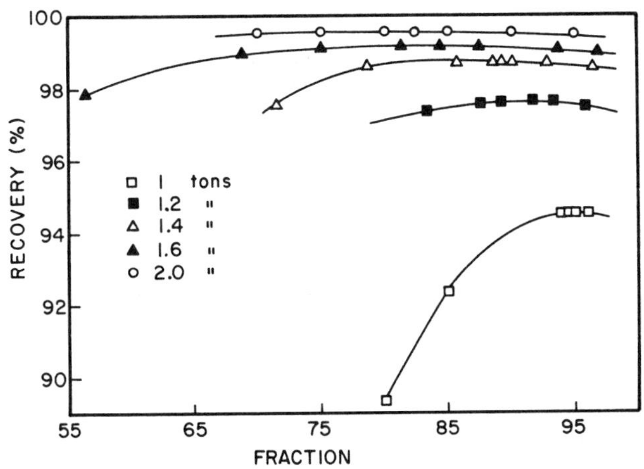

Fig. 9 Influence on gold recovery of the carbon transferred fraction for differents carbon inventories.

Transfer pumping rate

The effect of an increase of the carbon pump capacity (from 40 m^3/h to 100 m^3/h) was tested by simulation, assuming a constant pumping time or a constant transferred fraction of the carbon inventory. In the first case, Fig. 10 shows the improvement of the gold recovery with the pumping rate. The higher is the pumping rate, the higher is the amount of transferred carbon and the results are coherent with the effects observed in the previous section. In the second case, where the amount of transferred carbon is maintain constant, the gold recovery is also improved, because, with high pumping rates, the disturbance due to the carbon transfer is shorter.

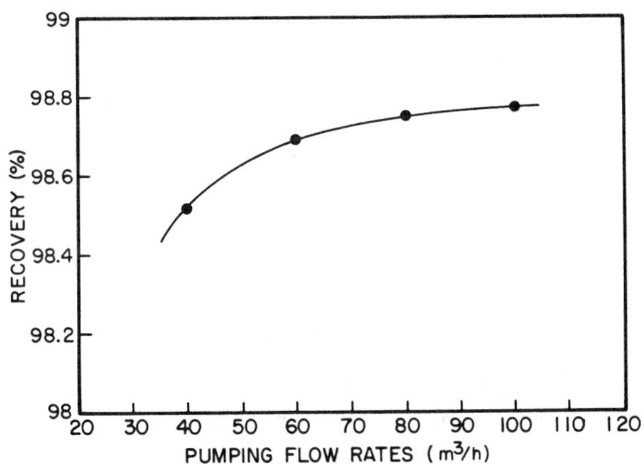

Fig. 10 Influence of pumping rate on gold recovery.

Carbon inventory in the last tank

The previous simulations were performed assuming that the carbon inventory is constant along the series of tanks. Since most of the gold losses are related to the increase of gold content in the solution of the last tank during the transfer periods, additional amount of carbon in this last tank only should minimize gold losses. Simulations have been made to evaluate this point by increasing the carbon inventory in the last tank, while maintaining the amount transferred to the previous tank. It can be seen in Fig. 11 that the gold recovery is enhanced with almost no effect on the gold concentration of the carbon going to the elution stage.

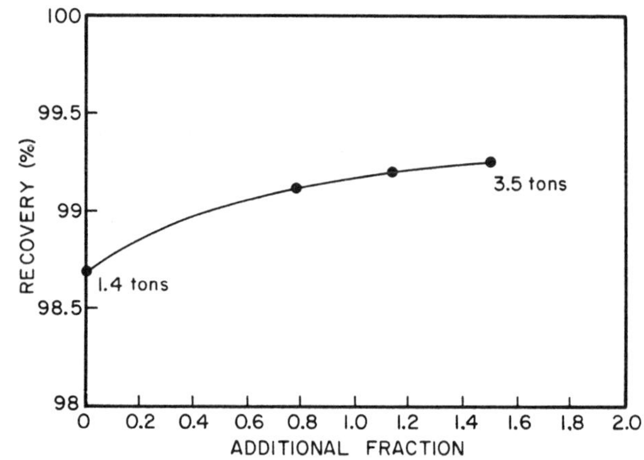

Fig. 11 Influence on gold recovery of an additional fraction of carbon in the last tank.

Tank size and transfer strategy

Dynamic simulation can also be very helpful to study the parameters which are considered in the design of a CIP unit: the number of tanks in series, the tank size and the type of transfer. The selection of the number of stages, which is related to the tonnage and the grade of the ore which has to be treated, is not studied here. The selection of the tank size is related to investment cost considerations, to carbon concentration, which is usually 25 g carbon per liter, and to the mean residence time for the pulp, which is usually about one hour. Recent developments (Kyffin, Charles, Wilson 1983) on the CIP circuit design show that, with a suitable tank/screen configuration, carbon concentration of 125 g/l and mean residence time of 12 minutes can lead to the same efficiency as the traditional design. To evaluate this aspect, simulations for different tank size have been performed, keeping constant all the other operating variables. The result show (Fig. 12) that small tank performances are better. This is mainly due to the behaviour of the gold content in the solution during the transfer periods. For higher carbon concentration (70 g/l for a 20 m^3 tank) the time required for transfering the same amount of carbon is reduced and the gold losses are consequently reduced.

The behaviour of the process has been also tested with a pumping rate proportionnal to the tank size and for three differents type of transfer (Table 2): 1- instantaneous transfer of the carbon fraction, 2- pumping and carbon screening, the pulp being immediatly returned to the tank which is pumped out, and 3- pumping of carbon and pulp. For the batch transfer there is no difference in the gold recovery between the two tank sizes tested, because of the absence of transient regimes during the transfer. For continuous counter-current operation or discontinuous operation with carbon batch transfer, gold recovery does not depend upon the tank size inasmuch as the carbon inventory is constant as well as the model parameters. This is based on the assumption that the hydrodynamics conditions which control the diffusion around the carbon particles are the same for higher carbon concentration in tanks. However, when the transient regimes are taken into account (2 last cases of Table 2), the tank size has an influence on the gold recovery as it is illustrated in Fig. 13 by the different solution gold content profiles in the last tank.

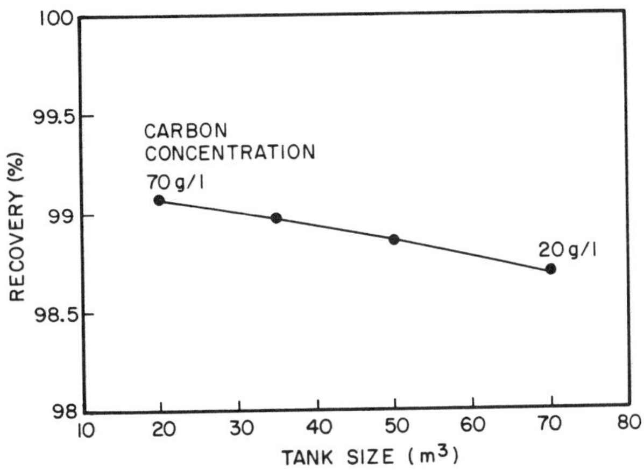

Fig. 12 Influence of tank size on gold recovery.

Table 2. Influence on gold recovery of tank size and type of transfer.

TYPE OF TRANSFER	RECOVERY (%)	
	V=20	V=70
BATCH	99.19	99.19
PUMP (CARBON ONLY)	98.91	98.84
PUMP (PULP AND CARBON)	98.87	98.69

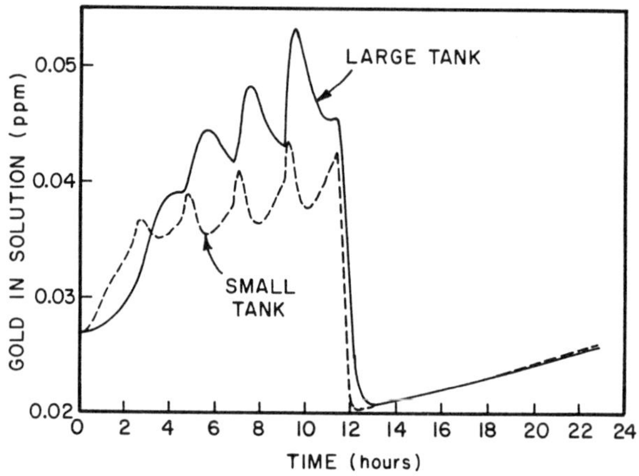

Fig. 13 Profile of gold content in solution in the last stage for differents tank sizes.

CONCLUSION

The analysis of the carbon-in-pulp process by dynamic simulation leads to the following concluding remarks:

- the CIP process, because of the intermittent transfer of carbon, is never in steady-state operation;

- the gold content of the solution is strongly disturbed during the transfer periods;

- for the tested circuit, half the gold losses can be explained by the transient increase of the gold content of the solution sent to the tailings;

- for a correct design and tunning of a CIP circuit, these transient regimes have to be taken into account;

- there is an optimal fraction of the carbon inventory to be transferred at each cycle;

- the transfer time must be minimized for a better gold recovery;

- small tanks with high carbon concentration is a valuable design to be explored;

- optimal carbon transfer strategy can be determined by dynamic simulation.

As for other mineral processes, the computer simulation of the CIP process gives valuable information on its behaviour and the manner to improve its efficiency. In future works, the simulated performances must be studied in connection to the cyanidation and the elution-regeneration stages.

ACKNOWLEDGMENTS

The authors are grateful to the Centre de Recherches minérales of the "Ministère de l'Energie et des Ressources du Gouvernement du Québec" for the financial support, the gold analyses and the permission to publish. The "Ministère de l'Enseignement Supérieur et de la Science du Gouvernement du Québec" is also thanked for its financial support though the "Programme d'Actions Structurantes".

BIBLIOGRAPHIE

Bazin, C., D. Hodouin (1987). Residence time distribution modelling for mineral processes evaluation, Accepted for publication in CIM Bulletin, 1987

Courchesne, M., G. Barbery (1987). Private Communication.

Dixon, S. N., S.N. Cho, C.H. Pitt,(1978). Fundamental Aspects of Hydrometallurgical Processes, Chapman, T.W. et al., (eds.). New York, AIChE. AIChE Symposium Series no. 173, vol. 74. pp 75-83.

Hodouin, D., C. Carrier, G. Barbery, M. Courchesne (1987). Evaluation of the performance of a CIP gold recovery process. Presented at the 19th Canadian Mineral Processors Operators Conference., Ottawa, Jan. 87

Hussey, S. J., H. B. Salisbury, and G. M. Potter (1979), Carbon-in-pulp gold adsorption from cyanide leach slurries,U.S. Bur. Mines. Report of Investigation no. 8368

Kyffin, D., W. D. Charles, and D. Wilson (1983). Gold recovery using activated carbon. Proc. 17 Int. Pre. Met. Inst. Conf. (San Fransisco, Cal.), pp 117-136

McDougall, G.J., R. D. Hancock, (1980). Activated carbon and gold- a literature survey. Minerals sci. Engng. vol. 12, no. 2, April

Menne, D. (1982). Predicting and assessing carbon-in-pulp circuit performance. CIM XIV International mineral processing conference. Toronto.

Peel, R.G., A. Benedek, and C. M. Crowe (1981). J. Environ. Eng. Div., Am. Soc. Civ. Eng., vol. 106, no. EE4. pp 797-813.

Skrivanek, J., J. Hostomsky (1976). A model of isothermal adsorption in a cascade of continuous flow stirred-tank reactors. Chem. Eng. Commun., Vol.2, pp. 109-114.

Van Deventer, J.S.J, (1984). Modelling of the adsorption of gold on activated carbon in batch and continuous systems. Presented: 1st International Symposium on precious metal recovery. Reno, Nevada.

Williams, D.F, D. Glasser (1985). The modelling and simulation of processes for the adsorption of gold by activated carbon. J. S. Afr. Inst. Min. Metall., vol. 85, no. 8, pp 237-243, august.

MICROWAVE IRRADIATION PRETREATMENT OF A REFRACTORY GOLD CONCENTRATE

by
K.E. Haque
CANMET, 555 Booth St., Ottawa, Ontario, K1A 0G1, Canada

ABSTRACT

Microwave energy was applied in the pretreatment of a typical refractory arsenopyritic gold concentrate. More than 80% of As and S were volatilized off as As_2O_3 and SO_2; whereas iron was oxidized mostly to hematite (Fe_2O_3). All the gold was present in the calcine which responded favourably to the conventional alkaline cyanidation process and yielded 98% of Au and 60% of Ag extractions respectively.

To avoid the oxidation of As and S, the concentrate was irradiated with microwave in an N_2 atmosphere. The majority of the sulphur and almost all of the arsenic were volatilized off as As_2S_3 and iron remained in the calcine as pyrrhotite (FeS). Alkaline cyanidation of this irradiated concentrated yielded only 89% of gold extraction. However, when the concentrate was irradiated after mixing with NaOH, no volatiles of As_2O_3 and SO_2 formed but instead Na_3AsO_4, Na_2SO_4, $FeSO_4$, $NaFeO_2$ and some Fe_3O_4 (magnetite). All of these products except Fe_3O_4 were removed from the calcine by dissolving in water. Alkaline cyanide leaching of the water washed calcine provided 99% Au and 70% of Ag extractions.

Various test data also confirmed that almost complete calcination was necessary for an efficient gold extraction by alkaline cyanidation of the concentrate. Finally, all the experimental observations lead to the conclusion that microwave energy has a sound technological potential in the pretreatment of the refractory arsenopyritic gold ores or concentrates.

KEYWORDS

Microwave, microwave pretreatment, microwave irradiation, microwave calcination, pretreatment of refractory gold concentrate, gold leaching from refractory concentrate

INTRODUCTION

Microwave energy, widely applied in communication, presently is finding increasing applications in cooking and heating of food products in the home as well as in industry. Apart from these uses, it is also finding commercial applications in the sterilization of agricultural products, vulcanization of rubber and densification of ceramic (Smith, 1984; Metaxas and Meredith, 1983; Pischner, 1966; Zavitsanos and coworkers, 1978; Dymock and Tuthill, 1986).

Microwaves are generated by a magnetron tube. It can propagate through space in continuous or in pulse mode to be received or absorbed by a substance at a distance from the origin. Figure 1 shows a simplistic diagram of a typical microwave heating system. Microwaves of frequencies 915 MHz (i.e. 32.8 Cm) and 2450 MHz (i.e. 12.8 Cm) have been set aside for industrial, scientific and medical (1SM) applications. However, 2450 MHz is the most commonly applied frequency for heating purposes (Smith, 1984; Metaxas and Meredith, 1983.). Microwave heating is due to ionic conduction and dipolar rotation of the molecules in the substance (Smith, 1984).

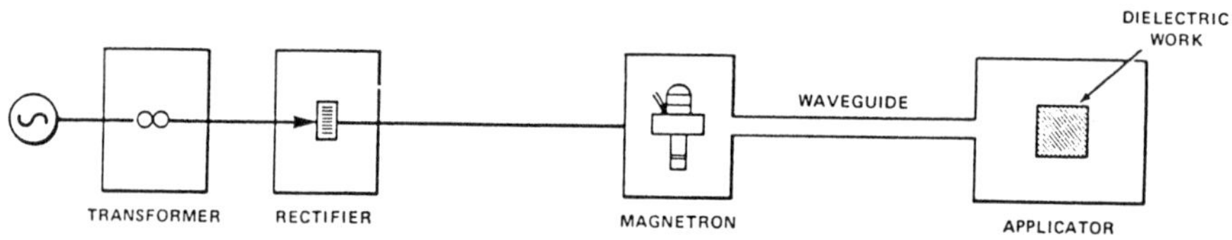

Fig. 1. Microwave Heating System

Microwave heating is material specific: that is, materials which absorb microwave energy generate heat (e.g., water, pyrite, assenopyrite, etc.) But when microwaves pass through a substance (e.g. plastic, paper, silica, carbonate, sulphur, etc.) or reflects from a material surface (e.g. metal), no heat is generated. Generally, microwave heating is rapid and instantaneously controllable but is dependent on the load factors, that is, the greater the mass, the more power or time required to bring it up to the desired temperature.

Because of material specificity, microwaves can heat or cause chemical or phase transformation of ore mineral in an ore without directly affecting other minerals (Smith, 1984; Chen and Coworkers, 1984). Consequently, a mineral which is difficult to process by conventional means may become amenable to processing by microwave irradiation. For example, pyrite, on microwave irradiation, transforms into pyrrhotite (eq. 1) which is chemically more reactive than pyrite and easily separable by magnetic separator; similarly, arsenopyrite transforms into pyrrhotite and arsenic (eq. 2) (Chen and coworkers, 1984).

$$FeS_2 \xrightarrow{hv} FeS + S^\circ \quad (1)$$

$$FeAsS \xrightarrow{hv} FeS + As^\circ \quad (2)$$

Presently the literature records a few examples of applications of microwave energy in the treatment of minerals. Of course, all these studies were limited to bench-scale investigations (Chen and Coworkers, 1984; Kruesi, W.H. and P.R.; 1984; Westbrook and Jefferson, 1986; Klein, 1985; Kreusi and Coworkers, 1984). More recently, Kruesi and Coworkers (1986) reported extractions of 70% of Ni and 85% of Co by 5 minutes of microwave irradiation of a lateritic ore mixed with 30% NH_4Cl. Besides all these, microwave energy is also finding effective applications in the field of chemical analysis of minerals (Nadkarni, 1984; Smith and Coworkers, 1985; Matthes and Coworkers, 1983; Abu Samura and Coworkers, 1975).

Objectives of Microwave Irradiation

The unique heating and decomposition mechanisms of various minerals caused by microwave energy can be applied in the pretreatment of refractory gold ore or concentrate. Generally, refractoriness of gold ore or concentrate is due to encapsultation of gold particles in sulphidic or arsenopyritic minerals. Because of encapsulation of gold particles in sulphidic minerals gold extraction by the conventional alkaline cyanidation process is always poor (Berezowsky and Weir, 1984). The pretreatment of refractory gold ores or concentrates can be conducted by a number of processes, for example, roasting (Wright, 1961), oxygen pressure leaching (Berezowsky and Weir, 1984), Arseno process (Vreugde, 1983), Caro's acid leaching (Lakshmanan and Childs, 1984), Nitrox process (Weert and Coworkers, 1986), bacterial oxidation (Bruynesteyn, 1984). Except for oxygen pressure leaching, none of these pretreatment processes have found any commercial application.

Because pyrite and arsenopyrite can be decomposed by microwave energy, then pretreatment of a typical refractory gold ore or concentrate with microwave energy is a possible option. As microwave energy decomposes these refractory minerals, gold would be released from this mineral matrix along with arsenic and sulphur. Thus freed gold would be extractable by conventional alkaline cyanidation.

To this end, a typical arsenopyritic gold concentrate was irradiated with microwaves, as a pretreatment process for subsequent extraction of gold by the alkaline cyanidation process. This paper represents the data on microwave pretreatment of an arsenopyritic refractory gold concentrate.

GENERAL PROCEDURES

A gold mine in Ontario, supplied the flotation concentrate of 85% minus 74 µm (i.e., 200 mesh). Pyrite and arsenopyrite are the major gold-bearing minerals in the concentrate. Tables 1 and 2 show the analyses of particle size and chemical composition of the concentrate.

X-ray diffraction analysis of the concentrate showed that the major elements Fe, As and S were present as FeS_2 and FeAsS. On the basis of chemical analyses the approximate relative concentrations of FeS_2 and FeAsS were 26.25% and 21.30% respectively.

Microwave Irradiation of the Concentrate

A microwave oven (model BPH-6000-P4-SP, manufactured by Cober Electronic Inc., Stanford, Conneticut) was employed for irradiation of the concentrate. This microwave oven is equipped with 6 x 1 kw power magnetrons, a turn table, an exhaust vent, and a timer with maximum 15 minutes duration. The microwave oven is also equipped with a variable power adjustment device for the generation of continuous or pulse mode microwaves. Microwaves of 2450 MHz frequencies are generated by magnetrons and delivered to the applicator placed on the turn table.

500 g. of the dry concentrate was evenly spread in a silica tray of 30 cm x 15 cm x 3 cm and irradiated with 6 kw power for a period of 15 minutes. The silica tray itself is transparent to microwaves. In fact, all the irradiations were conducted in stages of 5, 4, 3 and 2 minute periods, respectively. The stagewise irradiations were conducted in order to reduce arcing. Generally, within the initial 5 minutes of irradiation white fumes (As_2O_3 and SO_2) started to evolve. During the next 5-7 minutes of irradiation the evolution of white fumes essentially ceased and the whole mass in the tray became almost red hot. It was further observed that as soon as the microwave irradiation was stopped the evolution of the fume also stopped essentially immediately (e.g. less than a minute for 500.0 g scale test). This observation obviously indicated that microwave does not necessarily initiate an autogeneous roasting

process. Towards the end of calcination frequent arcing took place which, however, could be prevented by applying low power or short irradiation time and frequent stirring of the hot mass. Temperatures of the hot mass were measured by turning off microwave power and almost immediately a thermocouple was brought in contact with the red hot mass. The temperature was found to be in the range of 420°C to 550°C during fuming period.

TABLE 1. Size distribution of the flotation concentrate

Mesh	Micron	% wt.
+ 48	+ 295	0.08
+ 65	+ 208	0.25
+ 100	+ 147	1.33
+ 150	+ 104	4.0
+ 200	+ 74	9.0
- 200	- 74	85.0

TABLE 2. Chemical analysis of the flotation concentrate

Element	% Wt	Elements	% Wt
Au	8.95 oz/t	Al	1.4
Ag	1.25 oz/t	SiO_2	29.83
As	9.81	Sb	0.18
Fe	24.79	Cu	0.14
S	18.27	C	0.94
Ca	1.58	Mg	0.13
Ni	0.16	Zn	0.41

At the end of calcination the hot mass was allowed to cool. The calcine thus obtained was light brown in colour and agreed quite well chemically as well as mineralogically with a similar roaster calcine. Average weight loss during microwave irradiation was 20% which agreed well with complete volatilization of As and S and simultaneous oxidation of iron to Fe_2O_3 (hematite). Table 3 shows mineralogical and chemical analyses of the microwave calcine. For the sake of comparison the analyses of the roaster clacine are also included in this table. A portion of the white fumes which evolved during microwave irradiation of the concentrate condensed as sublimate on oven walls and was found to be arsenious oxide (As_2O_3) with almost 97% purity.

Alkaline Cyanide Leaching of the Microwave Calcine

The microwave calcine without further grinding and sizing was leached with alkaline sodium cyanide in a reaction kettle equipped with a mechanical stirrer, air or oxygen sparging device and a pH electrode. Calculated amount of NaCN, NaOH and water were added to a known weight of the calcine in order to maintain the predetermined concentration of NaCN, pH and pulp density. Generally pH of the leach slurry was maintained in the range of 10.0-11.0. After solid liquid separation the filter cake was washed with water three times with approximate solid to liquid ratio of 1:1. The filtrates and the dry residues were chemically analyzed.

TABLE 3. Analyses of Microwave and Roaster Calcines

Calcine	Minerals	Elements (wt %)							
		Au (oz/t)	Ag (oz/t)	Fe	S	As	Sb	SiO_2	Cu
Microwave	Hematite, Quartz, Magnetite, and glassy material of ($FeSiO_3$ type)	10.64	1.47	31.9	1.20	1.50	0.04	35.90	0.18
Roaster	Hematite, Magnetite, Quartz	5.45	0.86	28.1	1.23	1.25	0.05	40.75	0.13

RESULTS AND DISCÜSSIONS

Microwave Irradiation of a Gold Flotation Concentrate

Microwave irradiation decomposed pyrite and arsenopyrite of the concentrate, but SO_2 and AS_2O_3 were detected in the volatiles and Fe_2O_3 (hematite), Fe_3O_4 and SiO_2 solids. The formation of SO_2 and As_2O_3 was undesirable but was unavoidable because of the built-in air cooling system of the oven necessitating constant circulation of air through the oven. Some of the fumes which condensed on the oven walls as sublimate were identified by chemical and X-ray diffraction analyses as As_2O_3 (97%) and some As_2S_3.

Chen and Coworkers (1984) confirmed that microwaves decomposed arsenopyrite into pyrrhotite and arsenic) (equ. 2). However, they conducted microwave irradiation on 0.5 - 1.0 g sample placed in a long silica tube (1-15 cm x d-1 cm) with an open end which kept closed during microwave irradiation. In reality, the hot solids in this silica tube was not fully exposed to hot air. However, in the present case, 500.0 g concentrate were irradiated in an open tray in the oven where air was in constant circulationa. Therefore, hot air was in direct contact with the hot solids and the volatiles which initiated the oxidation of As, S and Fe into As_2O_3, SO_2 and Fe_2O_3 (eq. 3,4,5).

$$2\ As + 3O = As_2O_3 \quad (3)$$

$$S + 2O = SO_2 \quad (4)$$

$$2\ Fe + 3O = Fe_2O_3 \quad (5)$$

These products are also formed during conventional heating of arsenopyrite at 550°C in the presence of oxygen (Strathedec and Pidgeon, 1961; Mackenzie, 1970).

Furthermore, DTA studies showned this gold concentrate started to decompose at 550°C and completed at 570°C, whereas decomposition by microwave irradiation of the same concentrate apparently started at 420°C and essentially completed at 550° C (temperature measurements were not precise - (please see general procedure). Therefore, it is obvious that microwave irradiation and thermal treatment of this gold concentrate essentially follow the same decomposition made and form similar kinds of final products.

If microwave irradiation of this concentrate could be conducted in the absence of oxygen (i.e. in vacuum or in an inert atmosphere), then the formation of these oxides could be avoided. Although the design of this model of oven offers very little scope to conduct irradiation under such a modified condition, nevertheless, several irradiation tests were conducted in N_2 atmosphere in a silica tube.

1.0 - 1.5 g of the concentrate was placed in a silica tube (l-30 cm x d-1 cm) with one closed end. Nitrogen gas was flushed through this tube. After 5-7 minutes of flushing, the open end was closed with a rubber stopper while N_2 gas was flushing through the tube. The closed tube was irradiated with 6 kW microwave energy for 5 minutes. An orange coloured sublimate desposite on the cooler surface in the tube. X-ray diffraction analysis confirmed the presence of pyrrhotite (FeS) in the dark calcine and As_2S_3 and S in the orange sublimate.

Furthermore, as microwave energy can penetrate most of the minerals to an appreciable extent, this property may be utilized in the calcination of particles larger than -200 mesh sizes. To this end, pyrite and arsenopyrite minerals of various particle sizes were irradiated. Particles of -10 mesh sizes underwent calcination easily with almost complete removal of As and S. Whereas particles larger than -10 mesh (or 2 mm) sizes suffered incomplete calcination. It was observed that during microwave irradiation of the larger particles (i.e. > 2 mm) S and As oozed out of the particles and cracks appeared in them. These test data indicated that size reducing below -10 mesh size are not always necessary for microwave calcination.

Current practice necessitates fine grinding (i.e. -200 mesh or below) in order to liberate gold particles. Size reduction to -200 mesh or below is a costly operation. Since microwave energy can be applied in the calcination of larger particles (i.e. -10 mesh size); therefore, there is a definite prospect of cost savings specially in the grinding operation.

Gold Leaching from the Microwave Calcine

The Conventional alkaline cyanide leaching was conducted on the calcine obtained by microwave irradiation of the refractory arsenopyritic gold concentrate. Various leach parameters such as the concentration of NaCN, retention time, pulp density and temperature were optimized for the maximum extraction of gold. Figures 2 and 3 show the gold and silver extraction profiles as a function of NaCN and the retention time respecitvely, whereas Tables 4 and 5 represent the similar data as a function of temperature and pulp density respectively. The highest percent extraction of gold (98%) and silver (60%) took place when the leach conditions were maintained at NaCN (8.0 kg/t), retention (24 h), temperature (30° C), pulp density (30% solids), pH (10.5) and 2 hours of preparation.

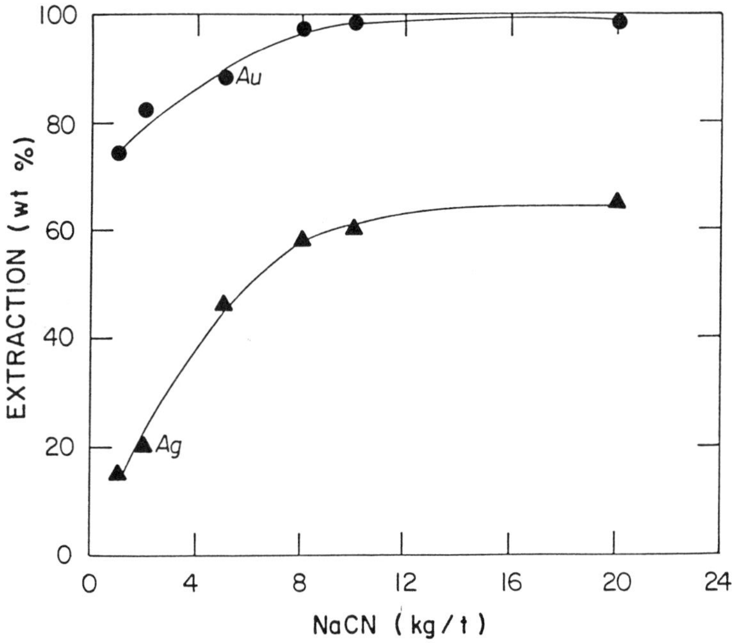

Fig. 2. Au and Ag extractions as a function of NaCN conc. (30°C), 30% solids, 24 hr. pH 10.5)

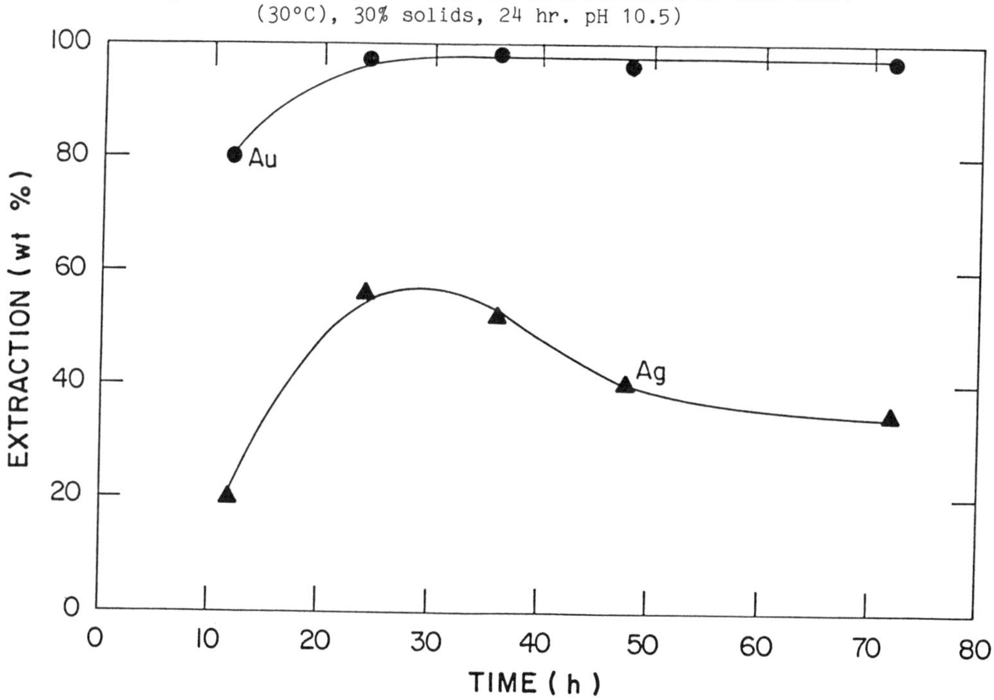

Fig. 3. Au and Ag extraction as as function of time. (NaCN - 8.0 kg/t, 30°C, 30% solids, pH 10.5)

TABLE 4. Leaching as a function of temperature
(NaCN - 8.0 kg/t, 30% solids, 24 h retention)

Temp (°C)	pH	Preaeration with O_2(g) (h)	Extraction (wt %)	
			Au	Ag
30	10.5	2	97	58
40	10.5	2	98	62
50	10.5	2	96	62

TABLE 5. Leaching as a function of pulp density
(NaCN - 8.0 kg/t, 30°C, 24 h retention)

Pulp density (% solids)	pH	Preaeration with O_2(g)	Extraction (wt %)	
			Au	Ag
20	10.5	2	97	60
30	10.5	2	98	59
40	10.5	2	97	61
50	10.5	2	93	62

Furthermore, an alkaline cyanide leach test was also conducted on the calcine obtained by microwave irradiation of the concentrate in a nitrogen atmosphere (test conditions were not optimized because of lack of sufficient feed material). However, the above leach conditions were applied and yielded 89% of gold extraction. This low gold extraction was most probably due to the presence of sulphides in this calcine because the major component of the calcine was pyrrhotite along with some uncalcined arsenopyrite.

Furthermore, since chemical and mineralogical properties of the calcines obtained either by microwave irradiation or by conventional roasting of the concentrate were the same; optimum leach conditions for the microwave calcine would be effective for efficient extraction of gold and silver from the roaster calcine. Consequently, several alkaline cyanidation leaches were conducted on the roaster calcine at the optimum conditions obtained for the microwave calcine. Table 6 shows the leach results. Gold extractions from both the calcines were 98%, but silver extractions were 65% from the roaster calcine and 60% from the microwave calcine. Whereas flotation concentrate responded poorly to alkaline cyanidation in terms of gold and silver exractions. Table 7 shows the concentration of principal components of the leach liquors.

TABLE 6. Gold and Silver Extractions from the Calcines

Feed	NaCw (kg/t)	pH	Pre-areation (h)	Pulp Density (% solids)	Temp (°C)	Ret. Time (h)	Extraction (wt%)	
							Au	Ag
Microwave Calcine	8.0	10.5	2	30	30°	24	98	60
Roaster Calcine	"	"	"	"	"	"	98	65
Flotation Concentrate	"	"	"	"	"	"	98	15

TABLE 7. Leach liquors composition

Calcine Leach Liquor	Concentration (g/l)							
	Free CN-	Au	Ag	Fe	Cu	As	S	Sb
Micro-wave	1.50	0.131	0.01	0.037	0.044	0.016	0.6	0.026
Roaster	1.20	0.068	0.007	0.017	-	0.009	0.69	0.014

Cyanidation of the concentrate irradiated as function of microwave power

As it appears now, pretreatment of an arsenopyritic refractory gold concentrate is possible with microwave energy. However, the question arises regarding the extent of pretreatment necessary for an efficient extraction of gold from the refractory concentrate. Generally, autogenous roasting of a sulphide concentrate leads to complete calcination, whereas microwave irradiation can really provide an option to control the degree of calcination of such a concentrate. To this end, this test program was initiated. In this test program, each sample of 100.0 g of the refractory gold concentrate was irradiated for 15 minutes with different power levels. Figure 4 shows the extraction data for gold and silver as a function of power.

These extraction data obviously indicate that the highest percent extraction for gold (98%) and silver (59%) were obtained from concentrates irradiated with 6 and 5 kw power respectively. The samples of the concentrate irradiated with 5 kw or 6 kw were essentially fully calcined (vide wt. loss data). These observations lead to the conclusion that for maximum extraction of gold and silver, almost complete calcination is necessary.

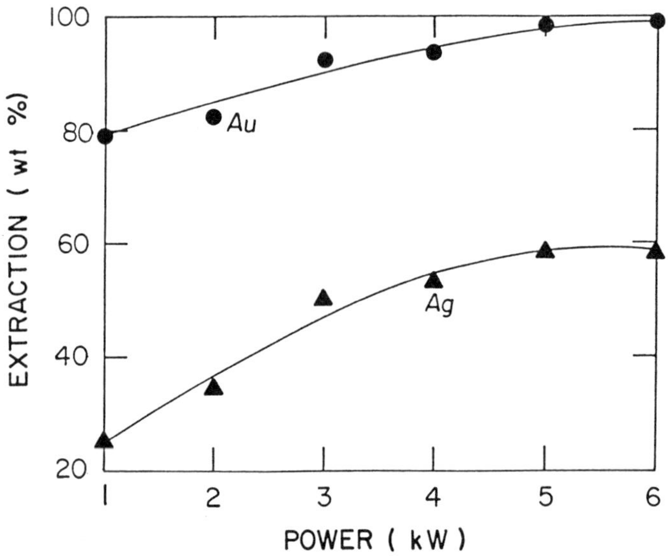

Fig. 4. Au and Ag extraction as a function of microwave power. (NaCN - 8.0 kg/t, 30°C, 30% solids, 24 hrs and pH 10.5).

As complete calcination was found to be necessary for an efficient extraction of gold and silver from an arsenopyritic concentrate, then it became necessary to determine how soon complete calcination could be obtained by employing full microwave power available in this oven. Consequently, five samples each weighing 100.0 g were irradiated with 6 kw power for 5, 8, 12, 13 and 16 minutes respectively. These time-prescribed irradiated calcines were leached separately with alkaline NaCN solution. Table 8 shows the extraction data for gold and silver and also the percent of weight loss for each of the samples.

TABLE 8. Leaching of Concentrates Irradiated for Different Times (NaCN - 8.0 kg/t, 30°C, 30% solids, pH 10.5, and 24 hours retention)

Irradiation time (min.)	Wt. Loss (wt%)	Preaeration (h)	Extraction (wt%) Au	Ag
5	1.6	2	76	34
8	10.2	"	89	45
12	13.2	"	97	56
13	17.0	"	98	59
16	20.1	"	98	60

The data in Table 8 clearly indicate that the highest percent extraction for gold (98%) and silver (59-60%) took place from the samples irradiated for 13 minutes. The weight loss data in Table 8 also indicate that more than 80% calcination took place during the first 13 minutes of irradiation.

As already stated in the introduction, the microwave power requirement is load dependent, that is, greater the load the more power or longer retention time is required. Therefore, the above data in Fig. 4 and in Table 8 do not represent any power or time optimization for complete calcination of the concentrate; rather, they indicated the fact that complete calcination is needed for efficient extractions of gold and silver from an arsenopyritic refractory gold concentrate.

Microwave Irradiation of the Concentrate in Presence of NaOH

The above discussions fully confirm that arsenopyrritic refractory gold concentrate can be pretreated with microwave energy. It was further observed that the overall chemistry of both microwave calcination and the conventional thermal roasting of the concentrate was essentially the same. For example, off-gases from both of these processes contain mainly arseious oxide (As_2O_3) and sulphur dioxide (SO_2), and the calcines contain hematite, quartz and some magnetite.

If As_2O_3 and SO_2 could be scavenged during microwave calcination of the concentrate then environmental concerns related to these oxides in off-gases would be eliminated. Accordingly, microwave irradiation was conducted on the concentrate in the presence of NaOH which was added in excess of stoichiometric requirement for the formation of Na_2SO_4 and Na_3ASO_4.

Typically, a mixture of 100.0 g concentrate and 73.0 g NaOH was irradiated with 5 kw power, but only for 4 minutes. During this time, the whole mixture essentially fused X-ray diffraction analysis of this calcine indicated the presence of Na_3AsO_4, Na_2SO_4, $FeSO_4$ and $NaFeO_2$. The formation of these products may have followed sulphating roasting mechanism (equs. 6, 8).

$$2\ FeAsS + 12\ NaOH + 7\ O_2 = 2\ NaFeO_2 + 2\ Na_3AsO_4 + 2\ Na_2SO_4 + 6\ H_2O \quad (6)$$

$$FeS_2 + 2\ NaOH + 3.50\ O_2 = Na_2Fe(SO_4)_2 + H_2O \quad (7)$$

$$FeS + 2\ O_2 = FeSO_4 \quad (8)$$

This irradiated calcine was pulverized and digested with water at 75°C. After solid-liquid separation, the filter cake was washed with water and dried. Net weight loss was 47% which agreed well with over 90% removal of aresenic and sulphur from the concentrate. Conventional alkaline cyanidation of the calcine yielded 99% of Au and 70% of Ag extractions.

During hot water digestion of the above fused mass, all the soluble species such as Na_3AsO_4, Na_2SO_4, $FeSO_4$ and $NaFeO_2$ went into solution. Arsenate (AsO_4^{-3}) can be removed by precipitating as $FeAsO_4$ under acidic contion either by adding Fe^{+3} to the solution or by oxidizing Fe^{+2} present in the solution with a suitable oxidant. After the removal of $FeAsO_4$ the solution contains Na_2SO_4 and some $Fe_2(SO_4)_3$ which are less toxic than arsenate.

The results of calcination of the concentrate with the addition of NaOH have some merits, but the optimization of NaOH requirement and the disposition of the liquors containing various sodium compounds, or possibly, the application of an alternative scavenger such as lime are yet to be studied.

Critical Assessment of Microwave Heating

Experimental observations clearly demonstrate that microwave energy can be applied in the pretreatment of refractory arsenopyritic gold concentrate. However, the overall results of microwave heating were the same with those of conventional roasting. Generally, microwave heating is rapid, highly efficient, immediately controllable, material-selective, requires no additional fuel nor a critical concentration of sulphur in the concentrate. Besides, microwave heating can be conducted on a suitable moving belt, on coarse particles (e.g. 2 mm size) and in an inert atmosphere.

Microwave heating of an arsenopyritic refractory gold concentrate yielded volatiles containing mainly As_2O_3 and SO_2 with essentially no gold particulates in it. However, the formation of these oxides could be avoided by conducting microwave irradiation in an inert atmosphere or in the presence of NaOH. Furthermore, microwave heating is much faster than conventional roasting; for example, 2-2.5 hours are required for conventional roaasting of a tonne of arsenopyritic gold concentrate (private communication with an operating gold mine in Ontario) as against ≈ 15 mins. by microwave heating.

The microwave heating system is still handicapped with the lack of reliable temperaturing device, because the metallic thermocouple does not respond to temperature measurements during microwave irradiation. At present, because of the type and nature of this test program (i.e. small scale batch operation and essentially preliminary in nature), and also the lack of data of actual power utilization during calcination, it is premature to estimate the capital and operating cost of a microwave heating system. However, the readers may find interesting information in the reports of Pei (1985), Smith (1984) and Vankoughnett (1970).

CONCLUSIONS

These preliminary test results lead to the following conclusions.

(1) The microwave energy has a positive technological potential in the pretreatment of the arsenopyritic refractory gold concentrate.

(2) Particles of 2 mm of sizes can be easily calcined with microwave energy.

(3) The formation of As_2O_3 and SO_2 could be prevented by conducting microwave irradiation of the concentrate in an inert atmosphere or in the presence of NaOH.

ACKNOWLEDGEMENT

The author expresses his sincere thanks to J.J. Laliberté for conducting the leach tests and the staff of the Chemical Laboratory for the chemical analysis. Finally, this author extends his sincere thanks to Dr. T.T. Chen of Mineral Processing Laboratory for mineralogical identifications and Dr. S.A. Mikhail of Extractive Metallurgy Laboratory for conducting the DTA studies on the arsenopyrite gold concentrate.

REFERENCES

Abu-Samara, A., J. S. Morris., and S. R. Kointyohann (1975). Wet Ashing of some biological samples in microwave oven. Anal. Chem., 47, (8), 1475-1477.

Berezowsky, R. M. G. S., and R. Weir (1984). Gold extraction from refractory concentrate. Presented in the CIM 14th annual hydrometallurgical meeting, Oct. 14-17, Timmins, Ontario, Canada

Bruynesteyn, A. (1984). Bioleaching of refractory gold/silver ores and concentrate. Presented in the CIM 14th Annual hydrometallurgical meeting, Oct. 14-17, Timmins, Ontario, Canada

Chen, T. T., J. E. Dutrizac, K.E. Haque, W. Wyslouzil, and S. Kashyap (1984).

The relative transparency of minerals to microwave radiation. Can. Metall. Quart., 23, (3), 349-351.

Dymoek, R. and H. Tuthill (1986). Microwave ninhydrin development - an evaluation. Identification Canada, 9, (3), p 10.

Hamilton, E. H. (1920). Chapters I and II in Manual of Cyanidation. McGraw-Hill Book Company, Inc.

Klein, A. J. (1985). Processing with microwaves. Advanced Materials and Processes, 1, (4), 36-39.

Kruesi, W. H., and P. R. Kruesi (1986). Microwave in laterite processing. Presented in the CIM 25th annual conference of metallurgists. August 17-20, Toronto, Canada

Lakshmanan, V. I., and A. M. Childs (1984). Treatment of refractory gold ores. Presented in the CIM 14th annual hydrometallurgical meeting, Oct. 14-17, Timmins, Ontario, Canada

Matthes, S. A., R. F. Ferrel and A. J. Mackie (1983). A microwave system for the acid dissolution of metals and minerals. U.S.B.M. TPR 120, Albany Res. Cent., U.S.A.

Metaxas, A. C. and R. J. Meredith (1983). In Industrial microwave heating. Peter Peregrinus Ltd., London, U.K.

Mackenzie, R. C. (1970). In Differential Thermal Analysis, Vol. I, p 198

Nadkarni, R. A. (1984). Applications of microwave oven sample dissolution in analysis. Anal. Chem., 56, (12), 2233-2237.

Pei, C. T. D. (1985). Applications of electrical microwave heating in food industries; and Applications of electrical microwave heating in Chemical industries. Canadian Electrical Association Report, CEA. No. 411, V. 481, Department of Chemical Engineering, University of Waterloo, Canada

Püschner, H. (1966). In Heating with microwaves, fundamental components and circuit techniques, Philips Library.

Smith, F. B. Cousins, J. Bozi, and W. Flora (1985). The acid dissolution of sulphide mineral samples under pressure in a microwave oven. Anal. Chim. Acta, 177, 243-245.

Smith, R. D. (1984). In Microwave Power in Industries, Report ERPI-EM 3645, Electrical Power Research Institute, Palo Alto, Calif., U.S.A.

Strathdee, B. A. and L. M. Pidgeon (1961). Thermal decomposition and vapour pressure measurements on arsenopyrite and an arsenical ore. CIM Bulletin, 59, 883-887.

Vaankoughnett, A. L. (1970). Unpublished data on fundamentals of microwave heating. Radio and Electrical Engineering Division, NRC, Canada, K1A-OR8.

Vreugde, M. J. A. (1983). The arsenoprocess for treating gold ores. Presented at the annual meeting of Alberta and British Columbia section of mineral processors, Vancouver, B.C., Canada

Weert, V. G., K. J. Fair and J. C. Schneider (1986). Prochem's Nitrox Process, CIM Bulletin, 79, (895), 84-85.

Westbrook, W. T. and R. H. Jefferson (1986). Dissolution of samples by heating with microwave oven in a teflon vessel for instrumental analysis. J. Microwave Power, 21, (1), 25-32.

Wright, K. P. (1961). Fluid bed roasting practice in the Red Lake Camp. CIM Bulletin, 59, (592), 595-600.

ACID PRESSURE OXIDATION OF REFRACTORY GOLD MINERAL CARRIERS

G.P. Demopoulos and V.G. Papangelakis

Department of Mining and Metallurgical Engineering
McGill University, Montreal, Quebec, Canada H3A 2A7

ABSTRACT

Acid pressure oxidation is currently strongly advocated as the most effective pretreatment method for refractory gold ores and concentrates. Arsenopyrite and pyrite are the most common mineral carriers of refractory gold. Pressure oxidation of these refractory gold mineral carriers is a very complex reaction system involving several oxidation and hydrolytic reactions which occur in series and/or parallel. Moreover these oxidation reactions are highly exothermic. Since the nature and composition of the reaction products can have a considerable impact on the downstream operations of gold and silver recovery, an understanding of the prevailing reaction regime is very important from the standpoint of process design and operation. Moreover the task of process optimization can be further facilitated with the aid of models simulating the performance of various pressure reactors operated isothermally or adiabatically. In this paper the chemistry and modelling of the acid (H_2SO_4) pressure oxidation of pyrite and arsenopyrite is reviewed on the basis of some recent research findings along with information obtained from literature. Finally some preliminary data on a novel acid pressure oxidation system (O_2-H_2SO_4-NaCl) for direct extraction of Au and Ag from the refractory materials is reported.

KEYWORDS

Pressure oxidation; pyrite; arsenopyrite; gold recovery; refractory gold; process chemistry; process modelling.

INTRODUCTION

Many of the gold deposits now being developed contain gold finely disseminated in refractory sulphide minerals such as pyrite (FeS_2) and arsenopyrite (FeAsS) - the two most common gold carriers (Boyle, 1979). In several of these deposits a significant portion of gold occurs in a submicroscopic form, probably as a lattice constituent of pyrite and arsenopyrite, so that even fine grinding fails to liberate the metal for recovery by cyanide leaching. The exact nature of this "invisible" gold within the structure of the pyrite and arsenopyrite is currently the subject of intensive mineralogical investigations (Mainwaring, 1985 and 1987; Chryssoulis, Cabri and Salter, 1987). These ores are the ultimate type of refractory gold since complete breakdown of the crystal structure of the host mineral by oxidation is necessary to liberate gold. The factors affecting gold extraction from refractory ores in general and the available processing options for their treatment were recently discussed by Demopoulos (1987). The classical commercial practice applied for the treatment of this type of refractory gold ores usually involves an oxidation roast (single stage or double stage depending on the As content) of a flotation concentrate to produce a porous calcine which is more amenable to conventional cyanidation. However this technology despite its established practice is not considered particularly attractive with today's standards for high product recovery and clean environment. Roasting is known to be very sensitive to heating parameters and feed composition (interfering impurities) and deviation from optimum conditions can result in reduction in permeability which is translated to incomplete gold liberation.

Thus a considerable amount of R & D work has been conducted in recent years with the purpose of developing hydrometallurgical oxidation processes instead for the treatment of refractory gold ores. Basically three aqueous oxidation approaches have been proposed: (a) pressure oxidation; (b) biological oxidation and (c) chemical oxidation. In pressure oxidation (Weir and Berezowsky 1984) oxygen and autoclaves are used at high temperatures and pressures; in biological oxidation (Lawrence and Bruynesteyn, 1983; Bruynesteyn, 1984) autotrophic bacteria (thiobacilli ferrooxidans) and either heaps (for ores) or agitated tanks (for concentrates) are employed, while in chemical oxidation strong oxidants such as chlorine gas (Guay, 1981), Caro's acid (Lakshmanan and coworkers, 1984) and nitric acid (Van Weert and coworkers, 1987) used in conjunction with agitated tanks operated under atmospheric pressure. Among these hydrometallurgical processes only pressure oxidation has gained commercial status so far, with the others being mostly at the (small or large) pilot plant scale. Homestake's McLaughlin gold plant (Guinevere, 1984) was the first example of industrial application of pressure oxidation to the treatment of refractory gold while recently a second plant in Brazil was due to come on stream (Weir and Berezowsky, 1986a). Pressure oxidation is also considered for the Porgera gold deposit in Papua, New Guinea (Weir and coworkers, 1986a and 1986b).

Pressure oxidation can be conducted in either alkaline or acidic media but it is only the latter approach that has been promoted to the level of commercial operation. Studies on alkaline pressure oxidation using sodium hydroxide and oxygen for the treatment of arsenical concentrates in particular have been reported in the Soviet Union (Khryashchev and coworkers, 1969; 1970). Alkaline pressure oxidation yields a residue basically composed of Fe_2O_3 while the sulphur and arsenic contents of the refractory minerals (i.e. FeS_2 and FeAsS) are solubilized completely as sodium sulphate and arsenate salts. At some point alkaline pressure oxidation was considered for the treatment of the Mercur gold ore body - a highly acid consuming ore (Mason and coworkers, 1984; 1986) but it was later abandoned when geostatistical studies revealed that only a portion of the total ore body was of refractory nature. Possibly further development of the alkaline pressure leach route has been hampered due to the high cost of reagents used (NaOH), the problem of arsenic isolation (Na_3AsO_4) and the release of undesirable contaminants such as mercury and thallium.

Thus it is the acid (H_2SO_4) pressure oxidation route that it is mostly advocated today as the preferred pretreatment method for the refractory pyritic type gold ores and/or concentrates. This route has been vigorously pursued by Sherritt Gordon Mines Ltd. (Weir and Berezowsky, 1986b and 1986c; Weir 1986a). Temperatures above the melting point of sulphur (118°C) and preferably above 170°C are employed to obtain complete oxidation of FeS_2 and FeAsS (and similarly of any other sulphide mineral present) with simultaneously minimal elemental sulphur formation. The latter is mostly undesirable since it results in occlusion of gold and unreacted sulphides during pressure oxidation and increased consumption of reagents during subsequent cyanidation. Work on similar lines has been reported by Muir and coworkers (1984) and Yen and Wyslouzil (1986). Others have instead proposed low temperature (< 118°C) pressure oxidation procedures to liberate gold from the refractory mineral structure (Wyslouzil and coworkers, 1980; Beattie and coworkers, 1985). From these two latter works only the method of Beattie and coworkers (the so-called ARSENO process) appears to be the subject of considerable R and D effort presently. It is suspected though that its chemistry differs radically from that of straightforward H_2SO_4 pressure oxidation since a non-disclosed "catalyst" is used.

In view therefore of the importance of high temperature acid pressure oxidation in extracting gold from refractory materials it was decided to study in detail the chemistry of the process in an effort to understand and optimize the whole operation. To this end the chemical reactions involved in the oxidation of arsenopyrite and pyrite (the two principal refractory gold carriers) are studied and chemically-based models are developed to be evaluated as an aid in reactor design and operation. Findings of this on-going research programme are discussed here in relation to the practice of H_2SO_4- pressure oxidation along with the results of some preliminary investigations on an alternative (H_2SO_4-NaCl) pressure oxidation system.

PRESSURE OXIDATION IN H_2SO_4 SOLUTIONS

Process Chemistry

Pyrite along with arsenopyrite are the most common host minerals of refractory gold. It is important from a process optimization standpoint to know the behaviour of each of these minerals during acid pressure oxidation. Although the aqueous oxidation of pyrite by molecular oxygen has attracted scientific interest for more than 100 years and has been investigated extensively by various researchers the same is not true for arsenopyrite. First a brief review of the chemistry of oxidation of pyrite will be given using data obtained from literature and then the chemistry of arsenopyrite oxidation will be considered based on the results of a recently completed experimental study (Papangelakis and Demopoulos, 1987).

Pyrite. Pressure oxidation of pyrite involves a number of reactions yielding ferrous ion, ferric ion, sulphate ion, and elemental sulphur as products. In industrial practice, most of ferric ion produced precipitates as hematite, basic ferric sulphate and/or jarosite since high slurry densities and temperatures are employed. The relative abundance of the various products formed depends on the applied conditions, namely, temperature, time, partial pressure of oxygen, acidity, and inert sulphate salt concentration.

From the large body of experimental data reported in literature (Lowson, 1982) only data from those investigations conducted at temperatures and pressures approaching the industrial practice of acid pressure oxidation are considered here. Some important features of these investigations are summarized in Table 1.

It has been now well established that the oxidation of pyrite follows two parallel competing paths: the first yielding sulphate sulphur (eq. 1) and the second yielding elemental sulphur (eq. 2) (Bailey and Peters, 1976)

$$FeS_2 + 7/2\ O_2 + H_2O \rightarrow FeSO_4 + H_2SO_4 \quad (1)$$

$$FeS_2 + 2O_2 \rightarrow FeSO_4 + S^o \quad (2)$$

It appears that reaction (1) is the dominant path at temperatures much above the melting point of sulphur. This is better seen by considering Fig. 1 which is

Table 1. Laboratory Studies on the High Temperature Acid (H_2SO_4) Oxidation of Pyrite

Material	Conditions	O_2 pressure Dependence	E_A kJ/mol	Mechanism	Reference
Natural pyrite	130-210°C 2.7-14.0 atmO_2	0.5	83.7	Surface Controlled-Chemisorption	Warren (1956)
Upgraded pyrite concentrate	100-130°C 0-4.04 atmO_2	1.0	55.7	Surface Controlled-Chemisorption	McKay and Halpern (1958)
Upgraded pyrite concentrate	130-165°C 6.1-23.8 atmO_2	0.5	70.3-77.4	Chemical control	Cornelius and Woodcock (1958)
Natural pyrite	85-130°C 0-66.4 atmO_2	0.5	51.1	Electro-chemical-Controlled by the Mixed Potential	Bailey and Peters (1976)
Natural pyrite	25-160°C 0-15.5 atmO_2	1.0	54.8	Surface Controlled-Chemisorption	Gerlach and coworkers (1966a)

based on data reported by Gerlach and coworkers (1966a). It can be seen that S^0 formation is reduced down to 10% of the total sulphur oxidized after 1 hr leaching at 160°C. This reduction in S^0 formation cannot be explained solely on the grounds

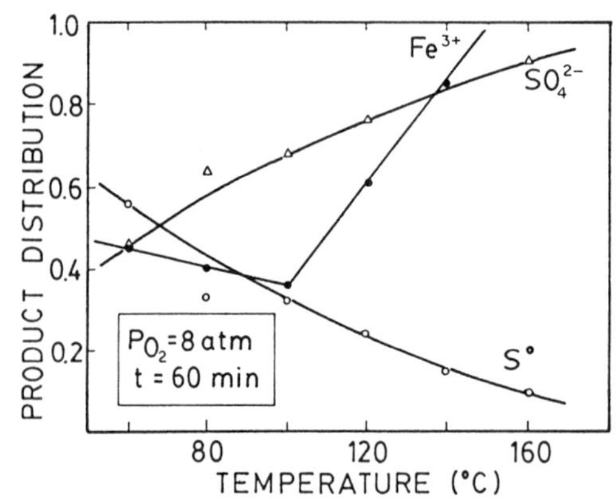

Fig. 1 Pyrite oxidation - the effect of temperature on product distribution (from Gerlach and coworkers, 1966a)

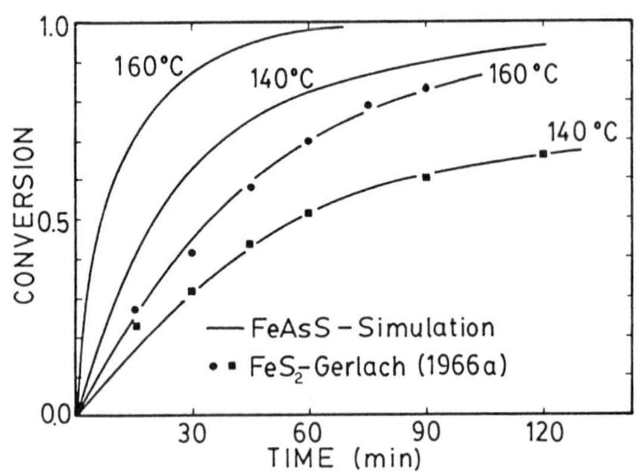

Fig. 2 Conversion vs time oxidation curves for pyrite and arsenopyrite concentrates (70% - 33μm) at 140 and 160°C.

of oxidation of the latter by O_2 as it was reported earlier (McKay and Halpern, 1958) since this reaction is known to proceed at very slow rates (Habashi and Bauer, 1960). Due to this kinetic stability of S^o, reactive minerals such as pyrrhotite which yield predominantly elemental sulphur could prove in practice even more "refractory" than FeS_2 in terms of gold recovery (Weir and Berezowsky, 1984). In addition to temperature, high O_2 pressures are also expected to suppress S^o formation by increasing the potential of the pressure leach slurry (Bailey and Peters, 1976), while increased acidities have the opposite effect (McKay and Halpern, 1958).

The oxidation of pyrite has been found to be chemically controlled (electrochemical and/or chemisorption control) exhibiting predominantly first order dependence at the lower O_2 pressure region (< 20 atm) and half order dependence at higher pressures. In addition to temperature and oxygen pressure fine particles also strongly favour the kinetics. Some typical oxidation curves for pyrite reported by Gerlach and coworkers (1966a) are reproduced in Fig. 2. It can be seen that for a concentrate of 70% - 33 μm size ∿ 82% of pyrite is oxidized at 160°C, 8 atm O_2 and 2 hrs. On the same graph the corresponding conversion curves for FeAsS of similar size distribution determined using a model developed by Papangelakis, Berk and Demopoulos (1986) are shown for comparison reasons. Despite however the extensive investigations performed by several researchers in the past no reliable rate equation representing the pressure oxidation of pyrite has been developed. This of course is a serious limitation from the standpoint of process modelling. The rate equations reported by McKay and Halpern (1958) and Gerlach and coworkers (1966a) are only partly valid since the surface area term (or equivalently the particle size effect) of these equations has not been correctly determined. Moreover all these previous investigations were limited to temperatures much below those applied in pressure oxidation of refractory gold ores. Kinetic investigations are currently in progress in the authors' laboratory dealing with these aspects of pyrite oxidation chemistry.

Ferrous ion produced by reactions (1) and (2) is subsequently oxidized to ferric ion by the following reaction:

$$2\ FeSO_4 + 1/2\ O_2 + H_2SO_4 \rightarrow Fe_2(SO_4)_3 + H_2O \qquad (3)$$

The role of ferric ion in catalysing the kinetics of pyrite decomposition during pressure oxidation has not been clearly determined. For example Gerlach and coworkers (1966a) reported nil effect while King and Lewis (1980) instead found ferric ion to have an accelerating effect. Weir and Berezowsky (1986b) on the other hand claim that high iron concentrations especially during the early stages of autoclaving favour the oxidation process.

Upon formation ferric ion undergoes hydrolysis reactions, the extent of which depends on the prevailing solution conditions (i.e. Fe^{3+} and H_2SO_4 concentrations, temperature, etc.). For temperatures above 150°C the following two principal hydrolysis reactions occur (Tozawa and Sasaki, 1986):

$$Fe_2(SO_4)_3 + 3H_2O \rightarrow \underbrace{Fe_2O_3}_{\text{hematite}} + 3H_2SO_4, \text{ at lower acidity} \qquad (4)$$

$$Fe_2(SO_4)_3 + 2H_2O \rightarrow \underbrace{2FeOHSO_4}_{\text{basic ferric sulphate}} + H_2SO_4, \text{at higher acidity} \qquad (5)$$

Precipitation of jarosite-type compounds is also known to occur (Weir and Berezowsky, 1986a) but to a lesser degree than Fe_2O_3 and $FeOHSO_4$:

$$3Fe_2(SO_4)_3 + 14H_2O \rightarrow 2H_3OFe_3(SO_4)_2(OH)_6 + 5H_2SO_4 \qquad (6)$$

(↑ hydronium jarosite)

From the process economics standpoint, hematite is the preferred form of iron precipitation (Weir and Berezowsky, 1986b) since it facilitates the subsequent neutralization and gold recovery operations. The upper limit of H_2SO_4 concentration allowing Fe_2O_3 precipitation is 55 g/l at 170°C and 70 g/l at 200°C. Maintenance in the leach solution of a high concentration of an additional inert salt such as $MgSO_4$ helps to extend this upper limit of hematite precipitation up to 100 g/l H_2SO_4 (Tozawa and Sasaki, 1986). The mode of action of $MgSO_4$ is to reduce the activity of protons thereby permitting higher concentrations of H_2SO_4 to be tolerated during hematite formation (Demopoulos, Kondos and Papangelakis, 1987). This beneficial effect of $MgSO_4$ on hematite precipitation during pressure oxidation of refractory gold ores is exploited by Sherritt Gordon Mines Ltd. (Weir and Berezowsky, 1986b).

Before this section on pyrite oxidation is closed it is considered important to remind the reader of the existence of two forms of iron disulphide, pyrite and marcasite. The latter is considerably more reactive than pyrite. Thus according to Gerlach and coworkers (1966b) more than 50% of marcasite (72% - 33 μm) oxidation occurs at 80°C and 8 atm O_2 pressure after 2 hours as opposed to only 25% of pyrite oxidation under the same conditions. Moreover, pyrite is known to exist into at least two distinct morphological forms, the well-developed crystal form, cubic (or euhedral) form - refractory in terms of reactivity - and the framboidal form which is rather poorly crystalline and possesses high specific surface area - non-refractory in terms of reactivity (Lowson, 1982). This latter form responds well to relatively mild oxidation methods as it happens with the Carling gold ore which is subjected to the Double Oxidation process (Guay, 1981). Thus it is mandatory an extensive mineralogical study to be performed on the refractory gold ore being under testing in order to identify the exact nature of the gold mineral carriers and thereby select the most appropriate pretreatment method.

Arsenopyrite. Berezowsky and Weir (1984) appear to be the only authors who have reported some information on the chemistry of acid pressure oxidation of arsenopyrite as a complement to Sherritt Gordon's development work for the treatment of refractory gold ores. These investigators postulated that arsenopyrite is first oxidized to yield trivalent arsenic (H_3AsO_3) and divalent iron which are subsequently oxidized further to pentavalent (H_3AsO_4) and trivalent states respectively. The same workers recognized that elemental sulphur also forms to some extent (in addition to sulphate sulphur), under relatively mild conditions (100-160°C) and in the presence of high acid concentrations. No postulation as to the mechanism of elemental sulphur production, i.e. intermediate product or product of a parallel reaction was made, nor kinetic data was reported. In addition to the above oxidation products the formation of ferric arsenate as a result of iron(III) hydrolysis was also reported.

The chemistry (product identification and kinetics) of acid pressure oxidation of arsenopyrite has been recently studied in detail by the present authors. Details of this study are given elsewhere (Papangelakis and Demopoulos, 1987). The distribution of reaction products was determined as a function of time, temperature, acid concentration and slurry density. Here the temperature effect results were chosen to be presented (Fig. 3). The top graph gives the effect of temperature on the conversion of arsenopyrite in terms of oxidation and precipitation. The bottom graph gives the respective product distribution in the form of ratios of Fe(II) to total iron extracted, S^0 to sulphur extracted and As(III) to total arsenic

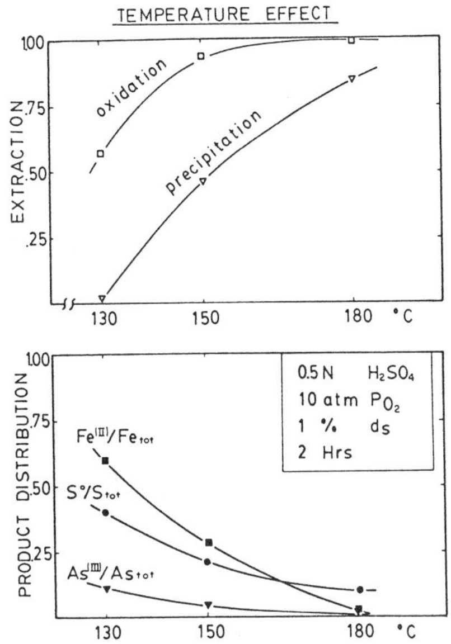

Fig. 3 Arsenopyrite oxidation - the effect of temperature on product distribution (from Papangelakis and Demopoulos, 1987).

extracted. It can be seen that all arsenic reports in solution as As(V) while the relative abundance of Fe^{2+} decreases with temperature. Similar effects on Fe^{2+} abundance had also an increase in slurry density, and reaction time. This behaviour of Fe^{2+} distribution is in agreement with the well established rate equation for oxidation of Fe^{2+} to Fe^{3+} by molecular O_2 in H_2SO_4 solution (reaction 3) (McKay and Halpern, 1958):

$$-\frac{dN_j}{dt}\frac{1}{V}\frac{1}{\nu_j} = 3.10^7\ e^{-69kJ/RT}|Fe^{2+}|^2 \cdot P_{O_2}\ (\text{mol } L^{-1}\ \text{min}^{-1}) \qquad (7)$$

where N_j is the number of moles of Fe^{2+}, ν_j is the stoichiometric coefficient of Fe^{2+}, t is the time (min), V the volume of the solution (L) and P_{O_2} the partial oxygen pressure (atm). On the other hand the formation of elemental sulphur was found to be suppressed by increasing temperature as well by increasing time and acidity in analogy, in other words, to the behaviour exhibited by pyrite (McKay and Halpern, 1958). It is of interest to note that even at 180°C still a 10% S^0 formation is observed.

Based on the experimental findings of this work the following reactions are proposed to account for the oxidation of arsenopyrite:

$$2FeAsS + 13/2\ O_2 + 3H_2O \rightarrow 2H_3AsO_4 + 2FeSO_4 \qquad (8)$$

$$2FeAsS + 7/2\ O_2 + 2H_2SO_4 + H_2O \rightarrow 2H_3AsO_4 + 2FeSO_4 + 2S^0 \qquad (9)$$

In other words it is postulated, that elemental sulphur formation is not an intermediate product but rather the result of a competing parallel reaction. Apparently, the different kinetic characteristics of the two competing reaction paths lead to variation in the distribution of S^0 and SO_4^{2-} products. The results of Fig. 3 support Sherritt Gordon's strategy of applying very high oxidation temperatures to prevent the formation of elemental sulphur. Moreover it is postulated that H_3AsO_4 is the initial form with which oxidized arsenic reports in solution. In other words H_3AsO_3 in analogy to S^0 is considered as the product of a parallel competing reaction rather than as an intermediate product. This view is supported by the results of a complemental study (Kondos and Demopoulos, 1987) in which molecular oxygen under comparable conditions (150°C, 10 atm O_2, 0.5 N H_2SO_4) failed to oxidize synthetically prepared H_3AsO_3 solutions. It is obvious that reaction (3) takes also place during pressure oxidation of arsenopyrite.

During these laboratory investigations, precipitation was found to occur under conditions of product build up in the solution (high slurry density, longer reaction times), high temperatures, and low acidities. X-ray diffraction analysis of the collected precipitates identified a single compound, that is scorodite, $FeAsO_4.2H_2O$. The respective formation reaction which is endothermic in nature (Papangelakis, Berk and Demopoulos, 1986) is given below:

$$Fe_2(SO_4)_3 + 2H_3AsO_4 + 4H_2O \rightarrow 2FeAsO_4.2H_2O + 3H_2SO_4 \qquad (10)$$

The precipitates were well crystallized exhibiting sharp diffraction peaks. A representative SEM picture of the produced precipitates is given in Fig. 4 which confirms the excellent crystallinity of ferric arsenate formed under pressure oxidation conditions (180°C; 10 atm O_2, 0.5 N H_2SO_4, 2h). This property is of

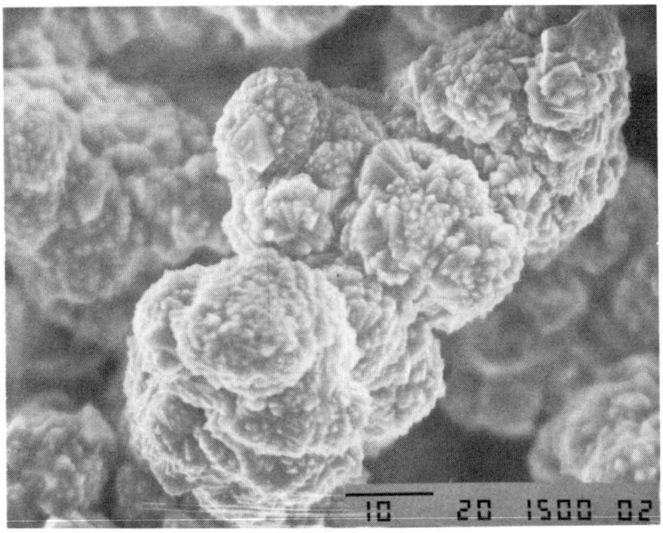

Fig. 4 Scanning electron microscopy photograph of scorodite ($FeAsO_4.2H_2O$) crystals produced during H_2SO_4-pressure oxidation of arsenopyrite at 180°C (from Papangelakis and Demopoulos, 1987).

critical importance in view of the current controversy surrounding the environmental stability of these compounds. Krause and Ettel (1987) have recently demonstrated that well crystallized scorodite exhibits two orders of magnitude lower solubility than reported previously in the literature (Robins, 1981). Thus it can be concluded that from the environmental standpoint pressure oxidation is expected to produce less troublesome arsenic residues than other hydrometallurgical processes - biological and chemical - operated under milder oxidation conditions.

In terms of kinetics, arsenopyrite oxidation is faster than pyrite oxidation as it is generally believed but not to the extent to allow preferential leaching to be feasible. Comparative oxidation curves for the two minerals are given in Fig. 2. Typical conversion curves for closely sized arsenopyrite fractions at various temperatures are plotted in Fig. 5. The reaction with the aid of shrinking core models (Papangelakis and Demopoulos, 1982) was found to be chemically controlled and the following intrinsic rate law was found to apply at the upper temperature region studied (i.e. T > 160^oC) where reaction (8) dominates:

$$- \frac{dN_i}{dt} \frac{1}{S} \frac{1}{\nu_i} = 12.385\ e^{-72.1\ kJ/RT}\ P_{O_2}\ \text{mol FeAsS min}^{-1}\ \text{cm}^{-2} \tag{11}$$

where N_i is the number of moles of arsenopyrite, ν_i is the stoichiometric coefficient of FeAsS, t is the time (min), S the total surface area (cm^2) of FeAsS particles and P_{O_2} the partial oxygen pressure (atm).

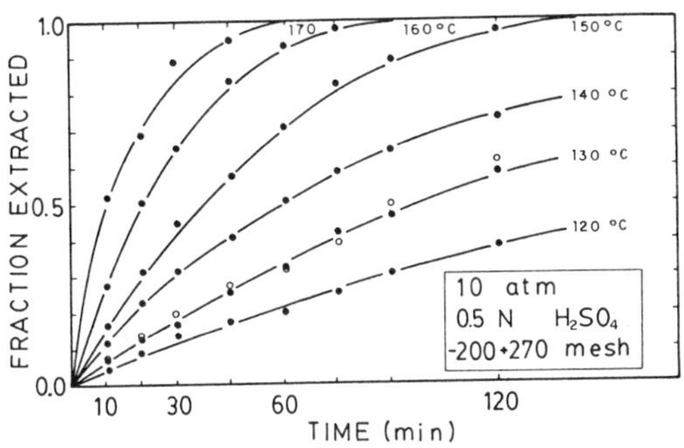

Fig. 5 Arsenopyrite oxidation - the effect of temperature on conversion (from Papangelakis and Demopoulos, 1987).

Process Modelling

Developing a global reaction model to represent all the reactions taking place during pressure oxidation could prove a very valuable tool from the standpoint of reactor design and process optimization. With the aid of these models computer-assisted simulation runs of alternative operational schemes can be executed in a relatively inexpensive way. This modelling and simulation approach is currently under development at McGill. The modelling procedure basically involves the set up of a series of mass balance and energy balance equations. The extent of each reaction is related either to the respective rate equation or to the equilibrium constant. For example in the case of arsenopyrite the extents of reactions (8) (3) and (10) were taken as the basis of the model formulation. For the first two reactions (8) and (3) their corresponding rate equations were used (i.e. eqs. (7) and (11) while reaction (10) was assumed to be always at equilibrium.

So far our modelling work has covered the acid pressure oxidation of arsenopyrite in a batch (Papangelakis, Berk and Demopoulos, 1986) and in a single CSTR reactor (Papangelakis, Berk and Demopoulos, 1987). Further work is currently underway to simulate the operation of a series of CSTR's which is the preferred mode of commercial practice (Weir, 1986b). Finally this work is to include eventually the pressure oxidation of predominantly pryite materials or mixed pyrite/arsenopyrite materials which of course represent more closely the composition of real concentrates.

Pressure oxidation of a high grade arsenopyrite concentrate (67% - 37 μmn) in a batch reactor was simulated both for isothermal and adiabatic operation. The pressure reactor was assumed to be a single vertical autoclave with 60 m^3 total volume and 2/3 static operating level. The model can predict at any time the solids and solution profiles in the reactor. The solids profile for three oxidation temperatures and a 20% solids slurry density is given for example in Fig. 6. The total amount of solids at 190^oC is also given. The corresponding solution profile for the same reactor operating conditions is shown in Fig. 7. Ferrous iron

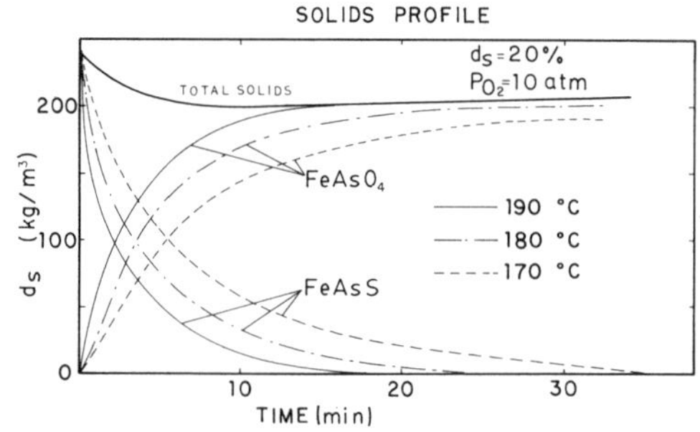

Fig. 6 Simulation of the batch pressure oxidation of FeAsS - the solids profile (from Papangelakis, Berk, and Demopoulos, 1986).

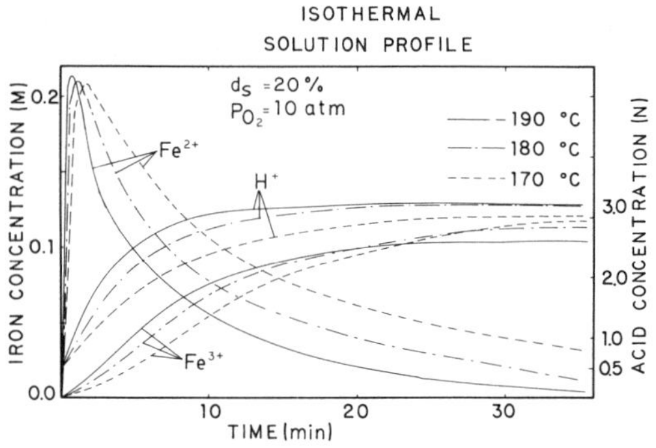

Fig. 7 Simulation of the batch pressure oxidation of FeAsS - the solution profile (from Papangelakis, Berk, and Demopoulos, 1986).

concentration is initially increasing rapidly followed by a gradual decay. The final ferric iron concentration level is lower at 190°C due to the increased $FeAsO_4$ precipitation at higher temperatures. The model of course at any time can predict the oxidation extent of a wide size distribution concentrate, as for example this is indicated with the simulation runs (FeAsS) plotted in Fig. 2 (70%-33 µm material).

In an oxygen pressure leaching system, two important process features that have to be critically considered are mass transfer and heat removal. There are limitations as to the maximum oxygen transfer rates in a given stirred tank reactor as well as to the heat removal capacity with a certain type of heat exchanger. The model permits the calculation of these two values which can subsequently aid the selection of industrial equipment. The reader is referred to the original research papers for further discussion of these features of the model (Papangelakis, Berk and Demopoulos, 1986; 1987).

Due to the highly exothermic nature of the oxidation of FeAsS and FeS_2 autogenous (or otherwise adiabatic) operation is preferred. The model allows to scan several combinations of starting temperatures and slurry densities in order to identify the conditions that would allow adiabatic operation at the desired final temperature. Some typical adiabatic curves are illustrated in Fig. 8. The temperature range 180 to 200°C was chosen as the preferred target final temperature. The graphs predict that some preheating of the concentrate will be required to reach the desired temperature region. A summary of the results of the batch adiabatic runs simulating the pressure oxidation of arsenopyrite are shown in Fig. 9. The top graph shows the final temperature obtained at different starting temperatures and slurry densities, while the bottom graph gives the time required to achieve 98% oxidation under the same conditions.

Some results from the simulation of the continuous pressure oxidation of arsenopyrite in a single CSTR are given in Figs. 10 and 11. In these graphs the rates of heat generated and heat removed of the system along with the average conversion as a function of the temperature of operation for a nominal retention time of 100 min (Fig. 10) and 50 min respectively (Fig. 11) are given. For these simulation runs two different slurry densities were considered (namely 5 and 10% solids) while the reactor working volume (40 m³) and the size distribution of the concentrate (67% - 37 µm) were the same as those of the batch simulation runs. The rate of heat removed lines are determined by the equation $R(T) = FCp|T - To|$ where F is the molal flowrate of H_2O, Cp is the heat capacity of H_2O, T is temperature of

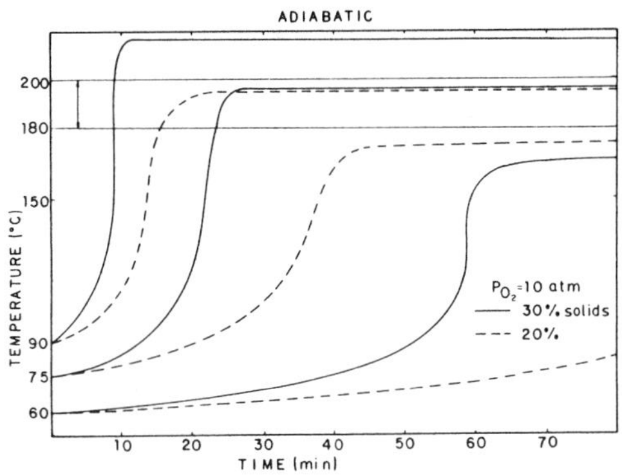

Fig. 8 Simulation of the adiabatic batch pressure oxidation of FeAsS - variation of reactor temperature with time at different slurry densities and starting temperatures (from Papangelakis, Berk and Demopoulos, 1986).

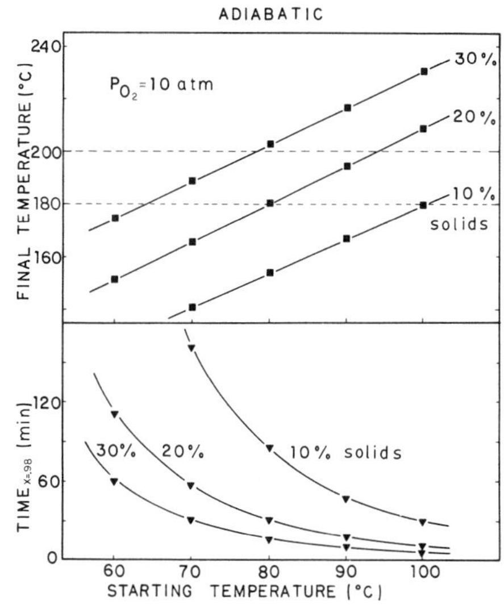

Fig. 9 Simulation of the adiabatic batch pressure oxidation of FeAsS - the effect of starting temperature for different slurry densities on the final temperature and retention time (from Papangelakis, Berk and Demopoulos, 1986).

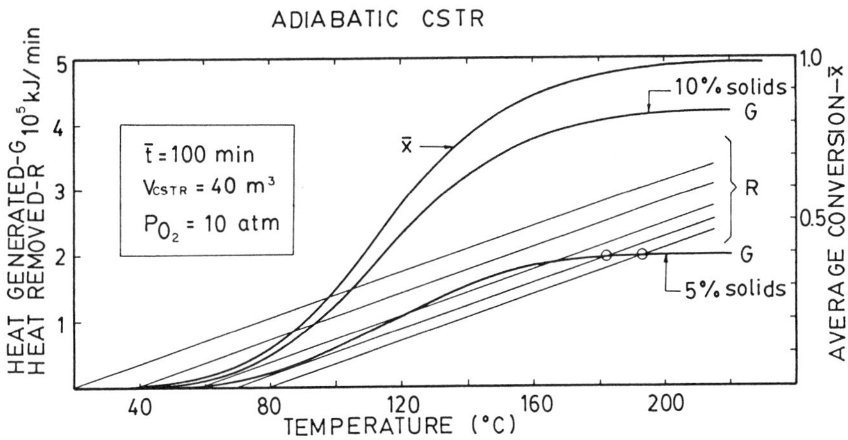

Fig. 10 Simulation of the adiabatic pressure oxidation of FeAsS in a CSTR- rates of heat generated and heat removed and conversion vs temperature curves for 100 min retention time (from Papangelakis, Berk, and Demopoulos, 1987).

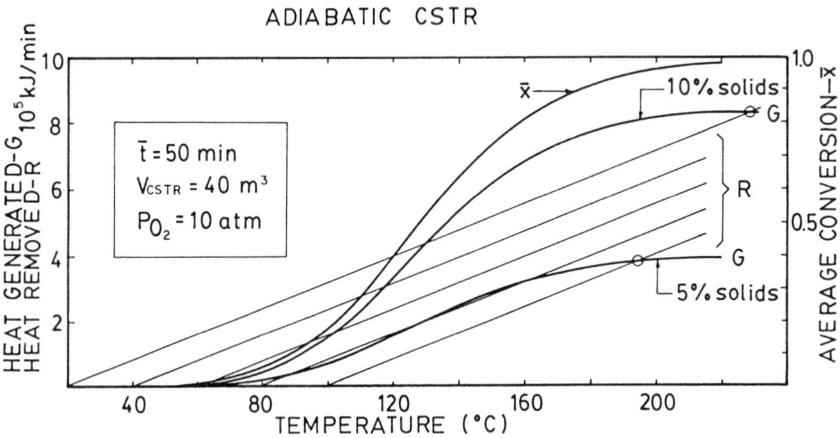

Fig. 11 Simulation of the adiabatic pressure oxidation of FeAsS in a CSTR - rates of heat generated and heat removed and conversions vs temp. curves for 50 min retention time (from Papangelakis, Berk, and Demopoulos, 1987).

operation and To is the inlet temperature. The capability of the system to remove heat depends solely on the flow rate of water which justifies the use of the term adiabatic in the present case. External cooling can of course be applied which would result in increasing the slopes of these lines and thus increasing the overall rate of heat removal. The importance of the diagrams of Figs. 10 and 11 is that they allow to identify conditions of steady state operation of the pressure reactor. The latter are defined by the crossing points of the rate of heat generated curves with the corresponding rates of heat removed lines. For example for the specific conditions of Fig. 10 steady-state adiabatic operation is possible only with the 5% solids slurry density and an inlet slurry temperature of 70°C or 80°C which respectively give 180 and 190°C operating temperatures and at least 95% conversion. Other steady-state points obtained at lower or higher temperatures are considered undesirable from an operational point of view. Obviously according to the model predictions of Fig. 10 provision for a cooling system is needed to process a concentrate of 10% solids or higher slurry density. A reduction in the nominal retention time to 50min by increasing the flow rate and using the same volume reactor makes, according to the model predictions of Fig. 11, possible the adiabatic operation of the single autoclave with 10% solids at a temperature slightly above 220°C, at 20°C inlet slurry temperature and 98% conversion.

Although aonsiderably more work is needed before these models are capable of simulating the actual industrial processing of real concentrates, it is the opinion of these authors that the process hydrometallurgist can be aided greatly in performing his difficult tasks of reactor design and process optimization.

Au and Ag Recovery

Following pressure oxidation the oxidized solids containing the precious metals are subjected to several stages of thorough washing and neutralization prior to cyanidation while the generated acidic solution is neutralized and the precipitated base metal compounds are disposed. A portion of the solution might be recycled to the pressure oxidation and the preceding feed preparation stages. Provided that elemental sulphur formation was limited (recycling of oxidized solids to the autoclave has been found to be an effective means to overcome some of the problems associated with S^0 formation (Weir 1986a)) and no significant slimes were generated during neutralization, very high (> 95%) gold recoveries are realized in the subsequent cyanidation stage.

Recently it was suggested (Yen and Wyslouzil, 1986) the use of thiourea for the extraction of gold directly from the pressure leach slurry without pH adjustment and oxidant addition. Thiourea consumption was kept at a minimum by using SO_2 to control the emf during leaching according to Schulze's method (1984). Thiourea leach vs cyanidation of the pressure oxidized solids appears to offer a number of advantages: (i) it avoids the extensive washing and neutralization operations, thus reducing the formation of slimes, the viscosity of the slurry and the cost of lime addition; and (ii) it prevents cyanide consuming side reactions.

In terms of silver recovery pressure oxidation in H_2SO_4 solutions gives a rather poor performance. Silver is initially liberated from the refractory sulphide minerals but subsequently most of it is incorporated in the jarosite compounds precipitating during pressure oxidation thus rendering Ag refractory again. This loss in silver recovery due to jarosite formation appears to prevail even under conditions favouring hematite precipitation (Weir and Berezowsky, 1986a) contrary to the findings of a recent fundamental study (Bull and coworkers, 1985). To enhance silver recovery from these jarosite-type residues it has been proposed to incorporate in the flowsheet a hot lime treatment stage that would decompose the jarosites and thus liberate the silver values (Weir and Berezowsky, 1984; Weir, Robinson and King, 1986b). In view of the rather intensive character of this treatment (90°C; 20 - 100 kg CaO/ton conc, 2 hr, several tanks) it would be expected this approach to be feasible only in the case of high silver content ores. An additional concern arising from this treatment is the potential of releasing arsenic by destabilizing the ferric arsenate precipitates. Direct thiourea leaching instead of cyanidation of the oxidized solids is also expected to yield low silver recoveries as a result of the refractory nature of Ag-bearing jarosites.

PRESSURE OXIDATION IN H_2SO_4-NaCl SOLUTIONS

Based on thermodynamic (Fig. 12) and other considerations these authors decided to explore the possibility of modifying the present H_2SO_4 pressure oxidation process with the addition of chloride salts in order to achieve simultaneous solubilization of the gold and silver values. It was thought that maintaining a high concentration of Cl^- ions in the pressure leach reactor it would be possible to complex gold and silver (eq. 12 and 13) as soon as these values will be liberated and thus achieving both operations, i.e. decomposition of refractory minerals and Au/Ag recovery, in a single stage.

$$4Au + 3O_2 + 16Cl^- + 12H^+ \rightarrow 4AuCl_4^- + 6H_2O \quad (12)$$

$$4Ag + O_2 + 4Cl^- + 4H^+ \rightarrow 4AgCl^0(aq) + 2H_2O \quad (13)$$

For the initial exploratory investigations HCl-NaCl leach solutions were used although it was understood that eventually H_2SO_4-NaCl solutions would be preferable since H_2SO_4 is in-situ generated in the reactor by the oxidation and hydrolytic reactions and could be recycled as well.

The exploratory investigations involved the use of samples from an auriferous concentrate ("A") (75% - 74 μm) having the following chemical composition: 307 g/ton Au, 43 g/ton Ag, 25% Fe, 18.3% S and 9.8% As and containing primarily pyrite (~30% of the total concentrate mass) and arsenopyrite (~20%) as the major sulphide minerals and silica (~30%). Direct cyanidation of this concentrate yields

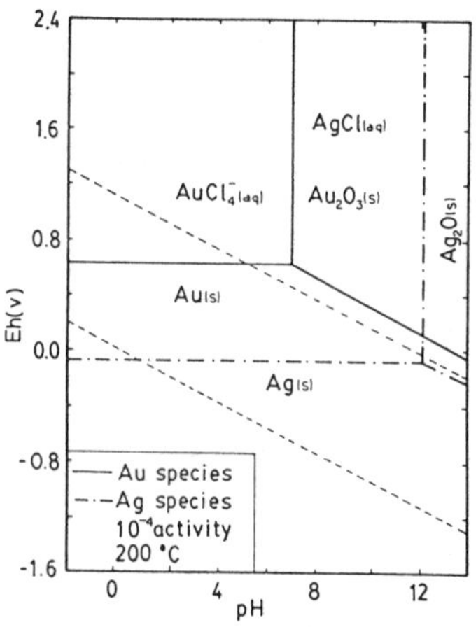

Fig. 12 E_h-pH diagram for the Au-Cl-H_2O and Ag-Cl-H_2O systems at 200°C- activities of all aqueous species 10^{-4} (F*A*C*T, McGill University).

Table 2. Co-extraction of Au and Ag from Concentrate "A" by Pressure Oxidation in a Mixed Sulphate-Chloride Acidic Medium (10% solids; 4N NaCl; 20 atm O_2).

Temp. (°C)	Time (h)	|HCl| (N)	% Extraction Au	% Extraction Ag
170	2	1.0	88.2	94.2
185	2	0.5	37.3	56.6
185	2	1.5	98.9	99.5
200	2	1.0	90.2	95.0
200	4	1.5	99.5	98.0

approximately 75% gold extraction. The parameters investigated were temperature (170 to 200°C), time (2 to 6 hrs.), HCl concentration (0.5 to 1.5 N), and type of chloride addition (4N NaCl or $CaCl_2$) while slurry density and oxygen pressure were kept constant at 10% solids and 20 atm respectively. These tests revealed that NaCl is a better source of complexing Cl^- ions than $CaCl_2$ since the latter results in gypsum precipitation which interferes with the gold solubilization reaction. Other investigators who advocate the use of $CaCl_2$ for similar type of experiments (Langhans and Lei, 1986) apparently have worked with sulphate-free solutions and thus have not observed this difficulty.

Typical results obtained from a series of statistically designed experiments are shown in Table 2. This set of results shows that gold and silver extraction with one-step pressure leach is technically feasible and recoveries as high as 99% can be obtained by maintaining a relatively high acid strength (1.5 N) in the solution to avoid iron compound precipitation. The experiments with 0.5 N HCl which resulted in poor precious metal extraction were associated with the production of redish color (hematite) precipitates. Some elemental sulphur observed to form under high acid concentrations (1.5 N) did not appear to hinder the extraction of the precious metals.

Following this first successful series of experiments a second auriferous concentrate ("B") of a different composition was tested to evaluate its amenability to the new chloride-based pressure oxidation approach. Concentrate "B" was somewhat coarser (53% - 74 μm) than conc "A" and had the following composition: 34.6 g/ton Au, 27.6 g/ton Ag, 40.6% Fe, 42.1% S and 8.5% As. In terms of mineral content this concentrate contained much more pyrite (∿73%) than concentrate "A" while arsenopyrite represented ∿18% of the concentrate mass. This concentrate under conventional cyanidation testing yields only 5% Au extraction. Some of the obtained results with concentrate "B" along with the applied pressure leaching conditions are given in Table 3.

Considering the predominantly pyritic nature of this concentrate (pyrite being a more refractory mineral than arsenopyrite, see Fig. 2) and the short reaction time (2 hrs) applied for most of the tests of Table 3 the obtained recovery values are considered good. Moreover the results obtained with H_2SO_4 are particularly encouraging from a practical standpoint. It is postulated that the action of acid

Table 3. Co-extraction of Au and Ag from Concentrate "B" by Pressure Oxidation in a Mixed Sulphate-Chloride Acidic Medium (10% solids, 20 atm O_2).

Temp. (°C)	Time (h)	Acid (N)	NaCl (N)	% Extraction Au	% Extraction Ag
170	2	1.5 HCl	2.5	58.7	82.1
185	2	1.5 HCl	2.5	76.1	95.2
200	2	0.5 HCl	2.5	53.9	96.8
200	2	1.5 HCl	2.5	93.8	98.1
200	4	1.5 H_2SO_4	4.0	60.7	98.8
200	4	3.0 H_2SO_4	4.0	99.0	99.5

is mainly to prevent extensive precipitation of iron compounds (such as Fe_2O_3, $FeOHSO_4$, or $FeAsO_4$) which appear to passivate the liberated gold values and thus hinder its solubilization. On the other hand it is the presence of NaCl which is responsible for the complexation and thereby solubilization of gold. It appears silver extraction to be unaffected by the precipitated solids due to its apparent different mode of occurrence in the ore. Moreover losses of silver by jarosite and/or AgCl precipitation are avoided in the present system due to the complexation of silver with Cl^- ions (Dutrizac, 1985; Dinardo and Dutrizac, 1985). Work is presently underway to investigate further this novel pressure oxidation approach in recovering gold and silver from refractory mineral carriers.

SUMMARY

Gold production from refractory ores is expected to become increasingly significant in the future as gold prices rise and fewer "free milling" deposits become available. Currently major technological efforts are underway to develop effective and environmentally acceptable methods for the treatment of these ores. Hydrometallurgical processes appear to attract most of this R & D activity. Among those only acid pressure oxidation has been demonstrated successfully at the commercial level, and the use of this technology is expanding.

As is often the case, the development of the technology has outpaced our understanding of the reactions involved. In this paper a review of the available information on the high temperature aqueous oxidation chemistry of pyrite and arsenopyrite - the two most important carriers of refractory gold - was made and the chemical factors affecting the metallurgical performance of pressure oxidation were identified. Areas warranting further fundamental research studies were indicated. Also, the potential benefits of developing mathematical models for computer-assisted simulations of alternative reactor operational schemes as a process optimization tool were outlined.

Finally some recently initiated investigations involving direct oxidation of the refractory minerals and simultaneous extraction of gold and silver by employing a mixed sulphate-chloride acidic medium were described. Potential advantages of this new pressure oxidation concept are flowsheet simplification and high silver recoveries.

REFERENCES

Bailey, L.K. and E. Peters (1976). Can. Metall. Quarterly, 15(4), 333-344.
Berezowsky, R.M.G.S. and D.R. Weir (1984). Minerals and Metallurgical Processing, 1, 1-4.
Boyle, R.W. (1979). Geological Survey of Canada Bulletin, no. 280.
Bruynesteyn, A (1984). In Leaching-Unit Operations and Processes. Proc 14th CIM Ann. Hydromet. Meet., paper no. 4, Timmins, Ontario.

Bull, W.R., D.J. Spottiswood and J-M. Tourre (1985). In A.D. Zunkel, R.S. Boorman, A.E. Morris and R.J. Wesely (Eds.), Complex Sulfides:Processing of Ores, Concentrates and By-Products. TMS-AIME, Warrendale, Pa., p. 141.

Chryssoulis, S.L., L.J. Cabri and R.S. Salter. This Symposium.

Cornelius, R.G. and J.J. Woodcock (1958). Proc. Australasian Inst. Mining Metall., 185, 65-107.

Demopoulos, G.P. (1987). Min. Proc. Extr. Metall. Review, 3(1), (In press).

Demopoulos, G.P., P.D. Kondos and V.G. Papangelakis (1987). In Inter. Symp. on Crystallization and Precipitation (ISCAP), Proc. 17th CIM Ann. Hydromet. Meet, Saskatoon, Sask.

Dinardo, O. and J.E. Dutrizac (1985). Hydrometallurgy, 13, 345-363.

Dutrizac, J.E. (1985). In L.F. Haughton (Ed.), MINTEK 50, Vol. 1. The Council for Mineral Technology, Randburg, S.A., pp. 39-62.

Gerlach, J., H. Hähne and F. Pawlek (1966a). Erzmetall, 14(2), 66-74.

Gerlach, J., H. Hähne and F. Pawlek (1966b). Erzmetall, 14(4), 173-180.

Guay, W.J. (1981). In W.J. Schlitt, W.C.Larson and J.B. Hiskey (Eds.), Gold and Silver-Leaching, Recovery and Economics. SME-AIME, Littleton, Co., pp. 17-22.

Guinevere, R. (1984). First Intern. Symp. on Precious Metals Recovery, Reno, NV, Paper no. 14.

Habashi, F. and E.L. Bauer (1966). Ind. and Eng. Chem. Fund, 5(4), 469-471.

Khryashchev, S.V., O.P. Berezkin, V.G. Sirotinin and T.A. Lovavova (1969; 1970). The Soviet Journal of non-Ferrous Metals, 7, 15-17; 2, 85-86.

King, W.E. and J.A. Lewis (1980). Ind. Eng. Chem. Proc., 19(4), 719-722.

Kondos, P.D. and G.P. Demopoulos (1987). Hydrometallurgy '87, Soc. Chem. Ind., Manchester, U.K., (in press).

Krause, E. and V.A. Ettel (1987). To be published in Hydrometallurgy.

Lakshmanan, V.I., D.W. Busch and A.M. Childs (1984). In Leaching-Unit Operations and Processes. Proc. 14th CIM Ann. Hydromet. Meet., paper no. 13, Timmins, Ontario.

Langhans, J.W. and K.P.V. Lei (1986). In D.R. Gaskell, J.P. Hager, J.E. Hoffman, and P.J. Mackey (Eds.), The Reinhardt Schuhmann International Symposium on Innovative Technology and Reactor Design in Extraction Metallurgy. TMS-AIME, Warrendale, Pa., pp. 259-271.

Lawrence, R.W. and A. Bruynesteyn (1983). CIM Bulletin, 76(857), 107-110.

Lowson, R.T. (1982). Chemical Reviews, 82(5), 461-497.

Mainwaring, P.R. (1985). Paper presented at the 24th CIM Ann. Conf. of Metallurgists, Vancouver, B.C.

Mainwaring, P.R. (1987). This Symposium.

Mason, P.G., R. Pendreigh, F.D. Wicks and L.D. Kornze (1984). In V. Kudryk, D.A. Corrigan, and W.W. Liang (Eds.), Precious Metals: Mining, Extraction and Processing. TMS-AIME, Warrendale, Pa., pp. 435-445.

Mason, P.G., F.D. Wicks and J.C. Gathje (1985). U.S. Patent No. 4,552,589.

McKay, D.R. and J. Halpern (1958). Trans. TMS-AIME, 6, 301-309.

Papangelakis, V.G., D. Berk and G.P. Demopoulos (1986). In R.G. Bautista, R.J. Wesely and G.W. Warren (Eds.), Hydrometallurgical Reactor Design and Kinetics. TMS-AIME, Warrendale, Pa., pp. 209-225.

Papangelakis, V.G., D. Berk and G.P. Demopoulos (1987). To be published in Met. Trans. B.

Papangelakis, V.G. and G.P. Demopoulos (1987). To be published in Can. Metall. Quart.

Robins, R.G. (1981). Met. Trans. B, 12, 103-109.

Schulze, R.G. (1984). Journal of Metals, 36(6), 62-65.

Tozawa, K. and K. Sasaki (1986). In J.E. Dutrizac, and A.J. Monhemius (Eds.), Iron Control in Hydrometallurgy. Ellis Horwood, Chichester, England, pp. 454-476.

Van Weert, G., K.J. Fair and J.C. Schneider (1987). Paper presented at the 116th Annual TMS/AIME Meeting, Denver, Co.

Warren, I.H. (1956). Austr. Jour. of Appl. Sci., 7, 346-358.

Weir, D.R. (1986a). U.S. Patent no. 4,605,439.
Weir, D.R. (1986b). U.S. Patent no. 4,606,763.
Weir, D.R. and R.M.G.S. Berezowsky (1984). In Leaching - Unit Operation and Processes. Proc. 14th CIM. Ann. Hydromet. Meet., paper no. 12.
Weir, D.R. and R.M.G.S. Berezowsky (1986a). In GOLD 100, Vol. 2, SAIMM, Johannesburg, S.A., pp. 275-285.
Weir, D.R. and R.M.G.S. Berezowsky (1986b). U.S. Patent no. 4,571,263.
Weir, D.R. and R.M.G.S. Berezowsky (1986c). U.S. Patent no. 4,571,264.
Weir, D.R., J.A. King and P.C. Robinson (1986a). Minerals and Metallurgical Processing, 3(4), 201-208.
Weir, D.R., P.C. Robinson and J.A. King (1986b). Minerals and Metallurgical Processing, 3(4), 208-218.

INVESTIGATION ON THE POTENTIAL TECHNIQUES TO RECOVER GOLD FROM THIOUREA SOLUTION

G. Deschênes

Mineral Sciences Laboratories,
CANMET, Energy, Mines and Resources Canada,
555 Booth Street, Ottawa, Canada, K1A 0G1

ABSTRACT

Recovery of gold from thiourea solution by techniques such as activated carbon, cementation with lead and aluminum powder, resins, solvents and pressure reduction with hydrogen was investigated. Activated carbon recovers gold with 99% extraction but high loadings of thiourea occur. Cementation with lead powder was found to be inefficient in sulfuric media. Aluminum powder produces an extraction comparable to activated carbon.

Strong acid cationic resins show selectivity for the gold thiourea complex and provide good recovery. Thiosulphate and cyanide solutions have shown the best performance to elute gold from the resins tested. Solvent extraction using D2EHPA 40% - TBP 10% was performed. Extraction and stripping are functions of the acidity.

Pressure reduction under hydrogen was successful to strip gold and also silver from thiourea solution. Barren solutions containing only 0.2 ppm gold were produced. The effects of temperature, pressure, composition of solution and nature of catalysts were studied. No decomposition of thiourea was reported using this method.

KEYWORDS

Thiourea, gold, recovery, cementation, carbon, resin, solvent, hydrogen.

INTRODUCTION

Cyanide leaching is the usual method used throughout the world to extract gold from ores. The development of new processes and techniques could increase the possibility for recovery of gold from refractory materials. These methods include roasting, oxidative leaching, bioleaching, pressure leaching, and the use of thiosulfate and thiourea as lixiviants.

All industries have to take sufficient precautions for environmental protection. Hydrometallurgical operations in the mineral industry are concerned with potential seepage of reagents from large tailings ponds. More specifically, EPA (USA) forces industries to provide options in non-pollution methods for gold leaching. Thiourea leaching is one of the more promising approaches in these new techniques. Strict regulations in water quality will be the reason for gradual replacement of toxic processing, such as cyanidation, by less toxic technology.

Leaching with thiourea has been successfully tested on different materials (Bodson, 1981; Chen, Lung and Wang, 1980; Deschênes and Ghali, 1986; Gabra, 1984; Moussoulos, Potamianos and Kontapoulos, 1984; Van Lierde, Ollivier and Lesoille, 1982; Yen and Wyslouzil, 1985a). Commercial application in the extraction of gold and silver is performed in some areas (Hisshion and Waller, 1984; Randol, 1981) or plan to do so (Anonymous, 1984). Various techniques to recover gold and silver from this solution have been investigated. Cementation with aluminum (Randol, 1981; Simpson, Peterson and Sandberg, 1984; Van Lierde, Ollivier and Lesoille, 1982), lead and iron (Tataru, 1968; Yen and Wyslouzil, 1985b; Wen, 1982), activated carbon, (Gabra, 1984; Hisshion and Waller, 1984; Huyhua and Gündiler, 1986; Moussoulos, Potamianos and Kontapoulos, 1984; Pesic and Wey, 1986; Schulze, 1985; Yen and Wyslouzil, 1985b), ion exchange (Becker, Knothe and Lobel, 1983; Gabra, 1984; Schulze, 1984b; Schulze, 1985; Simpson, Peterson and Sandberg, 1984; Yen and Wyslouzil, 1985b) and electrowinning (Deschênes, 1986) are the main avenues explored.

Pressure reduction is an attractive avenue to recover gold and silver from such solutions because of its selectivity and its ability to precipitate precious metals. In aqueous solutions, pressure reduction under hydrogen atmostphere was previously performed to recover platinum, rhodium, palladium (Findlay, 1983) and silver (Webster and Halpern, 1957). Researchers found that chloride, acetate and ethylene complexes of the metals are reduced but it is not possible to precipitate precious metals from cyanide solutions because of the stability of these complexes. In organic media, gold and palladium are recovered by the same technique (Pouskouleli and Demopoulos, 1984).

This progress report presents results of tests using cementation, activated carbon, ion exchange resins, solvent extraction and pressure reduction under hydrogen to recover gold from thiourea solution.

MATERIAL, EQUIPMENT AND PROCEDURES

Thiourea Solution

Experimental solution was prepared by leaching Campbell Red Lake gold concentrate. Composition of the gold concentrate, leaching conditions and composition of the leach liquor are listed in Tables 1, 2 and 3.

Reagent grade chemicals and distilled water were used in experimental tests. Thiourea analysis was performed by mercuric nitrate titration or potentiometric titration with potassium iodate. Metal ion analyses were performed by atomic adsorption spectroscopy (AAS). Gold was also analysed by AAS or by inductively coupled plasma (ICP).

Table 1. Analysis of the Campbell Red Lake gold concentrate.

Element	Analysis
Au	307 g/t
Ag	43 g/t
Fe	24.8 %
S	18.3 %
SiO_2	29.3 %
As	9.8 %
Ca	1.6 %
Al	1.0 %
C	0.8 %
Sb	0.2 %
Cu	0.1 %

Table 2. Leaching conditions.

Initial thiourea concentration	10.0 g/L
Initial $Fe_2(SO_4)_3$ concentration	5.0 g/L
Temperature	room
Pulp density (w/w)	30%
Stirring speed	800 rpm
pH control	1.0
Leaching time	60 minutes

Table 3. Analysis of the leach liquor.

Element	Analysis
Thiourea	7.6-8.9 g/L
Au	31-43 mg/L
Ag	1 mg/L
Fe	8.2-9.4 g/L
As	1.2-2.4 g/L
Cu	22-43 mg/L
Sb	19-45 mg/L

CEMENTATION

Equipment and Procedure

Metal cementation was performed in 250-mL reactors. Agitation of the system was provided by a mechanical stirrer set at 340 rpm and temperature control by a thermobath (Thermomix 1480, Braun).

Metallic powders used were 100% -38 μm. Aluminum powder was obtained from Anachemia and lead powder from Fisher.

Extraction of gold was evaluated by analysis of solutions with atomic absorption spertroscopy or with ICP.

Results and Discussion

Performance obtained with aluminum cementation is comparable with previous work (Simpson, Peterson and Sandberg, 1984; Van Lierde, Ollivier and Lesoille, 1982); providing gold recovery of 98.1%. Final cement has a gold content equivalent to the one mentioned by Van Lierde, i.e. 2%. Higher precious metals content in the cement can be obtained when there is high silver concentration in solution, resulting in a lower requirement of aluminum powder (Randol, 1981; Simpson, Peterson and Sandberg, 1984).

Tests have shown that zinc powder is not viable for such solutions because of the high hydrogen evolution and consequently the large acid consumption and also the possible formation of arsine.

Tests done with lead powder at -38 μm show that cementation with such a fine powder does not yield good recovery of gold. In fact, no gold recovery was possible with this powder, even with an increase of the temperature up to 60°C. The lead sulphate coating on particles was strong enough to prevent any gold deposition. The strength of the passivation coating decreases with the decrease of the contact surface. Yen and Wyslouzil (1985b) found that with lead shot 76% of gold was recovered in 4 hours. Gold refining of a lead cement is more attractive than treating aluminum cement using pyrometallurgy. Unfortunately, fine lead powder is not practical in sulfuric solutions. The decrease in gold content by using lead shot is thus not economical.

The relatively low gold content in aluminum cement is also a problem and the refining of such a cement becomes uneconomical. Increasing gold content or decreasing the aluminum consumption by decreasing the size of powder are the keys to the successful application of this practice. In solution containing silver, the aluminum cementation is applied to recover this metal efficiently (Randol, 1981). In such solutions, silver content is higher than the usual concentrations encountered in gold solutions. Aluminum requirement is thus reduced.

ACTIVATED CARBON

Equipment and Procedure

Norit activated coconut shell carbon was used in these tests. Carbon conditioning was performed by wet screening on 850 μm screen and reactivation by heating at 700°C for two hours. After being quenched and filtrated, the activated carbon was kept in a closed container for use in adsorption tests.

Carbon loading was carried out in an incubator shaker at 500 rpm and leach solution (100 mL) was contacted with different amounts of activated carbon (0.5, 2.0, 3.0 and 5.0 g) during different shaking times (15-60 minutes). Analysis of the solution was performed by AAS after filtration of the pulp at the end of each test.

Results and Discussion

Figure 1 shows that carbon loading increases with a decrease of gold extraction. In this case, a loading of 5.4 kg Au/t was loaded in 60 minutes with 5 g/L of carbon resulting in the lowest recovery (79%) but giving the maximum loading. Recovery of 99% of gold was obtained but with the lowest loading of 0.72 kg Au/t (50 g/L of carbon). Gold loading would have been higher if thiourea concentration was lower because loading increases when thiourea concentration decreases (Pesic and Wey, 1986). Dissolved oxygen has a similar effect to

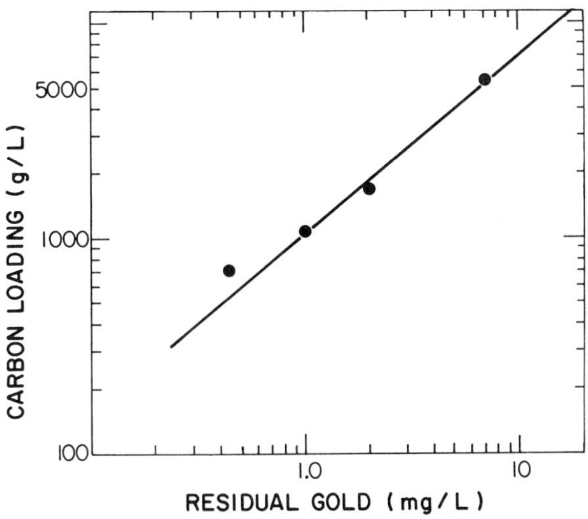

Fig. 1. Carbon loading with 60 minutes contact. Room temperature, feed: 32 mg Au/L.

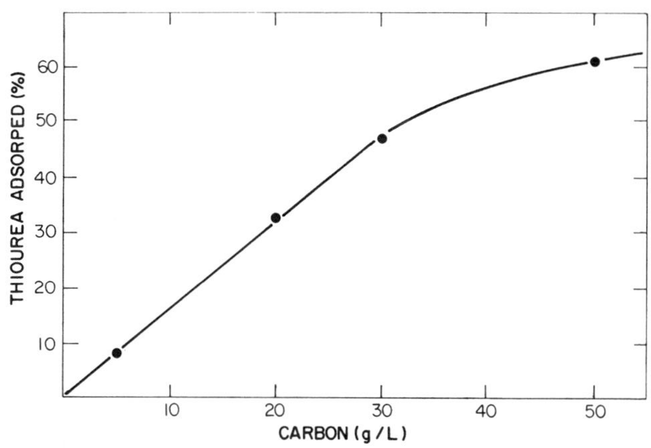

Fig. 2. Effect of activated carbon concentration on thiourea extraction from thiourea solution. Feed: 8.2 g/L thiourea, room temperature, 60 minutes contact.

thiourea concentration by interfering in the adsorption phenomena in acidic solutions.

High content of iron and arsenic in solution has resulted in high loading onto carbon, with up to 170 kg/t and 120 kg/t respectively at 0.5 g carbon/L solution. Such loadings of competing ions result in a drop in the activated carbon efficiency for gold. Gold and silver extraction were reported to be unaffected by ferrous ion in a synthetic solution having 0.2 to 1.8 g/L (Huyhua and Gündiler, 1986). Our leach solution having an higher concentration of Fe^{2+}, i.e., about 8.5 g/L, produces higher adsorption of this species.

Thiourea adsorption onto activated carbon is presented in Fig. 2. Extraction of thiourea increases with the concentration of activated carbon in a parabolic mode. With 50 g/L of carbon, 66% of thiourea was adsorbed in 60 minutes. These results confirm the results of Gabra (1984), Schulze (1984b) and Yen and Wyslouzil (1985b) and are opposed to Moussoulos' Potamianos' and Kontapoulos' results (1984) that indicated thiourea was not adsorbed by activated carbon.

Table 4 presents elution tests to determine if it is possible to recover thiourea from loaded activated carbon. An amount of 6.4 g of activated carbon loaded (11% thiourea) was contacted in a flask with 50 cc of solution for 60 minutes and stirred at 500 rpm.

Table 4. Elution of activated carbon to recover thiourea.

Eluant	Concentration of thiourea (g/L)	Thiourea elution (%)
Warmed water (40°C)	0.10	2.0
H_2SO_4, pH = 2.0	0.10	2.0
H_2SO_4, pH = 0	0.40	7.8
H_2SO_4, pH = -1.0	0.53	10.4

In these conditions, thiourea was not easily eluted from this material with warm water or even with acidic solution of H_2SO_4 at pH -1.0 (~50% V) at ambient atmosphere. In fact, maximum elution of 10.4% was obtained with a solution of H_2SO_4 at pH -1.0.

It is quite possible that oxygen is responsible for the difficulty in recovering thiourea. Schulze (1984a) said that it is possible to recover pure and concentrated thiourea solutions with a hot-water wash but this procedure must be performed in the absence of oxygen to avoid thiourea decomposition on carbon. It is reported that when activated carbon reaches saturation with thiourea, no further adsorption occurred (Yen and Wyslouzil, 1985b). This organic loading onto activated carbon represents a disadvantage and also affects the performance of the material in its gold loading (Pesic and Wey, 1986). Although activated carbon can provide high loadings of gold (200 kg/t (Huyhua and Gündiler, 1986)), competing ions like iron, copper and arsenic are also loaded onto this material. The prior removal of such elements would create a problem providing at least an additional step in the treatment. Elution of gold is also problematic. So far, only cyanide solution and thiourea solution with acetone (Yen and Wyslouzil, 1985b) can remove gold from the loaded carbon.

Because of the problems related to the use of activated carbon (thiourea adsorption, interfering elements, elution), further work was then oriented to other methods that could offer a simple, easy and economical way to recovery gold.

RESINS

Equipment and Procedure

Eight resins were examined in this investigation. Table 5 shows technical information about the materials.

Batch and column adsorption tests were performed on the as-received resins. Batch tests were run by contacting 15 mL of resin with 150 mL of solution in 250 mL flasks agitated with incubator shaker (orbital) for 60 minutes at 500 rpm and 25-27°C. Solution was filtered and analysed at the end of each experiment. The ion-exchange column set-up is shown in Fig. 3. Gold thiourea solution was pumped through a specific volume of wet-settled resin after being filtered to avoid any particles in the system. The volume of resin used was in the range of 2 mL-10 mL. The resin bed has been supported and retained by glass wool.

Table 5. Details on the resins used for the recovery of gold from thiourea solution.

Resin	Manufacturer	Type	Active group	Standard Mesh distribution
HCR-S	Dow Chemical	SAC*	sulfonic	20-50
50WX8	Dow Chemical	SAC	sulfonic	20-50
MSC-1	Dow Chemical	SAC	sulfonic	20-50
TG 650C	Dow Chemical	SAC	sulfonic	22-32
AG50W-X8	Bio-Rad	SAC	sulfonic	100-200
AG50W-X16	Bio-Rad	SAC	sulfonic	200-400
Lewatit C1014	Bayer	unknown†	polystyrene-DVB	12-48
Duolite GT73	Rohm & Haas	WAC[1]	polystyrene	14-48

* strong acid cationic: SAC; † thiourea in the resin; [1] weak acid cationic: WAC.

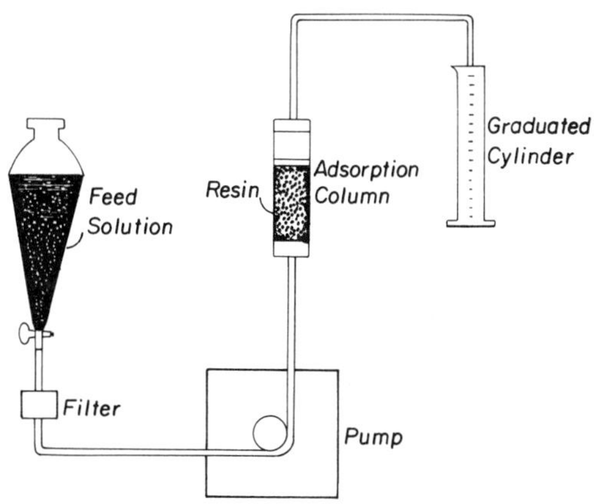

Fig. 3. Equipment set up for ion-exchange adsorption column.

Volumes of effluents were collected in graduated glass cylinders and analysed. The loadings of resins were evaluated by the difference in head and effluent analysis. Fire assay was conducted on the resin occasionally to check the results. Batch elution of resins was conducted with the same apparatus used in loading. Fifteen mL of loaded resin were contacted with 50 mL of various eluants during 120 minutes at room temperature. In some occasions a second contact was provided.

Results and Discussion

Loading Table 6 shows batch tests loading results obtained for 8 resins. With a ratio of 1/10 (volume of resin/volume of aqueous), all gold extractions were over 90%, i.e. 94% to 99%. The exchange mechanism involved is:

$$\bar{X}^+ + Au[SC(NH_2)_2]_2^+ \rightleftharpoons X^+ + \overline{Au[SC(NH_2)_2]_2}^+ \qquad (1)$$

where X^+ = ionic form of the resin. The highest gold recovery was achieved by the Bio-Rad AG50W-X16 with 98.6%. One reason is that this resin has the smallest beads. AG50W-X16 has produced the lowest gold loading. The best gold loading performance was achieved with the Bio-Rad AG50W-X8. This resin also loads easily the copper thiourea complexe. This adsorption of thiourea by ion-exchange resins is in agreement with the work of the U.S. Bureau of Mines (Simpson, Peterson and Sandberg, 1984). Species such as iron and copper are complexed with thiourea in solution and these complexes are adsorbed onto resin. Thiourea adsorption could then be explained by an exchange mechanism:

$$m\bar{X} + M[TU]_n^{m^+} \rightleftharpoons mX^+ + \overline{M[TU]}_n^{m^+} \qquad (2)$$

where $M[TU]_n^{m^+}$ = metal thiourea complex.

The Bayer resin, Lewatit OC1014 showed a low thiourea loading and provided a gold loading higher than the average (0.953 kg/t). In fact, the reason for the behavior of the Bayer resin with thiourea is the nature of the product. The Lewatit OC1014 is a chelating resin. The highest thiourea loading was achieved by the Duolite GT 73 (weak acid resin) with 44.7 Kg/t.

Table 6. Batch adsorption tests for the resins. 150 mL pregnant solution, 15 mL of resin, 60 minutes, 25°C.

Resin	Recovery (%)			Loading (kg/t)		
	Au	Cu	Thiourea	Au	Cu	Thiourea
Dowex 50WX8	94.6	71.8	11.1	0.819	0.62	19.52
Dowex HCR-S	93.8	79.4	11.4	0.875	0.70	19.89
Dowex TG 650C	95.1	69.2	11.1	0.890	0.65	21.10
Dowex MSC-1	94.6	71.8	11.1	0.867	0.66	20.66
Bio-Rad AG50W-X8	97.9	86.8	11.9	0.978	0.93	30.73
Bio-Rad AG50W-X16	98.6	84.4	13.1	0.760	0.70	27.08
Bayer Lewatit OC1014	94.4	97.9	2.5	0.953	1.07	5.66
Duolite GT 73	94.3	93.2	20.4	0.903	0.99	44.7

Exception of the Bio-Rad AG50W-X8, macroporous resins have a good efficiency to load gold compared to normal cationic type. In fact, Dowex MSC-1, Lewatit OC

1014 and Duolite GT 73 are both macroporous resins and have higher gold loading than the average. The Dowex HCR-S presents the lowest thiourea loading of the group of resins with a sulphonic active group.

Figure 4 shows the effect of the flow rate on equilibrium loading of gold onto the Dowex HCR-S at room temperature with a bottom flow ion exchange column. The decrease in the flow rate creates an increase in the gold loading. Plots 1 and 2 showing the isotherms at 12 and 6 bed volumes/h, result in respective maximum gold loading of 15.9 kg/t (7.0 g/L) and 22.6 kg/t (10.0 g/L) after being treated with 279 and 477 bed volumes of solution (Table 7). The adsorption at 3 bed volumes/h (plot 3) provides a maximum gold loading of 27.4 kg/t (12.1 g/L) after the treatment of the ion-exchanger with 675 bed volumes of pregnant solution. Analyses of the resin by fire assay show that the data that agree with calculated loadings are obtained by difference in head and effluent analysis. Gold loading onto HCR-S resin is a function of the gold in the effluent. A 95% extraction of gold (~1.6 mg/L in the effluent) results in a loading of 2.8 kg Au/t at a flow

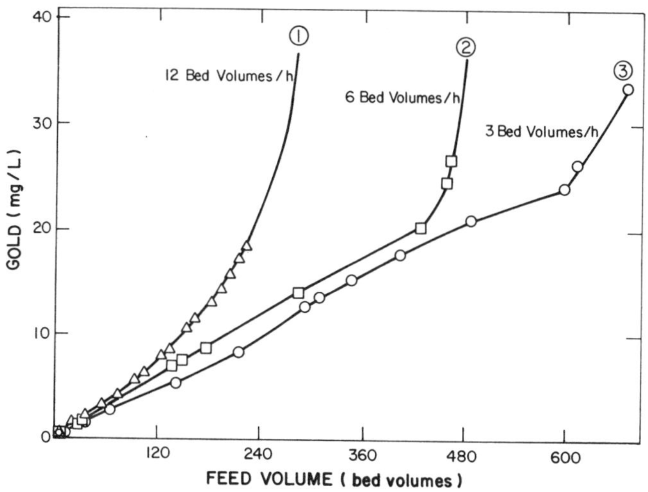

Fig. 4. Effect of the flow rate on equilibrium loading of gold onto strong acid cationic exchanger (HCR-S). Feed: 31 mg Au/L, room temperature.

rate of 6 bed volumes/h and 1.3 kg/t Au at 12 bed volumes/h. Higher loading will create lower extractions. These gold loadings then are lower than previous work (Yen and Wyslouzil, 1985b) and are related to thiourea concentration in the pregnant solution and also to competing ions. Higher thiourea concentrations decreases the loading capacity of ion exchangers (Simpson, Peterson and Sandberg, 1984). Competing ions in large concentrations such as iron and arsenic reduce the amount of sites available on the ion exchanger. Tests have shown that thiourea loading on the resin is a function of its concentation in solution. In fact, with the same experimental conditions, thiourea loading at equilibrium onto HCR-S was decreased by about 5 kg/t if the thiourea concentration decreases from 8 g/L to 5 g/L.

To optimize the resin inventory, lower extraction can be used providing a higher gold loading. In fact, at 6 bed volumes/h, a gold loading of 5 kg/t will result in about 90% extraction of gold. To minimize loading of base metals and other ions, we can limit the contact time of the resin. We see (Fig. 5) that resin selectively loads gold in presence of copper and antimony. The exchanger has more selectivity for gold than copper and antimony. However, copper has more facility than antimony to be loaded on the ion-exchanger. The selectivity of a resin for one ion over another is largely a function of the size of the ion. The literature (Sillen and Martell, 1964) presents the copper thiourea complex as a complex four, i.e., $Cu[SC(NH_2)_2]_4^{2+}$. This copper thiourea complex is possibly larger than the gold thiourea complex and it is less adsorbed onto the resin. According to these results, the antimony complex could also be larger than the gold thiourea ionic complex. So far, no information on the antimony thiourea complex is available in the literature. The copper loading at equilibrium is about three times less than the gold loading (Table 7). In spite of the fact that iron concentration in the feed is about 250 times the gold concentration, iron loading on the resin at equilibrium is not more than two times higher than gold loading.

Table 7. Equilibrium loading during column ion exchange adsorption test. HCR-S resin, room temperature. Feed: thiourea 8.2 g/L, 33 mg Au/L, 40 mg Cu/L, 8.5 g Fe/L.

Flow rate (bed volumes/h)	Loading (kg/t) Au	Fe	Cu	Thiourea
3	27.4	49.9	9.2	46.0
6	22.6	41.0	7.6	38.2
12	15.9	32.8	5.9	17.8

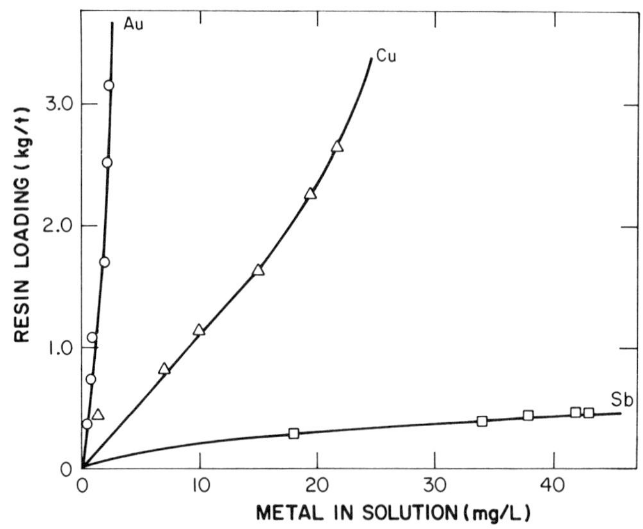

Fig. 5. Equilibrium loading of some metals onto a strong acid cationic ion exchanger (HCR-S). 12 bed volumes/h, room temperature.

Equilibrium loading of the resin with base metals such as iron can be modified after being reached. In fact, the resin continues to load gold but some of the loaded iron can be eluted by the thiourea solution at the same time and replaced by gold. Iron loading is thus changing with time.

Composition of the solution is the most important aspect for gold loading of a specific ion exchanger. Thiourea concentration, competing ions and also gold content will decide the equilibrium. In spite of the high concentrations of iron, thiourea and arsenic, the resin continues to load the gold thiourea complex after apparent saturation of these competing contaminants. Strong acid cationic ion exchanger shows selectivity for this gold thiourea complex but base metals concentration such as iron, being in large concentration affect this selectivity by a competing mechanism and create a problem.

Elution Gold desorption from the ion exchange resins was investigated with some eluants. These are sodium thiosulphate, HCl-Br, cyanide solution and acidic thiourea solutions. Table 8 shows the batch test results on 15 mL of resin (Dowex HCR-5, Bio-Rad AG50W-X8, Bio-Rad AG50W-X16 and Bayer Lewatit OC1014) contacted with 50 mL of eluant for 2 hours at room temperature.

Sodium thiosulphate, 5% solution at pH 9 gives the best performance with 85% of elution of gold on the resin AG50W-X8. Gold elution of the HCR-S shows also a good result with 74% but treatment of AG50W-X16 and Lewatit OC1014 resulted in elution below 20%. The mixture HCl 4N - Br 2% was found to elute with some efficiency on Dowex HCR-S and Bio-Rad resins but gave poor results on the Bayer resin. Acidic thiourea solutions reported in the literature to have some efficiency (Gabra, 1984; Yen and Wyslouzil, 1985b) or good efficiency (Schulze, 1984b; Schulze, 1985; Becker, Knothe and Lobel, 1983) were not as suitable in these tests as the cyanide solution reported previously to elute gold from the cationic Bio-Rad AG50W-X8 with success (Simpson, Peterson and Sandberg 1984). A second elution contact of resins, in the same conditions, is shown in Table 9.

Table 8. Batch elution tests for some resins. 50 mL eluant, 15 mL resin, 120 minutes, room temperature, gold loading ≃ 0.9 Kg/t.

Resin	Gold elution (%)			
Eluant	HCR-S	AG50W-X8	AG50W-X16	Lewatit OC1014
$Na_2S_2O_3$ 5%, pH=9	73.6	85.0	18.4	12.6
HCl 4N - Br 2%	23.7	40.6	33.6	2.0
NaCN 0.2 M-NaOH 0.2 M	2.6	9.5	2.0	4.1
HCl 4N-Thiourea 1%	8.3	5.5	2.8	2.0
H_2SO_4 0.5N-thiourea 10%	1.4	1.8	1.9	9.1

Table 9. Batch elution tests for resin. Second contact, 50 mL eluant, 15 mL resin, 120 minutes, room temperature.

Resin	Cumulative elution of gold (%)			
Eluant	HCR-S	AG50W-X8	AG50W-X16	Lewatit OC1014
$Na_2S_2O_3$ 5%, pH=9	97.0	90.9	34.9	21.6
HCl 4N - Br 2%	-	59.1	60.6	3.9
NaCN 0.2 M-NaOH 0.2 M	-	11.3	4.9	6.1
H_2SO_4 0.5N-thiourea 10%	3.2	3.6	3.7	16.8

The highest elution of gold was found to be 97% with the 5% solution of sodium thiosulphate at pH 9 on the HCR-S resin. Maximum elution achieved by the HCl-Br eluant was about 60% on Bio-Rad resins. The effect of sodium cyanide did not show promise. However, when the AG50W-X8 resin was previously contacted with a solution of HCl-thiourea, further gold elution with the solution NaCN 0.2 M-NaOH 0.2M has produced an extraction of gold up to 90% with few bed volumes. Thus, contacting the resin previously with an acidic thiourea solution provides successful elution of gold with a cyanide solution. The first treatment could be considered as a conditioning step; the acidic influent removing base metal complexes. The cyanide solution is efficient because less species remain to complex with cyanide and the elution is less dependant on cyanide concentration.

Elution with thiourea thus appears to be a function of the acidity of the solution and also a function of thiourea concentration. So far, HCR-S and Lewatit OC1014 show the easiest elution with this liquor. It is possible that the increase in acid concentration as mentioned in the literature (Gabra, 1984) and thiourea concentration and temperature (18) would improve this step. We have not performed stripping tests with 6M $HCl/NaClO_3$ mentioned in the literature (Becker, Knothe and Lobel, 1983), as an efficient eluant because of the chlorine formation and the high reagent cost resulting in the stripping and conditioning steps using this eluant.

We have analysed the effect of thiosulphate solution to see if this eluant could have some fouling effect on a strong acid cationic exchanger. Previous results show no drastic decrease in the loading capacity of the resin or in the efficiency of the stripping. Strong acid cationic resins show some selectivity to load gold but iron is a problem when present in high concentration (8-9 g/L). The ion exchanger does not provide the high loading of thiourea during the loading that occurs with activated carbon. The cationic resin also shows lower base metals loading than activated carbon.

SOLVENT EXTRACTION

Experimental

Batch "shake out" tests were conducted on some organic extractants with an aqueous organic ratio of one. These organic extractants are: 8-hydroxyquinoline derivative (LIX 26, Kelex 100), isodecanol, tri-octyl methylammonium chloride (Aliquat 336), tri-octylamine (Alamine 336), tri-isooctylamine (Adogen 381), tributyl phosphate (TBP) di-2-ethyl hexyl phosphoric acid (D2EHPA), mono-2-ethylhexyl phosphoric acid (M2EHPA), di-2,4,4-trimethylpentyl phosphinic acid (Cyanex 272), 2-ethylhexyl phosphonic acid mono-2-ethylhexyl ester (PC-88A: Daihachi, SME 418: Shell), tri-octyl phosphine oxide (TOPO) and methyl-isobutyl ketone (MIBK). After being separated, organic and aqueous phases were sampled to be analysed by AAS.

Stripping tests were performed with HCl, citric acid, sodium tartarate, NH_4HF_2 and water. Stripping isotherms were established with the most promising aqueous media at room temperature.

Results and Discussion

In the first batch of "shake out" tests, 9 organic extractants were tried. These were: 1) LIX 26 10% - isodecanol 10%, 2) isodecanol 100%, 3) TBP 50%, 4) Kelex 100 5% - isodecanol 5%, 5) Aliquot 336 5% - TBP 5%, 6) Alamine 336 0.05M - isodecanol 0.05M, 7) Adogen 381 5%, 8) D2EHPA 15% - TBP 5% and 9) MIBK 100%. The most promising extractants were Aliquat 336 5% - TBP 5% and D2EHPA 5% - TBP 5%; the former gave however phase precipitation in the organic. A mixture of D2EHPA and TBP was previously found effective to extract gold from a thiourea solution (Simpson, Peterson and Sandberg, 1984). The other tests provide poor gold extraction, or formation of precipitates or were totally inefficient like MIBK 100%.

In the second batch of "shake out" tests, different concentrations of D2EHPA and TBP were tried. Other forms of phosphorus acids, i.e. phosphonic and phosphinic, were studied. Additional tests with Aliquat 336 and MIBK were performed to improve the gold extraction.

With 10% TBP, the increase of D2EHPA concentration from 25% to 60% improves the extraction from 47.6% to 77.1%. The use of M2EHPA with TBP 10% slightly decreases the gold recovery. Phosphonic acid (PC-88A and SME 418) and phosphinic acid (Cyanex 272) resulted in a substantial decrease in extraction of gold. Precipitation often occurred in the aqueous phase or in the organic phase during these tests. Only high concentration of TBP alone, i.e. 100%, shows efficiency in the gold extraction with 61.9% but the precipitation causes a problem.

The addition of D2EHPA 15% in TBP 50% improves the gold extraction to 77.4%, however, the precipitation still occurs with 50% of TBP. Tests with Aliquat 336 - TBP or Aliquat 336 - TOPO or MIBK with D2EHPA did not improve the gold recovery, and precipitation also resulted.

Kinetic tests for the extraction of gold by D2EHPA 40% - TBP 10% have shown that equilibrium is reached in less than 10 seconds (see Table 10). The extension of the contact time of the organic with the aqueous phase causes an increase in iron extraction. If the contact time is less than 5 minutes, iron concentration in the organic could be kept below 200 mg/L with an A/O of 2. The increase of the temperature decreased the gold extraction but increased the iron recovery as shown in Table 11. There is then no advantage to increasing the temperature of solvent extraction when using such reagents.

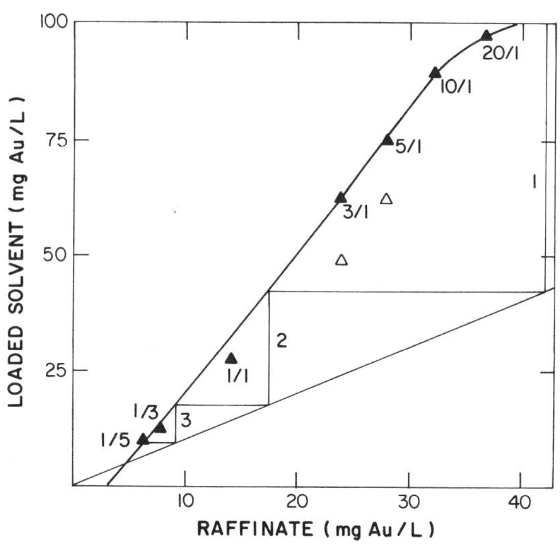

Fig. 6. Equilibrium isotherm for gold recovery with D2EHPA 40% - TBP 10%. Room temperature, feed: 43 mg Au/L.

Table 10. Kinetic extraction of gold from thiourea solution with D2EHPA 40% - TBP 10%. Feed: 33 mg Au/L, room temperature, phase ratio A/O = 2.

Test	Time (s)	Organic composition Au (mg/L)	Fe (mg/L)
1	10	32	86
2	30	32	100
3	60	33	110
4	120	32	140
5	180	33	150
6	300	33	180
7	360	31	500
8	720	30	830

Table 11. Effect of the temperature on the gold extraction with D2EHPA 40% - TBP 10%. 3 minutes contact time. Feed: 41 mg Au/L.

Phase ratio A/O	Temperature (°C)	Extraction (%) Gold	Iron	Separation time (s)
3/1	room	42.4	15.0	120
1/3	room	75.8	8.5	100
3/1	50	12.1	24.0	30
1/3	50	51.5	12.8	15

The equilibrium isotherm of the extractant D2EHPA 40% with TBP 10% is given in Fig. 6. Equilibrium data were obtained by changing the phase ratio A/O from 1/5 to 20/1. This figure shows that gold extraction with this system will require 4 stages to reduce the precious metal concentration from 43 mg/L to 5 mg/L with a phase ratio O/A of 1. Successive extractions of gold from thiourea solutions are presented in Table 12.

The extraction with this solvent has some selectivity for gold. Most of the iron and the copper remain in the aqueous phase. Maximum loading of the organic appears to be around 85-90 mg Au/L. At the same time, the relative composition of iron and copper are respectively 560 mg/L and 0.5 mg/L. Iron and copper concentrations were then reduced by 16 times and 44 times while gold was concentrated by two times.

Table 12. Successive extractions of gold from thiourea solution with D2EHPA 40%, TBP 10%. A/O = 1, room temperature, contact time = 2 minutes. Feed: 41 mg Au/L, 9.02 g Fe/L and 22 mg Cu/L.

Contact	Organic phase composition (mg/L) Au	Fe	Cu	Extraction (%) Au
1	28	130	<0.5	68
2	48	220	<0.5	49
3	61	320	<0.5	39
4	72	390	<0.5	27
5	84	490	<0.5	29
6	85	560	<0.5	2

Table 13 shows the results of tests performed to strip the gold from the organic. Hydrochloric acid provides the best gold stripping. This solution was previously found to be effective in stripping gold (Simpson, Peterson and Sandberg, 1984). Almost 85% of gold was been recovered with HCl 20% and of ratio O/A equal to 1. Citric acid 10% strips gold at 27% extraction but some precipitation occurs in the organic phase. Sodium tartrate 10% and NH_4HF_2 5% do not appear to efficiently remove the gold from the organic phase.

Table 13. Screening of stripping solutions for gold from D2EHPA 40% - TBP 10%. O/A = 1, room temperature, 3 minutes, feed: 51 mg Au/L.

Reagents	Gold extraction (%)
Water	33
HCl 5%	62
HCl 10%	82
HCl 20%	84
HCl 30%	80
HCl 40%	67
Citric acid 10%	27
Sodium tartarate 10%	-
NH_4HF_2 5%	-

Optimum gold stripping of about 85% occurs between 10% and 20% of HCl at a phase ratio O/A = 1. However, the iron stripping increases with the increased acidity. It is therefore advantageous to strip gold at low acidity. The decrease of the phase ratio O/A to 1/10 has shown gold stripping of 98% with HCl 10%.

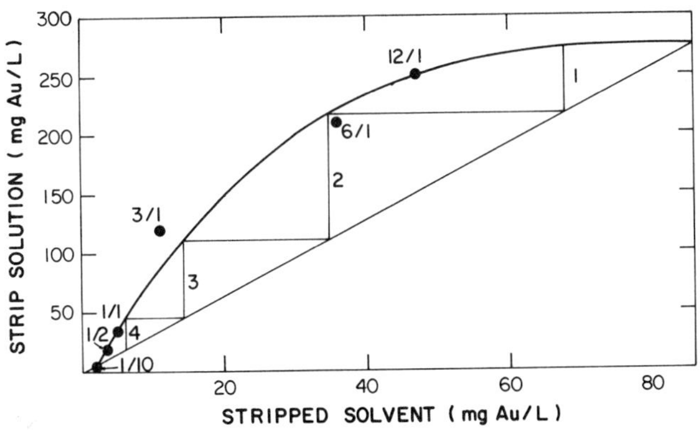

Fig. 7. Stripping isotherm for gold with HCl 10%. Contact time = 3 minutes, room temperature, feed: 86 mg Au/L.

A room temperature stripping isotherm is shown in Fig. 7 for the removal of 68 mg Au/L from the loaded organic. A solution of HCl 10% readily strips the loaded solvent containing 68 mg Au/L to 4 mg Au/L in 4 stages with a phase ratio O/A of ~3.16.

Solvent extraction thus shows selectivity for the recovery of gold from a thiourea solution. The stripping stage also provides an increase in the gold concentration, other ions remaining in the organic phase.

HYDROGEN REDUCTION

Experiment

A standard 1 litre titanium Parr autoclave was used to carry out batch reduction tests for aqueous solutions. A glass liner was used and metallic immersed parts were either titanium grade 12 or unalloyed titanium with an oxidized surface to prevent corrosion. The solution was filtered just before experiments using a Millipore filter (0.45 μm). The autoclave was purged with high purity nitrogen and heated to the appropriate temperature. Introduced through an inlet tube immersed in solution, high purity hydrogen was used to purged nitrogen when temperature level was reached and then hydrogen pressure was set at the desired level and stirring was started. The pressure was maintained constant throughout the test. Samples were drawn at convenient time intervals, filtered with Millipore filters (0.45 μm) and analysed by ICP for gold and occasionally for silver. Atomic absorption spectrometry (AAS) was used for other metals analysis.

Results and Discussion

The following Table 14 shows the hydrogen reduction of 500 mL of Campbell Red Lake leach liquor at 2068 KPa, 95°C during 6 hours. Nickel powder was added to activate the reduction (40 mg/L).

No detectable thiourea decomposition occurred during the test. Approximate composition of the solid residue was obtained by mass balance using the solutions analysis and also X-ray fluorescence spectroscopy of the residue. Estimate of the composition shows 55% Au, 1% Ag, 28% Cu, 13% S about 1% Fe and 1% Ni. We see that is is possible to reduce selectively gold and silver in an acidic thiourea

Table 14. Hydrogen reduction of a gold thiourea solution, 300 PSI, 95°C, 5 hours, 450 rpm, 40 mg/L Ni powder.

	Starting solution	Final solution	Recovery
Thiourea	7.0	7.0	N A
pH	1.1	1.1	N A
Au (mg/L)	40	1.2	97.0
Ag (mg/L)	1.0	0.1	90.0
Fe (g/L)	9.2	9.1	1.1
Cu (mg/L)	40	20.1	49.7

solution. The main impurities in the residue obtained from the CRL leach solution are copper and sulphur. Most of the nickel powder added went in solution during the experiment.

To study the effects of pressure and temperature on the recovery of gold by hydrogen reduction, a synthetic gold thiourea solution was prepared using double distilled water. The pH of the solution was adjusted by addition of concentrated sulphuric acid from BDH (ACS grade). Thiourea from Anachemia (ACS grade) was dissolved in the acidic media. The gold-thiourea complex was prepared by adding gold chloride (Johnson Matthey Ltd.) to this solution.

Tables 15 and 16 shows the effect of pressure and temperature on the recovery of gold.

Table 15. Effect of temperature on the recovery of gold in thiourea solution by hydrogen reduction; feed: 110 mg Au/L, 2068 KPa, 60 minutes, 450 rpm, stainless steel impeller.

Temperature (°C)	Gold in solution (mg/L)
80	85
95	0.25

Table 16. Effect of pressure on the recovery of gold in thiourea solution by hydrogen reduction; feed: 110 mg Au/L, 95°C, 30 minutes, 450 rpm, stainless steel impeller.

Pressure (KPa)	Gold in solution (mg/L)
2068	55
2500	49

A stainless steel impeller was used to replace nickel powder to activate the reduction. We see that the decrease of temperature (Table 15) and pressure (Table 16) creates a decrease in the reduction rate of gold. At temperatures under 95°C at 2068 KPa, the reduction is slow. At 95°C the concentration of gold goes under 0.5 ppm at either 2068 KPa or 2500 KPa after 60 minutes. No decomposition of thiourea was noticed during these tests and no elemental sulphur was detected. The gold powder recovered was almost pure with traces of chromium/nickel sulphide.

CONCLUSIONS

Recovery of gold by aluminum cementation provides extraction up to 98%. However the precious metal content of the cement is poor (~2% gold) rendering the process uneconomical. Cementation with fine lead powder is totally inefficient in sulphuric media and the increase of the surface particle will increase the gold recovery but such practice is unattractive because of the decrease in precious metal content.

Activated carbon recovers gold from thiourea solution with 99% extraction. However, the extraction occurs with high loadings of thiourea which are poorly stripped from the carbon in presence of air.

Strong acid cationic ion exchange resins show some selectivity to load the gold-thiourea complex in a solution containing iron, arsenic copper and antimony. Equilibrium loading is a function of the solution composition. Important iron loading occurs during contact. Gold is not easily eluted from the strong acid cationic ion exchanger. Thiosulfate solution has been shown so far to have the better performance to elute gold from a strong acid cationic ion exchange resin removing almost all gold with a limited volume of solution. Solvent extraction of gold with D2EHPA 40% - TBP 10% has shown a selective extraction of gold from a thiourea solution. Effective stripping is possible with HCl 10% and selective stripping is a function of the acidity.

Pressure reduction with hydrogen selectively reduces the gold (1) complex with thiourea. Barren solutions having only 0.2 ppm gold were obtained. Reduction at 95°C under 2500 KPa is fast with the use of metallic nickel.

ACKNOWLEDGEMENTS

The author gratefully acknowledges the work of P.J. Prud'homme, E.W. Wong and B. Taylor, who performed the tests and analyses. The author also wishes to thank G.M. Ritcey and V. McNamara for the useful discussions in solvent extraction and ion exchange and the Chemical Laboratory for the excellent analyses received.

REFERENCES

Anonymous (1984). "Sonora's gold mine looks major", The Northern Miner, 70, 35, 1.

Becker, E., Knothe, M. and Lobel, J. (1983). "Gold recovery from non-metallic secondary raw materials by leaching with thiourea and adsorption on ion exchangers"; Hydrometal. 11, 265-75.

Bodson, F.J.J. (1981). "Traitement des matières sulfureuses en vue de récupérer les métaux y contenus"; Brev. Fran. 2476137.

Chen, C.K., Lung, T.N. and Wang, C.C. (1980). "A study of the leaching of gold and silver by acidothioureation"; J. Hydromet, 5:207-12.

Deschênes, G. and Ghali, E. (1986) "Gold leaching by thiourea in chalcopyrite concentrate and chalcopyrite residue"; 25th Ann. Conf. Metall, CIM, Toronto.

Deschênes G. (1986). "Literature survey on the recovery of gold from thiourea solutions and the comparison with cyanidation". CIM Bull., 19, 895, 76-83.

Findlay, M. (1983). "The use of hydrogen to recover precious metals"; Precious Metals 1982, M.I. El Guindy, Pergamon press, Toronto, Ont., 477-501.

Gabra, G. (1984). "Leaching of gold from pyrite and chalcopyrite concentrate using acidified thiourea"; 23rd Ann.Conf.Metall., CIM, Québec.

Hisshion, R.J. and Waller, C.G. (1984). "Recovering gold with thiourea". Mining Mag, 237-42.

Huyhua, J.C. and Gündiler, I.H. (1986). "Recovery of gold and silver from acidic thiourea solutions"; TMS/SME Ann. Meet., New Orleans, Louisiana.

Moussoulos, L., Potamianos, N. and Kontapoulos, A. (1984). Recovery of gold and silver from arseniferrous pyrite cinders by acidic thiourea leaching; Proc. TMS-AIME Int. Symp, L.A., Cal.

Pesic, B. and Wey J.E. (1986). "Carbon adsorption of silver from acidified thiourea solutions"; TMS/SME Ann. Meet., New Orleans, Louisiana.

Pouskouleli, G. and Demopoulos, G.P. (1984). "Direct recovery of precious metals by integratin of solvent extraction and hydrogen reduction techniques"; Proc. 8th Int. Prec. Met. Inst. Conf., Toronto, Ont..

Randol, Int. Ltd. (1981). "Vivier Plant: Ag flotation and thiourea leach"; Gold and Silver Recovery Innovations, Vol. 1, 497-502.

Schulze, R.G. (1984a). "New aspects in thiourea leaching of precious metals"; J. Met., 6, 62-65.

Schulze, R. (1984b). "Hydrometallurgical extraction of precious metals using thiourea"; Germ. pat. DE 3401961A1.

Schulze, R. (1985). "Method for hydrometallurgical recovery of noble metals"; Germ. pat. DE 3407049A1.
Sillen, L.G. and Martell, E. (1964). Stability constants of metal ion complexes, The Chem. Soc., London.
Simpson, W.W., Peterson, L. and Sandberg, R.G. (1984). "Gold and silver recovery from thiourea leach solutions"; Pac. North. Met. Min. Conf., Portland, Or..
Tataru, S. (1968). "Précipitation par cémentation de l'or en solutions acides"; Rev. Roumaine de Chimie, 13, 1043-49.
Van Lierde, A., Ollivier, P. and Lesoille, M.V. (1982). "Développement du nouveau procédé de traitement pour le minerai de Salsigne"; Ind. Min., Les Tech. 1a, 399-410.
Webster, A.H. and Halpern, J.V. (1957). "Homogeneous catalytic activation of molecular hydrogen in aqueous solutions by silver salts. III. Precipitation of metallic silver from solutions of various silver salts"; J. phy. chem., 61, 1245-1248.
Yen, W.T. and Wyslouzil, D.M. (1985a). "Gold extraction from refractory ore by pressure oxidation and thiourea leach"; 17th C.M.P. Conf, Ottawa, Ontario.
Yen, W.T. and Wyslouzil, O.M. (1985b). "Gold recovery from thiourea solution"; Gold and Silver Recovery Forum 85, Sante-Fe, N.M.; Oct. 1985.
Wen, C.D. (1982). "Studies and prospects of gold extraction from carbon bearing clayey gold ore by the thiourea process"; Proc. XIV Int. Min. Proc. Cong. Can., Toronto, Ontario.

ION EXCHANGE RECOVERY OF GOLD AND SILVER FROM CYANIDE SOLUTIONS

Patricio A. Riveros and W. Charles Cooper

Department of Metallurgical Engineering
Queen's University
Kingston, Ontario, CANADA K7L 3N6

ABSTRACT

The use of activated carbon and carbon-in-pulp in the recovery of gold from cyanide solutions is well established. However, ion exchange resins present a number of advantages over carbon such as higher loading, the regeneration of cyanide ions for recycling to the leach circuit and the elimination of periodic re-activation and high temperature stripping. These advantages have prompted increased research on weak-base, strong-base, and chelating and adsorbent resins for the recovery of gold (and silver). Studies on the equilibrium and kinetic behaviour of gold, silver, copper, iron and zinc cyano complexes point to an interesting potential for strong-base resins in gold and silver recovery. This potential may be related to the marked differences in diffusivity in the resin phase on the part of gold and silver cyano complexes as compared with the base metal complexes. Also of interest are the development of superior weak-base resins and improved elution procedures. These advances are reviewed and discussed in the light of their application to the recovery of precious metals by ion exchange from cyanide solutions on an industrial scale.

KEYWORDS

Strong-base resins; weak-base resins; elution; thiocyanate; ion exchange, cyano complexes.

INTRODUCTION

Although carbon adsorption and in particular carbon-in-pulp are well established in gold recovery technology, the use of ion exchange resins appears to provide a number of advantages over activated carbon. These advantages include faster kinetics, higher loadings, and a wider scope of application. Thus cyanide solutions emanating from the heap leaching of low-grade gold ores, the leaching of calcines and of flotation tailings are all amenable to treatment using ion exchange. This versatility is related to the ability of resins to tolerate significant amounts of calcium salts and organic matter. In addition resins do not require periodic re-activation treatment or high temperature stripping which are standard operations with activated carbon. An important feature that is often overlooked is the fact that resins can also extract free cyanide ions for recycling and pollution control.

In spite of these advantages ion exchange resins have not been used on a commercial scale in the Western World for the recovery of precious metals. However, in recent years there has been a renewed interest in this application prompted in part by the successful industrial use of ion exchange for gold recovery in the U.S.S.R. and by the intensive research including pilot-plant operations in South Africa and elsewhere. The synthesis of new resins and studies on the kinetics of the ion exchange behaviour of metal cyano complexes have also contributed to advancing gold ion exchange technology.

Since in a cyanide leach liquor the noble and base metals are found as anionic cyano complexes of various compositions (Riveros and Cooper, 1985b), the most appropriate ion exchangers are weak- and strong-base resins. Chelating and adsorbent resins present further possibilities. This paper examines the potential which is offered by each of these types of resins for the recovery of gold and silver and the ion exchange technology applicable to gold-bearing cyanide leach solutions.

WEAK-BASE RESINS

By definition, weak-base resins are those having primary, secondary, or tertiary amine groups attached to the hydrocarbon matrix. Their most typical characteristic is that above a certain pH, known as the pKa, the amine groups become deactivated and incapable of functioning as ion exchangers. Consequently, weak-base resins can be easily eluted with an alkali solution. However, all commercial weak-base resins have a pKa lower than the normal pH of 10-11 of a cyanide leach liquor. Therefore a lowering of the solution basicity is necessary to promote the extraction of cyano complexes. In recent years there has been a significant emphasis on the development of weak-base resins having higher pKa values.

The loading reaction for a secondary amine weak-base resin may be represented by the following equations, where ⊢ indicates the polymeric resin matrix.

$$\vdash NRH + H^+ + X^- = \vdash NRH\,H^+X^-$$

$$\vdash NRH\,H^+X^- + Au(CN)_2^- = \vdash NRH\,H^+\,Au(CN)_2^- + X^-$$

A series of weak-base resins with phenol-formaldehyde condensate matrix and polyamine functionality capable of extracting gold from cyanide solutions has been developed. Of these resins DUOLITE A-7 has been studied the most extensively for gold recovery. The maximum loading and selectivity for gold on the part of this resin are strongly dependant on pH. At neutral or slightly acidic pH, there is a significant loss in selectivity (Fleming and Cromberge, 1984a). The selectivity of weak-base resins for gold and silver cyano complexes over base metal complexes is enhanced considerably at high vs low pH (e.g. pH 11 vs 6). Thus there is a real advantage to developing weak-base resins with higher pKa values.

Mehmet and Te Riele (1984) carried out pilot-plant testwork with DUOLITE A-7 in a NIMCIX column. The solution used was a bleed from the pregnant feed to a conventional gold plant using the Merrill-Crowe process. The gold adsorption was increased by reducing the pH of the solution from 11 to 6.7 with 2% sulphuric acid. At this pH the gold adsorption was highly satisfactory. For a gold concentration in the feed of 1.47-1.95 ppm, more than 99.85% of the gold was retained by the resin while the barren solution had <0.002 ppm Au. The resin loading was between 1.38 to 1.95 g Au/kg. However, as shown in Table 1, the selectivity at this pH is almost completely lost.

TABLE 1. Adsorption of Metals by DUOLITE A-7 at pH 6.7

Element	Feed (ppm)	Barren (ppm)	Adsorption (%)
Au	1.5	<0.002	>99.87
Ag	0.2	<0.05	>76.2
Ca	733	723	1.4
Co	0.75	<0.2	>72.4
Cu	8.1	<0.2	>97.6
Fe	0.6	<0.2	>64.3
Ni	4.9	<0.2	>95.9
Pb	0.7	0.3	57
Si	7.8	7.7	1.3
Zn	11.9	6.0	49.6

Table 2 shows the composition of the loaded resin. In agreement with the data in Table 1, at the pH used, no selectivity for gold and silver was found, which fact can be seen from the fraction of the total resin capacity used by each metal adsorbed.

Table 2. Composition of the Loaded Resin

Element	Resin loading (g/kg)	Fraction of total capacity (%)
Au	1.6	0.10
Ag	0.4	0.05
Ca	1.3	----
Co	0.8	0.49
Cu	8.0	3.11
Fe	0.7	0.54
Ni	4.0	1.71
Pb	0.5	----
Si	0.9	----
Zn	2.4	0.89

It will be noted that only 0.10 percent of the nominal resin capacity was used by the gold and only 6.89 percent by all of the species listed in Table 2. The resin was readily eluted with a caustic solution having a pH of 13.

More recently, Hiskey and Ette (1986) studied the characteristics of a new weak-base resin, PAZ-4, developed by SELA Inc., New York. This resin which is reported to be an improved version of DUOLITE A-7, is synthesized in an ultrasonic chamber which imparts a high degree of surface porosity and a large internal capillary network. PAZ-4 can extract significant amounts of gold even at pH 12 and can be eluted at pH 13 or higher. Figure 1 shows a comparison between DUOLITE A-7 and PAZ-4. No data were presented on the behaviour of cyanide leach solutions.

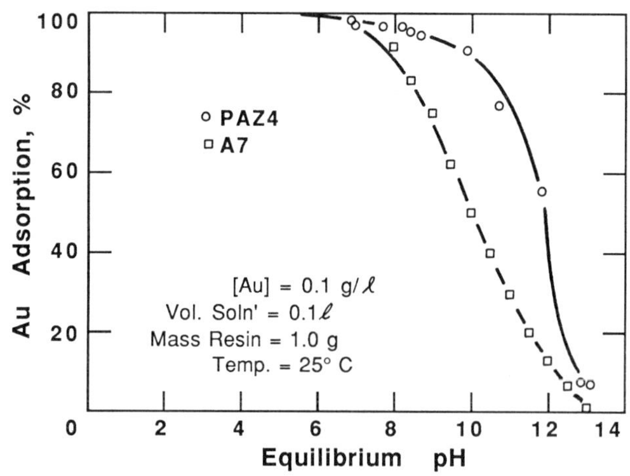

Fig. 1. Equilibrium adsorption curves for aurocyanide loading onto weak-base resins PAZ-4 and DUOLITE A-7.

The U.S. Bureau of Mines (Palmer, 1986) has investigated a number of weak-base resins. Six commercial resins were studied, vis., DUOLITE A-7, SYBRON A-305, DUOLITE A-340, DUOLITE A-561, DOWEX MWA-1, and DOWEX WGR-2. In addition, five experimental resins were tested: DOWEX XU-40138, DOWEX XU-40139, DOWEX XFS-40114, DOWEX XFS-43309, and DOWEX XFS-43356. The structure of these resins was not disclosed. A series of preliminary tests with synthetic and actual cyanide solutions showed that three of the experimental resins can function at higher pH (9-11) than commercial resins while retaining a selectivity for gold (Fig. 2). Testing was expanded to determine the effectiveness of the three promising resins with actual leach solutions in a flow-through contact column. The leach solution contained 1.7 ppm Au, 3.0 ppm Ag, and 0.3 ppm Hg, with a pH of approximately 10.5 and a free cyanide concentration of 0.5 g/L. Approximately 5 litres of solution were pumped through 1 gram of resin in a 1-cm (ID) glass column at a flowrate of 0.5 bed volume per minute. Following the adsorption phase, the loaded resin was eluted with 200 ml of 1 M NaOH at a flowrate of 0.17 bed volume per minute. These three resins were generally selective for gold and mercury, with minimal silver and other base metals being adsorbed. The resin DOWEX XU-40138 appeared to show the best loading and elution characteristics. This resin loaded to about 4.80 g Au/kg (140 tr.oz./st) and produced an eluate having a peak concentration of 35 ppm Au.

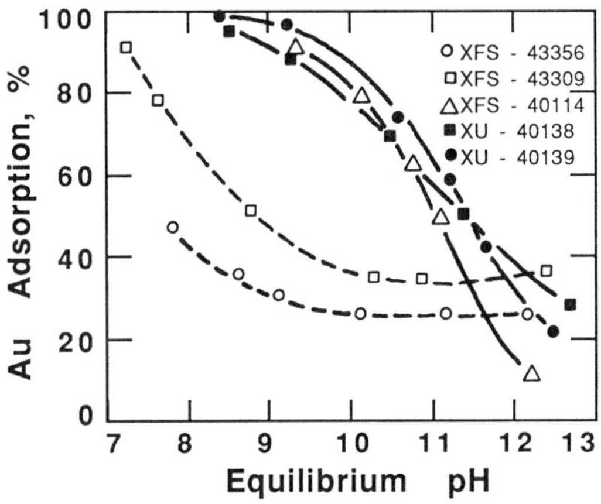

Fig. 2. Equilibrium adsorption curves for aurocyanide with experimental weak-base resins.

Green and Potgeiter (1984) have reported the synthesis of polyvinyl-based resins with imidazoline and tetrahydropyrimidine structures having apparent pKa's of 10.2 and 9.7 respectively. These values are approximately 2 pKa units higher than those for commercial weak-base resins. The imidazoline resin loaded gold from a cyanide solution containing 100 ppm gold to a maximum of 47.5 g Au/kg at a pH of 9.8. At this same pH the loading on DUOLITE A-7 was 22 g Au/kg. From their work on a number of experimental resins, the authors concluded that favourable loading and elution properties are related not only to the pKa value of the resin but also quite strongly to the physical nature of the resin matrix.

The extraction of silver with a series of weak-base resins, including DUOLITE A-7, DOWEX WGR-2, DOWEX MWA-1, DOWEX XFS-4195, and AMBERLITE IRA-35, has been reported by Riveros and Cooper (1985a). As shown in Fig. 3 the extraction of silver is significant only when the pH is lower than 8. A comparison of these resins shows that DUOLITE A-7 and DOWEX MWA-1 have the best characteristics in terms of capacity and elution efficiency. However, it is apparent that the extraction of silver requires a lower pH than does the extraction of gold.

STRONG-BASE RESINS

Unlike weak-base resins, strong-base resins adsorb metal cyanoions over a broad pH range. Although they have a higher loading capacity than weak-base resins, there is no particular selectivity for gold. Thus as shown by Fleming and Cromberge (1984a), the cyano complexes of copper, nickel, and zinc are more strongly adsorbed than gold on the resin IRA-400 whereas cobalt, silver, and iron are less strongly adsorbed. An important conclusion from this adsorption behaviour is that if gold is to be recovered efficiently from a cyanide leach solution by a strong-base resin, all of the copper, nickel and zinc present in the cyanide solution must be adsorbed.

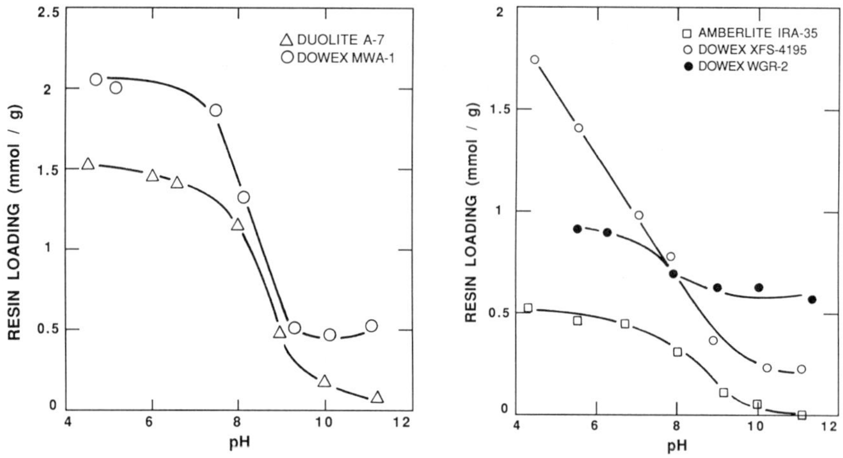

Fig. 3. Silver extraction as a function of the equilibrium pH: Aqueous feed solution = 1 mmol/L Ag and 10.2 mmol/L free cyanide.

Strong-base resins have quaternary ammonium groups which do not become deactivated at high pH as is the case with weak-base resins. Consequently, they are more difficult to elute since the attached gold cyano complex must be either displaced or decomposed.

It has been recognized for some time that the main problem with strong-base resins is the difficulty of elution. In fact, the elution has been found so difficult in the past that destruction (Waitz, 1982) and wet ashing (Kalocsai and Fabics, 1983) of the resin have been proposed as viable methods of recovering the adsorbed gold. Naturally, such procedures can only be applied in the case of very pure gold solutions since if other base metals are present they will be adsorbed and occupy a large fraction of the resin.

Although numerous eluants have been proposed for removing gold from strong-base resins, recently interest has focused on four eluants, viz., sodium hypochlorite, thiourea in acid solution, zinc cyano complex, and thiocyanate.

In a process developed by the U.S. Bureau of Mines (Palmer, 1986) elution is done sequentially with gold being eluted with NaClO which, being a strong oxidant, destroys the gold-cyanide complex and liberates the gold. During the tests, strong-base resins (AMBERLITE IRA-430, AMBERLITE IRA-900, DOWEX 21K, DOWEX MSA-1, and DOWEX SBR) were loaded with gold, silver, and mercury, then eluted sequentially with various eluants. Figure 4 shows a simplified flowsheet for the preferred eluting sequence. Generally, all five strong-base resins behaved similarly. With the first 3 hours, 100% of the mercury was eluted with 2N H_2SO_4. About 20% of the silver was eluted with the mercury in the case of the DOWEX SBR resin; however, no gold was eluted. A subsequent elution with 200 g/L NaCl in 1N HCl removed all of the remaining silver from the five resins in 6 hours with less than 10% of the gold being eluted. A final elution with 0.75% NaClO in 150 g/L NaCl plus 5% NaOH removed 93 to 98% of the gold in 9 hours.

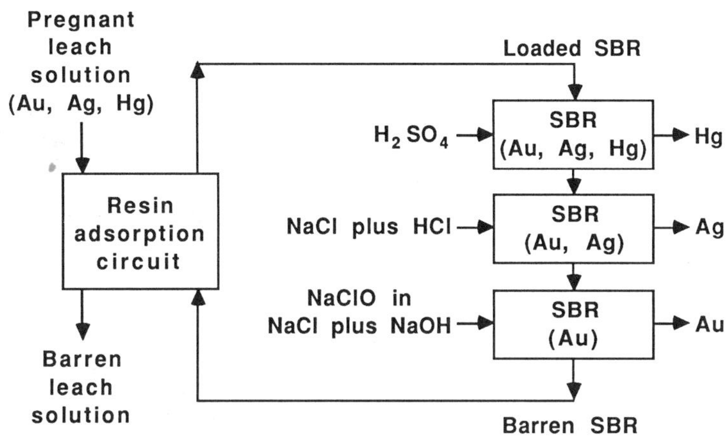

Fig. 4. Flowsheet for recovery of mercury, silver, and gold from a loaded strong-base resin (SBR) by sequential elution.

Thiourea, which appears to be the preferred eluant in the U.S.S.R., functions by forming a gold-thiourea complex which being positively charged, is not retained by the resin. However, the use of an acid solution of thiourea presents numerous operational problems. A recent pilot-plant comparison showed that the thiourea method does not compare favourably with the use of zinc cyanide or thiocyanate (Fleming and Cromberge, 1984c). The problems associated with this procedure include breakage of the resin beads by osmotic shock and poisoning due to the accumulation of cobalt and iron species on the resin. Also the cost of the reagent is prohibitive owing to the decomposition of thiourea in acid solution. In addition, the elution efficiency of base metals such as cobalt, iron, copper, and nickel is poor.

In the Russian process, non-selective adsorption is followed by selective elution (Fleming and Cromberge, 1984a). Zinc, nickel, and cyanide are eluted first by dilute mineral acid. This step is followed by the elution of copper with an ammoniacal solution of ammonium nitrate and, finally, gold and silver are eluted with acidified thiourea as the eluant and electrowon continuously.

The use of another metal cyano complex to displace the adsorbed gold and silver from the resin was first patented by Hazen (1957). Zinc cyanide complex, $Zn(CN)_4^{2-}$ was found to be particularly acceptable because: 1) it is preferred by the strong-base resin groups to gold and silver complexes; 2) it is readily decomposed by a mildly acidic solution so that the zinc and cyanide ions are removed while the resin is regenerated for further use; and 3) zinc is a metal which is compatible with most cyanide circuits using the Merrill-Crowe technology. The method patented by Hazen uses a saturated solution of sodium zinc cyanide to elute gold from the resin, after which the gold cyanide passes through a bed of zinc powder. The gold cements onto the zinc, and the zinc cyanide that is generated is recycled to elution.

One limitation of the use of the zinc cyano complex is the relatively low solubility of zinc cyanide. Fleming and Cromberge (1984b) overcame this problem by dissolving zinc sulphate heptahydrate in sodium cyanide solution so as to obtain a molar ratio of cyanide to zinc in solution of 4.5:1. Zinc concentrations as high as 50 g/L were realized. For solutions having a zinc concentration greater than 0.5 M, a temperature of 50°C was required to avoid the crystallization of $Na_2[Zn(CN)_4]$.

Fleming and Cromberge (1984c) evaluated all three eluants on a pilot-plant scale and concluded that the zinc cyanide method appears to be substantially cheaper than either the thiourea or the thiocyanate method. However, the rate of elution and the overall elution efficiency with thiocyanate are excellent and, if an efficient method could be found for recycling the thiocyanate from the eluted resin, this procedure could compete with the zinc cyanide method.

The efficiency of thiocyanate for eluting gold and silver cyano complexes from strong-base resins has been known for some time. However, due to its high affinity for the resin active groups, thiocyanate is difficult to remove once it has been adsorbed. Two interesting developments related to the use of thiocyanate as eluant have taken place in recent years. It has been discovered that the presence of dimethyl formamide greatly improves the elution kinetics of gold from AMBERLITE IRA-400 with NaSCN (Law, 1983). This behaviour was attributed to the solvating effect of dimethyl formamide on the quaternary ammonium groups of the strong-base resin. Another important development is the use of a solution ferric ions to elute the thiocyanate ions in the form of the red complex $FeSCN^{2+}$ (Fleming, 1985). This complex can be subsequently decomposed to produce more thiocyanate and ferric ions.

$$2\ \vdash\!NR_3SCN + Fe_2(SO_4)_3 = (\vdash\!NR_3)_2\ SO_4 + 2[FeSCN]SO_4$$

Thiocyanate is recovered from the strip liquor for recycling by the treatment of the solution with a solution of hydroxide, carbonate, bicarbonate, or any other suitable base. Iron precipitates from the solution as ferric hydroxide and, after filtration, the clear thiocyanate solution can be recycled to elution.

$$FeSCN^{2+} + 3OH^- \rightarrow Fe(OH)_3 + SCN^-$$

A proposed flowsheet for the gold recovery process is shown in Fig. 5.

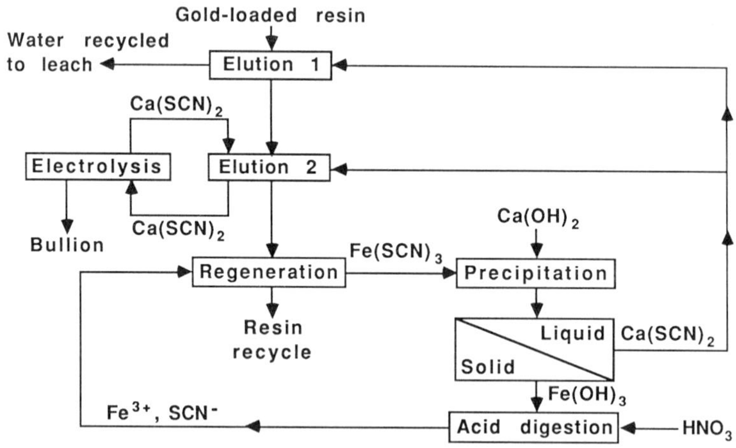

Fig. 5. Proposed flowsheet for the elution of aurocyanide from a strong-base resin with thiocyanate, followed by regeneration of the resin with ferric ions.

The elution of silver from strong-base resins has not been studied in detail. As noted previously in the U.S. Bureau of Mines process, silver is displaced, prior to the gold elution, by means of an acidified NaCl solution. Riveros (1985) found that silver is much easier to elute than gold. In fact, an efficient inexpensive eluant is sodium nitrate (Fig. 6). It can be seen that nitrate is almost as effective as thiocyanate or thiourea. Zinc cyanide can also be used for the elution of silver in concentrations that are considerably lower than those used for gold.

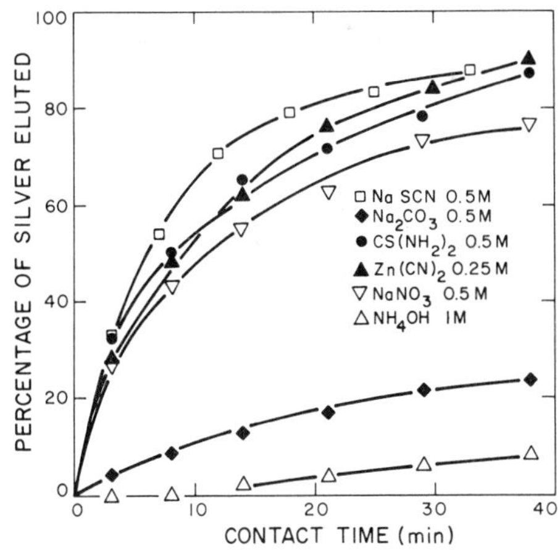

Fig. 6. Elution of silver cyanide by various reagents from DOWEX MSA-1 strong-base resin.

CHELATING AND ADSORBENT RESINS

The use of resins other than weak- or strong-base resins has received much less attention. Hodgkin and Eibl (1981) patented the use of linear or cross-linked resins synthesized from diallylamine. The preferred co-monomer is 1, 6-bis (diallylamine) hexane. The resins are easily and cheaply produced being synthesized directly from simple monomers. They are chelating resins and as such they were found to have a high selectivity and capacity for gold over other metals including silver.

In an evaluation of these resins, one gram of resin was agitated for 24 hours in 100 ml of a solution containing 100 ppm each of copper, cadmium, iron, nickel, silver and gold cyanide in excess KCN (pH 8) and then filtered and washed with 100 ml of dilute cyanide solution. Only gold was adsorbed by the resin, the concentration of the other metals in solution remaining at their initial value. The maximum gold capacity from a pure gold cyanide solution was 908 g/kg of resin (or 4.61 mmol/g).

Akser, Wan and Miller (1986) found that certain neutral polymeric adsorbents such as the polyacrylates AMBERLITE XAD-7, AMBERLITE XAD-8 exhibit modest capacities for loading gold from alkaline cyanide solutions (Fig. 7).

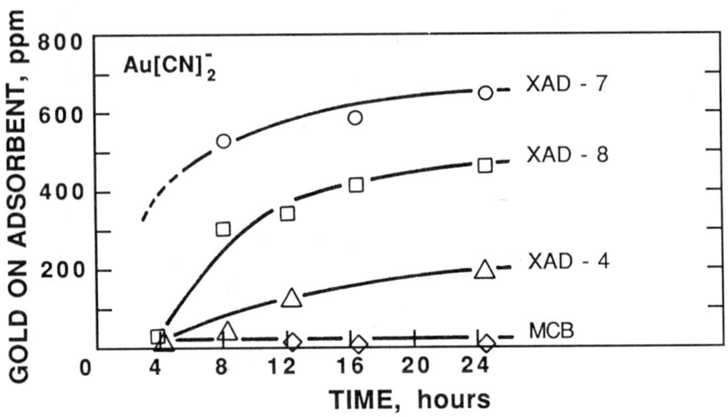

Fig. 7. Comparison of gold adsorption from $KAu(CN)_2$ solutions by different neutral polymeric adsorbents ($Au(CN)_2^-$ 5×10^{-5} M, Na_2SO_4 10^{-2} M, pH 10, 22°C, 5 gpl resin).

The development of selective adsorbents for gold from alkaline cyanide solution is continuing. Of interest are new phosphonate ester resins in which adsorbed gold can be readily eluted by a change in temperature and ionic strength of the eluant (Akser and co-workers, 1987).

ADSORPTION KINETICS OF METAL CYANO COMPLEXES

Little research has been conducted on the kinetics of the adsorption of cyano complexes onto ion exchange resins. Equilibrium has commonly been assumed to take place in less than 2 hours. Recently it has been shown that equilibrium on the strong-base resin Dowex MSA-1 is attained only after a period of several hours during which time the resin composition undergoes marked changes (Riveros and Cooper, 1987). Thus the loading of gold and silver passes through a maximum value whereas the leading of a metal such as zinc increases steadily as equilibrium is approached. This displacement of precious metals by the zinc cyanoion is the basis of the use of this species as an eluting agent. The kinetic behaviour of two-component systems involving the cyanoions of gold, silver, zinc, copper and iron is shown in Figs. 8 and 9.

Kinetic parameters obtained by applying mathematical models to the experimental data, indicate that there are significant differences in diffusivity amongst the various cyano complexes in both the Nernst diffusion layer and in the resin matrix (Riveros and Cooper, 1987).

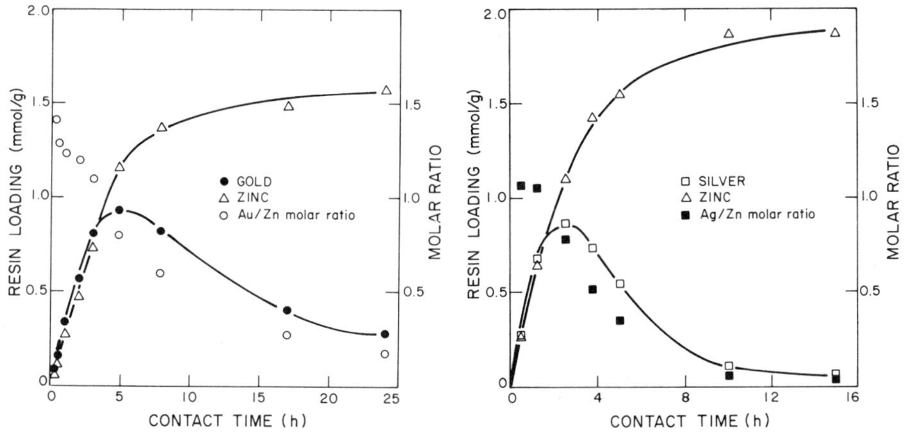

Fig. 8. Simultaneous extraction of Au or Ag (0.5 mmol/L) and Zn (0.5 mmol/L): DOWEX MSA-1, free cyanide 2.0 mmol/L, NaOH 2.5 mmol/L.

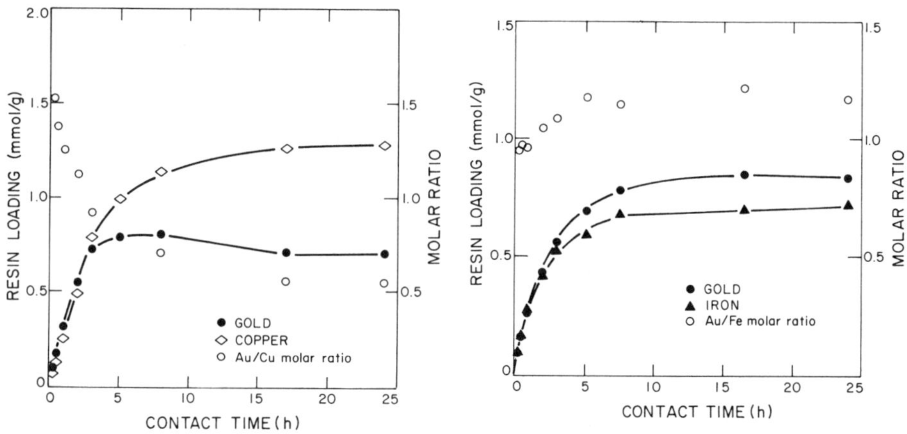

Fig. 9. Simultaneous extraction of Au (0.5 mmol/L) and Cu or Fe (0.5 mmol/L): DOWEX MSA-1, free cyanide 2.0 mmol/L, NaOH 2.5 mmol/L.

ION EXCHANGE TECHNOLOGY IN GOLD RECOVERY

In the U.S.S.R. the resin-in-pulp (RIP) process has been employed for some time for gold recovery from cyanide leach solutions on an industrial scale (Anon., 1978). In South Africa RIP has been applied successfully on a pilot-plant scale to several different types of pregnant cyanide feed solutions (Fleming and Cromberge, 1984c). The pilot plants ranged from a miniplant with 5-litre adsorption stages to a pilot plant with 400-litre stages. A flowsheet for the miniplant is shown in Fig. 10.

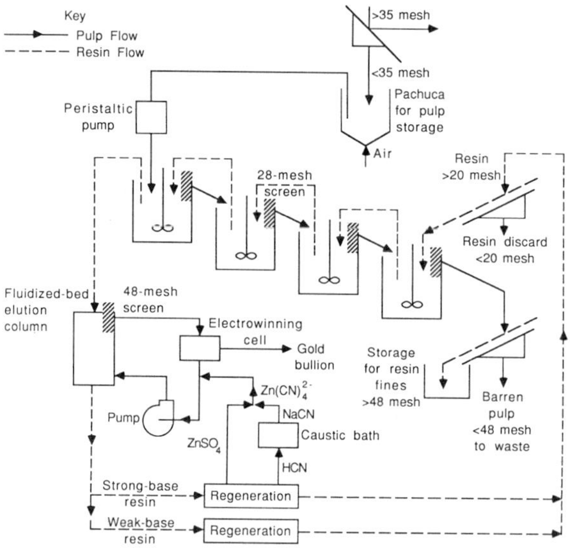

Fig. 10. Resin-in-pulp miniplant used in testwork in South Africa.

As pointed out by Fleming and Cromberge (1984c), the advantages of RIP in terms of faster reaction kinetics, higher loading capacities, and lower costs of elution and regeneration are particularly favourable to small operators. These authors indicate that while the advantages of resins are principally chemical their disadvantages are largely physical. These relate 1) to the smaller particle size of resins (-16 +20 mesh) compared with carbon (-6 +16 mesh), necessitating the use of smaller screens, 2) to resin attrition during adsorption and resin bead cracking due to osmotic shock during elution and regeneration, and 3) to the possible elutriation loss of resin due to the lower density of the resin compared with carbon.

An important advantage of ion exchange resins lies in their ability to effect a recycling of the cyanide solution and to eliminate the destruction of cyanide which is commonly practiced in the Merrill-Crowe and carbon-in-pulp processes. Ion exchange can be applied to the recovery of cyanide from process streams containing cyanide and base metals (Cy-Tech Inc., 1987). The cyanide is recovered from the loaded resin by elution with an acidic eluant. The eluant releases hydrogen cyanide which is stripped from the eluate using air and absorbed in a caustic solution which is then recycled to process. The base metals can be precipitated by neutralization of the stripped eluate or recovered.

It has been estimated that approximately 30 percent of mine-produced gold in the United States is realized through the heap leaching of low-grade ores (Thomas and Boyle, 1986). The indications are that this trend will continue. In the recovery of gold from the resulting leach liquors, an economic comparison of carbon-in-column and resin-in-column processes suggests lower capital and operating costs for the resin as opposed to carbon at a flowrate of 1000 U.S. gal. per min. for cyanide leach solutions containing either 1 ppm or 5 ppm gold (Mooiman and co-workers, 1983). The higher operating costs in the case of the carbon arise from the heating for elution and in the regeneration of the carbon. The principal operating cost in the resin-in-column process is the resin itself. It would seem that this cost could be reduced significantly given the availability of resins with the appropriate physical properties such as resistance to osmotic shock and to loss through elutriation. Of importance also is the nature of the resin transfer processes in the ion exchange technology which is employed.

CONCLUSIONS

Important developments have taken place in recent years in the application of ion exchange resins to the recovery of gold and silver. Both strong- and weak-base resins have potential in this field and have been evaluated on a pilot-plant scale. The use of other resins, such as chelating, polymeric adsorbents or impregnated resins, although in an early stage of development, has produced some interesting results.

Weak-base resins appear particularly suitable for the extraction of gold because of the selectivity found for this metal at high pH. With commercially available resins the gold loadings at high pH do not compare favourably with the loadings obtained with activated carbon. Unlike the latter, however, weak-base resins can be readily eluted with inexpensive alkaline solutions at room temperature and atmospheric pressure. Several promising new weak-base resins, capable of operating at higher pH, are under development.

The extraction of silver with weak-base resins is much less satisfactory than the extraction of gold, especially in the presence of base metals. In this case, no selectivity for silver over copper, zinc, iron, and other common metals is found. Also, silver is extracted at a significantly lower pH than is gold.

Several interesting advances have taken place in the field of strong-base resins as regards their elution behaviour towards cyanoions. These resins have a much higher capacity (although not selectivity) than weak-base resins. Since the loading is not affected by the pH, strong-base resins have been considered traditionally to be extremely difficult to elute. The new approaches to back-extract the gold from the strong-base resin include the oxidation of the gold cyano complex with NaClO, the displacement of the complex with zinc cyanide, and the elution with thiocyanate followed by recovery of the thiocyanate with ferric ions.

Silver has been found to be less difficult to elute from strong-base resins than gold. Therefore, these resins are particularly suitable for the recovery of this metal in the absence of contaminating base metal complexes.

The kinetic effects associated with the displacement of adsorbed gold and silver from a strong-base resin by base metal cyanions particularly the zinc cyanocomplex need to be taken into consideration in adsorption studies on actual cyanide leach solutions.

The development of new and improved resins, the advances in elution procedures, the recent successful pilot-plant testwork, and the inherent advantages of resins over activated carbon should in due course lead to the application of ion exchange for the recovery of gold on an industrial scale outside of the U.S.S.R.

ACKNOWLEDGEMENT

The authors would like to express their gratitude to the Natural Sciences and Engineering Research Council of Canada for financial assistance.

REFERENCES

Akser, M., R.Y. Wan, and J.D. Miller (1986). Gold adsorption from alkaline aurocyanide solution by neutral polymeric adsorbents. Solvent Extr. Ion Exch. 4 (3), 531-546.

Akser, M., R.Y. Wan, J.D. Miller, D.R. Quillan and S.D. Alexandratos (1987). Synthesis of new phosphonate ester resins for adsorption of gold from alkaline cyanide solution. Metall. Trans. B, submitted.

Anon. (1978). Gold from Russia's Muruntau deposit. Coal, Gold Base Miner., 26, 75.

Cy-Tech Inc., Mississauga, Ontario (1987). Communication.

Fleming, C.A. (1985). A novel process for the recovery of gold cyanide from strong base resins. In Proceedings of the Extraction Metallurgy '85 Conference. I.M.M., London, U.K.

Fleming, C.A., and G. Cromberge (1984a). The extraction of gold from cyanide solutions by strong- and weak-base anion exchange resins. J.S. Afr. Inst. Min. Metall., 84 (5), 125-137.

Fleming, C.A., and G. Cromberge (1984b). The elution of aurocyanide from strong- and weak-base resins. J.S. Afr. Inst. Min. Metall., 84 (9), 269-280.

Fleming, C.A., and G. Cromberge (1984c). Small scale pilot plant tests on the resin-in-pulp extraction of gold from cyanide media. J.S. Afr. Inst. Min. Metall., 84 (11), 369-378.

Green, B.R., and A.H. Potgeiter (1984). Unconventional weak-base anion exchange resins, useful for the extraction of metals, especially gold. In D. Naden and M. Streat (Eds.), Ion Exchange Technology, Horwood, Chichester, U.K. pp. 626-636.

Hazen, W.C. (1957). Method of eluting absorbed complex cyanides of gold and silver. U.S. Patent 2,810,638.

Hiskey, J.B., and A.O. Ette (1986). Kinetics of aurocyanide loading with an anion exchange resin. In D.R. Gaskell, J.P. Hager, J.E. Hoffmann, and P.J. Mackey (Eds.), Proceedings of the Reinhardt Schuhmann International Symposium, Fall Meeting of TMS-AIME, Colorado Springs, 295-309.

Hodgin, J.H., and R. Eibl (1981). Selective extraction of gold. European Patent Application 81303415.4 (Date of filing 24 July, 1981).

Kalocsai, G., and F. Fabics (1983). Recovery of gold and silver from ion exchange resins. Hung. Teljes HU 26,353. (Chem. Abs. 100:37678m).

Law, H.H. (1983). Recovery of gold from ion exchange resins with dimethylformamide/water mixture. In M.I. El Guindy (Ed.), Precious Metals 1982, Proceedings of the Sixth International Precious Metals Institute Conference, Newport Beach, California, June 1982. Pergamon Press Canada Ltd.

Mehmet, A., and W.A.M. Te Riele (1984). The recovery of gold from cyanide liquors in a countercurrent contactor using ion exchange resin. In D. Naden and M. Streat (Eds.), Ion Exchange Technology, Horwood, Chichester, U.K., pp. 637-652.

Mooiman, M.B., J.D. Miller, J.B. Hiskey, and A.R. Hendriksz (1983). Comparison of process alternatives for gold recovery from cyanide leach solutions. In Proceedings of the Soc. Min. Eng. of AIME, Fall Meeting, Salt Lake City, Utah.

Palmer, R.G. (1986). Ion exchange research in precious metals recovery. In Precious Metals Recovery from Low-Grade Resources. U.S. Bureau of Mines Information Circular IC 9059.

Riveros, P.A. (1985). The application of ion exchange resins to the extraction of silver and gold from cyanide solutions. Ph.D. Thesis, Queen's University, Kingston, Canada.

Riveros, P.A., and W. Charles Cooper (1985a). The extraction of silver from cyanide solutions with ion exchange resins. Solvent Extr. Ion Exch., 3, (3), 357-375.

Riveros, P.A., and W. Charles Cooper (1985b). Speciation study on the ion exchange extraction of silver, copper, iron, and zinc cyano complexes with an iterative slope analysis technique. Solvent Extr. Ion Exch., 3 (6), 909-930.

Riveros, P.A. and W. Charles Cooper (1987). Kinetic aspects of the ion exchange of gold, silver and base metal cyano complexes. Solvent Extr. Ion Exch., submitted.

Thomas, P.R., and E.H Boyle Jr. (1986). U.S. Bureau of Mines Information Circular IC 9070.

Waitz, W.H., Jr. (1982). Ion exchange for recovery of precious metals. Plat. Surf. Finish., 69 (4), 56-59.

AUTHORS INDEX